W9-AEC-438

Zerovalent Compounds of Metals

ORGANOMETALLIC CHEMISTRY

A Series of Monographs

BRIAN G. RAMSEY: Electronic Transitions in Organometalloids, 1969.
R. C. POLLER: The Chemistry of Organotin Compounds, 1970.
RUSSELL N. GRIMES: Carboranes, 1970.
PETER M. MAITLIS: The Organic Chemistry of Palladium, Volume I, Volume II – 1971.
DONALD S. MATTESON: Organometallic Reaction Mechanisms of the Nontransition Elements, 1974.
RICHARD F. HECK: Organotransition Metal Chemistry: A Mechanistic Approach, 1974.
P. W. JOLLY AND G. WILKE: The Organic Chemistry of Nickel, Volume I, Organonickel Complexes, 1974.
P. S. BRATERMAN: Metal Carbonyl Spectra, 1974.
P. C. WAILES, R. S. P. COUTTS, AND H. WEIGOLD: Organometallic Chemistry of Titanium, Zirconium, and Hafnium, 1974.
U. BELLUCO: Organometallic and Coordination Chemistry of Platinum, 1974.
L. MALATESTA AND S. CENINI: Zerovalent Compounds of Metals, 1974.

Zerovalent Compounds of Metals

L. Malatesta
and
S. Cenini
Istituto di Chimica Generale,
University of Milan,
Italy

1974 ***Academic Press*** London New York San Francisco
A Subsidiary of Harcourt Brace Jovanovich, Publishers

ACADEMIC PRESS INC. (LONDON) LTD.
24/28 Oval Road,
London NW1

United States Edition published by
ACADEMIC PRESS INC.
111 Fifth Avenue
New York, New York 10003

Library of Congress Catalog Card Number: 74-5650
ISBN: 0 12 466350 8

Printed in Great Britain by
William Clowes & Sons Limited
London, Colchester and Beccles

Preface

In spite of the growing number of books and reviews in the field of coordination and organometallic compounds, none has recently been published concerning the chemistry of zerovalent compounds of metals, an area interesting an increasing number of researchers, in the fields of both fundamental and applied chemistry. The widespread introduction in many laboratories of refined physico-chemical techniques such as i.r., n.m.r., e.s.r. and mass spectroscopy, and X-ray diffractometry on the one hand, and the interest of the development departments of many industries on the other, are the main reasons for the tremendous growth of this type of chemistry. This field, which is on the border line between organometallic and coordination chemistry, embraces many compounds of metals in low oxidation states. These are very stimulating not only for their unique reactivity, but also because they represent an ideal model for studying the chemical behaviour of a single centre which compares with a metal surface, a problem which is connected with the relation between homogeneous and heterogeneous catalysis. We will deal here with compounds having the metal in the zero oxidation state, but this review will not specifically include the carbonyl derivatives because they have recently been widely surveyed. However, since carbon monoxide is one of the commonest and most important ligands stabilizing metals in low oxidation states, many compounds where carbon monoxide is present together with other ligands will be reported here, particularly when they provide interesting examples for discussing the nature of the bonds and the reactivity of the compounds. A small section devoted to the isonitrile derivatives, which have been recently reviewed by one of us, will summarize only the very recent aspects of the chemistry of this type of compound. In Part 2, devoted to complexes having tricovalent P, As and Sb derivatives as ligands, an effort has been made to cover the literature to the end of 1972. Finally, complexes having only unsaturated hydrocarbons as ligands have not been considered since excellent books concerning this type of derivative have recently been published.

June, 1974

L. MALATESTA
S. CENINI

Contents

Part 1

The Stabilization of Low Oxidation States of Transition Metals

1.1. INTRODUCTION

The casual discovery of a volatile compound of nickel by C. Langer, during the study of the transformation of CO into CO_2 and carbon, catalysed by metallic nickel, led to the isolation in 1890 of the first coordination compound of a metal in a low oxidation state, $Ni(CO)_4$ [1]. The next member of this new class of compounds, $Fe(CO)_5$, was discovered in 1891 independently by Mond [2] and Berthelot [3]. At this time, the lack of any knowledge on the nature of chemical bonds did not allow any interpretation of the structure of these very unusual derivatives. Even in the first decade of the twentieth century, when Werner's theory, proposed in 1893, had been generally accepted as a new satisfactory key for interpreting "complex" compounds, nobody extended this theory to the metal carbonyls. They were in fact still regarded as anomalous organic compounds, with structures of the type:

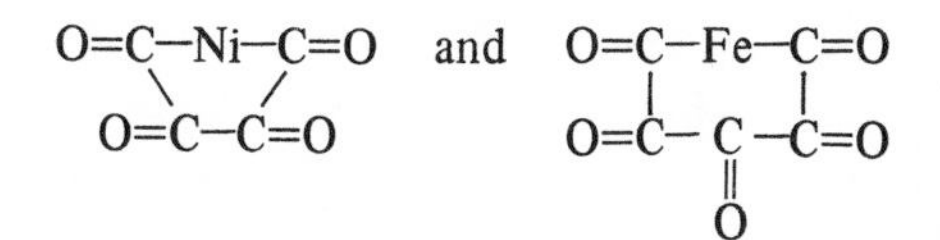

without any obvious relationship to other known inorganic substances.

Even when the right structure of metal carbonyls was recognized, they were not considered as derivatives of zerovalent metals because, before 1942, the idea of an oxidation number of zero was not taken into serious consideration. It was not until 1942, that $K_4[Ni(CN)_4]$ [4] and $K_4[Pd(CN)_4]$ [5] were obtained, chemists had to recognize the possibility that a metal in a compound might be in the oxidation state of zero [6]. This view was formally supported by the fact that $K_4[Ni(CN)_4]$ could be obtained by the additive reaction between elemental nickel and potassium cyanide:

$$Ni + 4\,KCN \longrightarrow K_4[Ni(CN)_4]$$

This reaction is analogous to that between nickel(II) cyanide and potassium cyanide:

$$Ni(CN)_2 + 2\,KCN \longrightarrow K_2[Ni(CN)_4]$$

in the course of which none of the atoms changes its oxidation number. At this point the realization that the cyanide ion $[C{\equiv}N:]^-$ is isoelectronic with carbon monoxide $:C{\equiv}O:$ and consequently that $Ni(CO)_4$ is isosteric with $[Ni(CN)_4]^{4-}$ wholly justified the consideration of metal carbonyls as neutral coordination compounds with the metal in the zero oxidation state.

The stability of the bond between carbon monoxide and the metal was attributed in valence bond terms, by Pauling, to resonance between single and double nickel–carbon bonds [7]:

$$Ni \leftarrow C{\equiv}O \longleftrightarrow Ni{=}C{=}O$$

The very weak basic character of the lone pair in carbon monoxide results in the dative σ bond involved not otherwise being strong enough. Moreover, the increase in negative charge on the metal as a consequence of a pure σ donation to the metal, if not compensated by an electron back-donation from nickel to the ligand, should increase the tendency of the metal to be oxidized instead of stabilizing the low oxidation state, as actually happens.

These same concepts were extended by Deasy to the complex cyanides [8]. The presence of the retrodative bond can be related to the so-called Pauling Electroneutrality Principle, which states in a qualitative way that in all stable compounds the charge is distributed as evenly as possible over all atoms [9].

The hypothesis of partial double bond character is in accordance with the low value of the nickel to carbon bond distance (1.82 Å), compared with the calculated distance for a single bond (2.0–2.2 Å). Considering the electronegativity of carbon in CO and CN^-, a zero charge on the metal corresponds to a double bond percentage of 75–78% for $Ni(CO)_4$ and about 67% for $K_4[Ni(CN)_4]$ [10, 11].

The concept of "zerovalent metal" or, better, of "metal in a zero oxidation state", requires some comment. It implies that in some way the charge left on the metal by complex formation with a ligand is nearly zero. This can be achieved in a coordination compound by considering resonance between single and double bonds, according to the valence bond theory. However, chemists now prefer to talk in terms of molecular orbital theory. According to this theory, bonds with little, if any, transfer of negative charge from the donor atom of the ligand to the metal can be accounted for by invoking a π-back donation from the metal into a suitable empty orbital of the ligand, superimposed on the σ donation from the ligand to the metal, giving place to a bond order greater than one. This second hypothesis is schematically illustrated in the case of a metal carbonyl complex in Fig. 1.1.

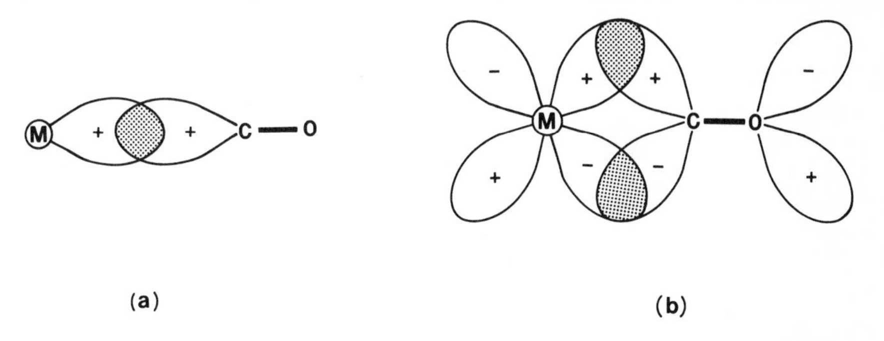

Fig. 1.1. (a) Dative σ bond from CO to the metal; (b) π-back donation form an occupied $d\pi$ metal orbital into the π^* antibonding orbital of CO.

The definition of oxidation number zero of a metal in a compound does not mean that the electron density on the metal atom is the same as in the fundamental metallic state, but rather it is only the result of a formal application of the rule by which we define oxidation number as the charge left on the central atom of a coordination compound when the ligands are removed in their closed-shell electronic configurations, i.e. CO, CN^-, I^-, NH_3, H_2O, OH^-, NH_2^-, etc.

As a consequence of the Electroneutrality Principle, the real amount of electronic charge on the metal atom corresponds more to the oxidation number of the atom when this is low than when it is high.

Although the concept of oxidation state in essentially covalent compounds has little physical meaning [12], it is useful to chemists for classifying compounds and discussing their stoichiometry and stereochemistry. However, it should be regarded as an over-simplification to consider that the charge distribution in a bond is in an extreme situation (assumed to be the most favourable), while in fact this exactly balanced electronic distribution is more or less perturbed. This point will be clarified when we discuss complexes such as the olefin derivatives of metals. In any case, the bonding in zerovalent compounds must be largely of the covalent type, since there can be no strong electrostatic attraction in the absence of a net charge on the metal. If we assume that a compound is "stable" when its lifetime is long enough for it to be isolated and characterized, then to discuss the "stability" of a zerovalent compound to a first approximation we must look at the factors affecting the strength of the σ and π bonds between the metal and the ligands.

We should recall that they are not simply two independent bonds. In fact the original σ-bond is strengthened by removal of the charge on the metal through the π-bond. On the other hand, a stronger σ-bond allows more extensive metal-to-ligand π-bonding. This produces what has been described as a "synergic" interaction between the two types of bonding [13]. This approach to the stability of a zerovalent compound is related only to thermodynamic factors, and the gain in energy following bond formation is the most important point to discuss. In fact whereas the formation of an organic compound is usually kinetically controlled, the product of an inorganic reaction is generally determined by the laws of thermodynamics. When considering the energy relations between an element, such as nickel, in the bulk metallic state, when its atoms have by definition an oxidation state of zero, and a zerovalent compound of the same metal, such as NiL_4, the change in heat content in the formation of the zerovalent compound from the metal can be expressed as follows:

$$Ni_{cryst} \longrightarrow Ni_{gas} + \Delta H_1 \quad (\Delta H_1 = 102.8 \text{ kcal mol}^{-1})$$

$$Ni_{gas} + 4\,L \longrightarrow NiL_{4\,gas} + \Delta H_2$$

ΔH_1 and ΔH_2 are very large in relation to the term $T\Delta S$, which to a first approximation can be ignored in determining the free energy of the reaction. We can thus state that the necessary condition for thermodynamic stability of the zerovalent compound is that ΔH_2 must have at least the same order of magnitude as ΔH_1 but the opposite sign. This rough thermodynamic calculation leads to the result that the Ni–L bond energy must be not less than 25 kcal mol^{-1}, in order to ensure the stability of the zerovalent compound with respect to its decomposition to metal.

We can now discuss the factors affecting ΔH_2, that is the σ-donor and π-acceptor properties of the ligands, and the ease with which electron density can be removed from the metal in back-donation to the ligands.

The problem of the existence of π-back donation from metal to ligands, will be discussed in detail. This point is in fact very important for the understanding of the physical-chemical behaviour of complexes of metals in low oxidation states, and it has stimulated a very large amount of research in this field. As we will see, in spite of a tremendous effort to reach a conclusion, no definite answer has been given to this problem. At this stage apparently a "suspension of hostilities" has taken place among researchers in the expectation that something new will give an unambiguous answer. This situation has at present discouraged speculation on this subject, although the existence of π-back-donation in specific cases is generally accepted. We feel that, although direct proof for the π-back donation has not yet been achieved, this hypothesis is still extremely valid and useful, and it has the advantage in respect of the rigid Valence Bond formalism, of offering a flexible approach in all the situations met with the different electronic interactions between metals and ligands. Anyhow, it is a matter of convenience to choose the most satisfactory approach, and the π-back-donation hypothesis seems in our opinion more adequate to interpret the properties of low oxidation state complexes.

Moreover, the importance of steric effects has recently been pointed out as one of the main factors which contributes to the instability of the complexes, and this point will also be considered.

1.2. THE σ-DONOR PROPERTIES OF THE LIGANDS

The most common ligands which allow the isolation of compounds of metals in low oxidation states are summarized in Table 1.I [6, 14]. An inspection of this Table shows that these compounds are not formed by the ligands most commonly found in Werner complexes, such as water, ammonia, halide ions, the nitro group, etc., that is with typical Lewis bases. In addition some of the ligands reported here, such as carbenes, cyclobutadiene, tetraazadienes, unstable as free molecules, can only be stabilized by coordination to a transition metal in a low oxidation state. Finally, Table 1.I includes N_2 as quite a common ligand, and this has been recognized only in the last few years.

The σ-donor properties of the various ligands can be estimated only in a rough qualitative way. For anionic ligands it appears to follow the electronegativity of the donor atom. For molecules we can measure the ionization potential (I.P.), that is the energy required to remove an electron from a given orbital. This can be taken as a measure of what has been defined as the "absolute" Lewis base strength [15], that is the tendency of a molecule to donate electrons as governed by the inductive and resonance effects, but uncomplicated by steric and solvent effects. Large I.P.s would correspond to poor σ-donor capacity. The I.P.s for ligands such as substituted phosphines PR_3, will reflect the influence of the groups R bonded to phosphorus on the energy of the lone pair on that atom. For these ligands it appears that the I.P.s are largely determined by partially mesomeric effects, measured by the Brown $\Sigma\sigma_p^+$ of the substituents R [16, 17]. The availability of the electron pairs of donor atoms will also be measured by their tendency to bind a proton. It has been shown that the pK_a for a large number of phosphines, as determined by potentiometric titration in nitromethane [18], linearly correlate with the sum of the Taft polar constants ($\Sigma\sigma^*$) [19] of the R groups bonded to phosphorus, which is a measure of the inductive power of the R substituents. Analogously, the I.P.s of aliphatic phosphines also correlate well with their $\Sigma\sigma^*$ values [17].

The importance of steric effects in determining the pK_a values has been pointed out, and a slight deviation from linear correlation was observed when R

= phenyl, for which a contribution due to mesomeric effects should be taken in account.

Although I.P.s and pK_a values are not directly comparable, one reaches similar conclusions concerning the basicity of ligands like phosphines, from both these experimental data. In Tables 1.II and 1.III the pK_a values [18] and I.P.s

TABLE 1.I

Most common ligands in compounds of metals in low oxidation states

CO	carbon monoxide
CN^-	cyanide ion
C_2R^-	acetylide ion
$C_5H_5^-$	cyclopentadienide ion
CNR	isocyanides
$RCOR$	alkylcarboxycarbenes
$RCNR_2$	alkyldialkylaminocarbenes
C_6H_6	benzene
C_4H_4	cyclobutadiene
C_3H_5	allyl
C_2R_2	alkynes
C_2R_4	alkenes
$C_nH_{(2n-2)}$	dienes
$C_nH_{(2n-2m)}$	polyenes
$C_6H_4O_2$	*para*-benzoquinone
N_2	nitrogen
NO	nitric oxide
RCN	nitriles
$C_{12}H_8N_2$	2,2′-bipyridyl
R_2N_4	tetraazadienes
PR_3	trihalogeno, triaryl, trialkyl phosphines
$P(OR)_3$	triaryl and trialkyl phosphites
AsR_3	triaryl and trialkyl arsines
SbR_3	triarylstibines
SR_2	dialkyl and diarylsulphides
SO_2	sulphur dioxide
CS_2	carbon disulphide

[17] of many phosphines are reported. For a number of phosphines different values of the ionization potentials can be found in literature, and they are reported in the Table.

We should recall that only if we consider I.P.s of homologous series of compounds, such as phosphines (Table 1.III) or aromatic hydrocarbons [20] (Table 1.IV), can we obtain a comparative measure of the donor ability of the electrons involved in the ionization process. In particular, the I.P.s for isocyanides [17] (Table 1.V) do not refer to ionization from the carbon lone pair. It appears in fact that this ionization process involve the π system of the CN

TABLE 1.II

pK_a values for tertiary, secondary and primary phosphines determined by potentiometric titrations. Data taken from ref. 18

	pK_a	$\Sigma\sigma^*$
Tertiary phosphines		
$P(cyclo\text{-}C_6H_{11})_3$	9.70	−0.450
$P(CH_3)_3$	8.65	0.000
$P(CH_3)_2(C_2H_5)$	8.62	−0.100
$P(CH_3)(C_2H_5)_2$	8.62	−0.200
$P(C_2H_5)_3$	8.69	−0.300
$P(n\text{-}C_3H_7)_3$	8.64	−0.345
$P(n\text{-}C_4H_9)_3$	8.43	−0.390
$P(n\text{-}C_5H_{11})_3$	8.33	−0.390
$P(i\text{-}C_4H_9)_3$	7.97	−0.375
$P(CH_3)_2(CH_2CH_2CN)$	6.35	+0.800
$P(CH_3)_2(C_6H_5)$	6.49	+0.600
$P(n\text{-}C_4H_9)_2(CH_2CH_2CN)$	6.49	+0.540
$P(n\text{-}C_8H_{17})_2(CH_2CH_2CN)$	6.27	+0.540
$P(C_2H_5)_2(C_6H_5)$	6.25	+0.400
$P(CH_2CH_2CN)_2CH_3)$	3.61	+1.600
$P(C_6H_5)_3$	2.73	+1.800
$P(CH_2CH_2CN)_3$	1.36	+2.400
Secondary phosphines		
$PH(cyclo\text{-}C_6H_{11})_2$	4.55	+0.190
$PH(n\text{-}C_4H_9)_2$	4.51	+0.230
$PH(n\text{-}C_8H_{17})_2$	4.41	+0.230
$PH(i\text{-}C_4H_9)_2$	4.11	+0.240
$PH(CH_3)_2$	3.91	+0.490
$PH(C_6H_5)_2$	0.03	+1.690
$PH(CH_2CH_2CN)_2$	0.41	+2.090
Primary phosphines		
$PH_2(n\text{-}C_8H_{17})$	0.43	+0.850
$PH_2(i\text{–}C_4H_9)$	−0.02	+0.855
PH_3	−14	+1.470

group. Although when we compare the I.P.s for different types of ligands [20, 21], (Table 1.VI), we can only have a qualitative idea of their relative σ-donor capacity, we get the general impression that ligands apt to stabilize low oxidation state complexes have a poor σ-donor capacity. Carbon monoxide, for instance, which is one of the less effective bases, as shown by its low dipole moment, by the size of the orbital containing the lone pair and by its very high I.P., is one of the ligands which more readily stabilizes complexes of metals in low oxidation states.

TABLE 1.III

Ionization Potentials of Phosphines (eV).
Data taken from ref. 17

Phosphine	I.P. (eV)
$P(C_4H_9)_3$	8.00
$P(C_2H_5)_3$	8.18–8.27
$P(C_6H_5)_3$	8.2
$P(OC_2H_5)_3$	8.40–8.63
$P(OC_4H_9)_3$	8.44
$P(iso\text{-}OC_3H_7)_3$	8.46
$P(C_6H_5)_2Br$	8.72
$P(C_6H_5)_2Cl$	8.75
$P(OCH_3)_3$	8.82–9.00
$P(CH_3)_3$	8.60–9.20
$P(C_6H_5)Cl_2$	9.45
$PH_2(C_2H_5)$	9.54
$PH(CH_3)_2$	9.7
$PH_2(CH_3)$	9.72
PBr_3	10.0
PH_3	10.2 –10.3
PCl_3	10.50–10.6–10.75
PF_3	13.00

TABLE 1.IV

Ionization Potentials (eV) for some aromatic hydrocarbons.
Data taken from ref. 20

Hydrocarbon	I.P. (eV)
Benzene	9.24 ± 0.01
Toluene	8.82 ± 0.01
iso-propylbenzene	8.69 ± 0.01
mesitylene	8.39 ± 0.02
naphthalene	8.18 ± 0.02

TABLE 1.V

Ionization Potentials (eV) of isonitriles.
Data taken from ref. 17

Isonitrile	I.P. (eV)
$C_6H_5CH_2NC$	9.61
$CH_3C_6H_4NC$	9.63
C_6H_5NC	9.70
$C_6H_{11}NC$	10.72
C_4H_9NC	11.71

TABLE 1.VI

Ionization Potentials (eV) of some common ligands. Data taken from ref. 20 and 21

Ligand	I.P. (eV)
N_2	15.6
CO	14.1
CH≡CH	11.41
$CH_2{=}CH_2$	10.52
NH_3	10.15
C_5H_5N	9.8
NO	9.40
C_5H_5	8.72
$(CH_3)_3N$	7.82

Interestingly, among the ligands above considered, PF_3 is one of the few molecules which has an I.P. near to that of carbon monoxide, and it is known that trifluorophosphine is one of the ligands that most easily replaces carbon monoxide in low oxidation state complexes.

1.3. THE π-ACCEPTOR PROPERTIES OF THE LIGANDS

We should now consider the π-acceptor ability of the ligands, which we can define as π-acidity. Figure 1.2 gives a diagrammatic representation of the orbitals which can be involved in the $M_{d\pi} \rightarrow L_{\pi}$ π-back donation, assuming that the metal orbitals are pure *d* orbitals.

Of the various orbitals which can overlap to form a π-bond, the d_π-d_π overlap (a), the $d_\pi-\pi_\pi^*$ (b), and the $d_\pi-\pi_\mu^*$ (c), are generally considered. As σ-bonds formed by overlap directed towards the centre of a multiple bond as shown in (c) are sometimes called μ-bonds, it is convenient to distinguish between $d_\pi-\pi_\pi^*$ and $d_\pi-\pi_\mu^*$ overlaps [22].

All the ligands reported in Table 1.I possess empty *d* orbitals (phosphines, arsines, disulphides, etc.) or π* antibonding orbitals (CO, CN^-, olefins, etc.) of low energy suitable for back-donation.

As we said before, the idea of π-back bonding was originally introduced in order to explain the stability of the complexes in low oxidation states. In fact one had on the one hand to devise a mechanism by which the negative charge accumulated on the metal by the σ-bond could be dissipated, and on the other hand, to explain how, in spite of the low basicity of the ligands, strong bonds could actually be formed in these compounds. This attractive picture of this

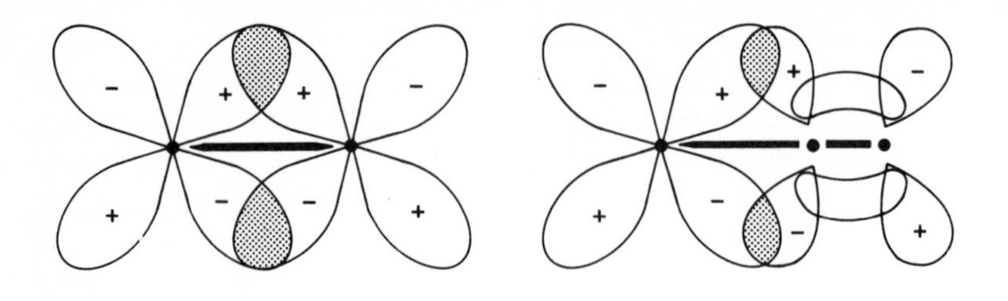

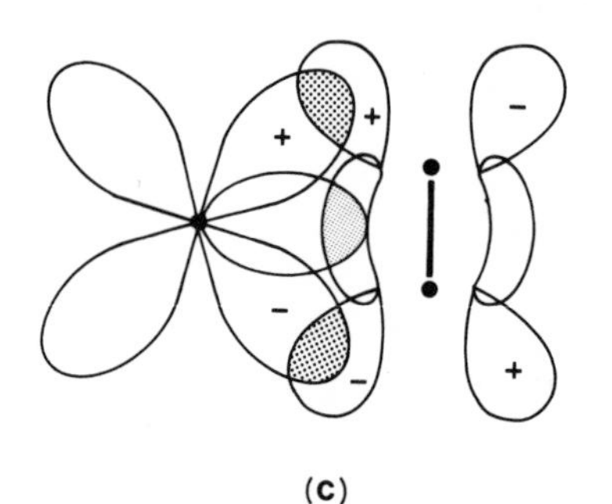

Fig. 1.2. (a) d_π-d_π overlap; (b) d_π-π_π^* overlap; (c) d_π-π_μ^* overlap.

type of bond has stimulated much research to find an experimental proof of back-donation, a problem which does not yet appear to be resolved. In fact, the same experimental facts that can be invoked in order to "prove" the presence of a π-bond, "the panacea of inorganic ills" [23], can also be given a different or even opposite interpretation [22]. As a consequence the only evidence for the existence of π-back donation is still indirect, that is: (1) in zerovalent compounds the interatomic distances between metal and ligand are generally shorter than expected for a single bond, while the geometry of the coordinated ligand can vary according to the amount of π-back-donation [24], and (2) bond moments are very low, as expected for nearly apolar bonds of the synergic type described above [25].

Other indirect evidence has been used to support this hypothesis, primarily the infrared spectra of the complexes. In considering the infrared spectra of complexes of metals in low oxidation states, we will also survey the nature of the most important metal–ligand bonding present in these compounds.

1.3.1. Infrared Spectroscopy and Theoretical Contributions

For ligands such as N_2, CO, NO, CN^-, O_2, alkenes, alkynes, etc., the stronger the π-back donation from the metal into their antibonding orbitals, the lower should be the respective N–N, C–O, N–O, C–N, O–O or C–C bond order. Experimentally one finds longer bonds and lower stretching frequencies in the ligands coordinated to the metal in comparison with the free ligands. This phenomenon has been widely used to estimate the π contribution in the metal–ligand bonds. However, for ligands such as alkenes and alkynes, an increase in C–C bond length does not prove back donation since a strong donation from the ligand to the metal would give a similar result. On the contrary, it is important to recall that for a ligand such as carbon monoxide, the σ-bond alone should slightly increase the values of the C–O stretching frequency [26]. The reversible adsorption of carbon monoxide on ZnO results in an increased ν(CO). Admittedly with this substrate π-bonding is relatively unimportant. In contrast, adsorption on *d* transition metals generally results in a lowering of ν(CO), and in these cases a π-bonding interaction with the metal *d* orbitals is considered to be operative.

Unfortunately, the metal–ligand absorption bands are difficult to assign in the i.r. spectra [27]. This does not allow any easy relation between the decreasing stretching frequency of the ligands, and the strength and the stretching frequencies of the metal–ligand bonds, to be made. This lack of information is particularly significant with ligands such as phosphines, as with the exception of fluorophosphines they do not contain any vibrations which can be related to the amount of π-back donation in the infrared spectra. A collection of i.r. data concerned with ligand and metal–ligand vibrations can be found in refs 27a, 27e, 35.

*1.3.1.1. The Metal–Phosphorus Bond**

Many vibrational studies involving the assignment of the ν(P–M) mode have been reported [27a, c] and a wide range of frequencies from *ca.* 150 to 550 cm^{-1} seems to be possible. However, several of these assignments appear to be doubtful [28], and it has been pointed out that only by using the recently developed metal isotope technique can a correct assignment be made [27c, 29].

For zerovalent compounds the range of frequencies is restricted from *ca.* 300–150 cm^{-1}, but a tentative assignment at 424 cm^{-1} for ν(Pt–P) in $Pt(PPh_3)_4$ has also been reported [27d]. The same type of problem exists for ν(M–As). It appears that more accurate assignments of the metal–phosphorus stretching frequencies should be made, in order to allow a discussion of these vibrations in terms of M–P multiple bond character.

Thus the existence of a metal–phosphine π-back donation can be inferred from the i.r. spectra mainly in an indirect way, for instance analysing the effects

* For a recent review on spectroscopic studies of metal-phosphorus bonding in coordination complexes, see: J. G. Verkade (1972). *Coord. Chem. Rev.*, 9, 1.

on the ν(CO) frequencies brought about by phosphine substitution in a metal–carbonyl complex. At present while some recent theoretical calculations seem to support the conclusion that the $d_\pi \rightarrow \pi_\pi^*$ bonding is relevant in pure metal-carbonyl compounds, the existence of a metal-phosphorus $d_\pi \rightarrow d_\pi$ bond is much more controversial.

By plotting the overlap integrals for the σ and π bonds in carbon monoxide versus the C–O internuclear distance, Kettle has shown that the σ interaction is relatively insensitive to vibration [30a]. Changes in the i.r. spectrum of carbon monoxide should be thus mainly attributed to effects within the π-electron system. This has been supported by a semi-empirical MO calculation in the series of compounds $[V(CO)_6]^-$, $Cr(CO)_6$ and $[Mn(CO)_6]^+$ [30b], where the σ(C–O) bond is approximately constant in each compound and the charge density on the ligand changes only in the π orbitals. It has also been found by means of an overlap population analysis [31], that the $d_\pi \rightarrow \pi_\pi^*$ back donation is strong in $[V(CO)_6]^-$, while it is almost negligible in $[Mn(CO)_6]^+$, as expected on the basis of the charge density on the metals. These conclusions have been confirmed by more recent calculations on the spectra of $M(CO)_6$ ($M = V^-$, Cr, W, Re^+) [32] by which the increased back-bonding in $[V(CO)_6]^-$, compared with $Cr(CO)_6$, has been attributed to the fact that the *d* orbitals are closer in energy to the π^* orbital of carbon monoxide. There are at present insufficient data to establish the expected magnitude of M–CO stretching constants in complexes where *d* orbitals are involved in the bonding [27b].

However, on the one hand from studies on platinum alkyls it seems that a metal–carbon single bond may be associated with a stretching force constant between 2 and 2.5 mdyn Å^{-1} [33], while, on the other hand, in hexacarbonyl molybdenum a Mo–C stretching constant of 1.8–1.9 mdyn Å^{-1} has been reported [34, 35], which is very low for a bond with a formal bond order of 1.5. The possibility that there is a mixing of various symmetry modes in the normal modes must be considered and this renders any conclusions on metal–ligand bond orders using vibrational spectral data uncertain.

Earlier results from a study of the carbonyl force constants in compounds of the type $LM(CO)_5$ (M = Cr, Mo, W; L = phosphine) [36, 37], were interpreted on the basis of an extensive π-bonding between metal and phosphines. Such a π-bond was also indicated by an analysis of the force constants of the carbonyl *cis* and *trans* to the ligand L in the compounds $LMo(CO)_5$ [38]. The inductive effect through the ligand–molybdenum σ bond brought about by changing L, should have the same influence on carbonyls in *cis* or *trans* to L. In contrast, a difference in the π-acidity of the ligands L affects the force constant of the *trans* carbonyl twice as much as the *cis*. Studies on series of compounds such as $Ni(CO)_{4-n}(PR_3)_n$ [39] and $LW(CO)_5$ [40] (L = amine, pyridine, phosphine) produced a different conclusion. From the linear correlation between the ν(C–O) and the Taft polar constants of the substituents R, Bigorgne [39] concluded that the $d_\pi \rightarrow L_\pi$ bonding is not important, or at least that it has a negligible value, nearly independent on the nature of R. Moreover, by studying the competition of ligands in a series of metal complexes, Bigorgne concluded that coordinated ligands cannot be characterized by definite donor-acceptor capacities, as these capacities are also dependent on the overall electron density of the complex to which they are bonded [41]. Similar results have been obtained by Angelici and Malone [40], who found a linear correlation between the C–O force constants and the basicity of the ligands L, given as their pK_a values in the series of compounds $LW(CO)_5$.

A method for determining electron donor–acceptor properties of tricovalent phosphorus ligands PR_3 based on the A_1 carbonyl stretching frequency in $Ni(CO)_3L$ has been published by Tolman [42]. On the basis of data for 70 ligands, an additivity rule is proposed, by assigning values to each R group, and this makes possible calculation of $\nu(CO)(A_1)$ for the various ligands with an average deviation of ± 0.3 cm^{-1}. From this study an ordering of the various ligands in terms of their electron-withdrawing properties can be derived. However, the main conclusion was that there does not appear to be sufficient evidence at this time to say to what extent the reduced electron density on nickel (as evidenced by an increase of the $\nu(CO)(A_1)$), can be attributed to reduced σ donation or enhanced π-acceptor behaviour of phosphorus in the complexes.

In favour of the first hypothesis seems to be the fact that the above substituent parameters are in good correlation with Kabachnik's parameters based on acid ionization constants of phosphinic acids of the type $R'R''POOH$ in water, or with phosphonium pK_a s.

These conflicting results are difficult to explain. If one considers the metal–ligand bond as a synergic σ–π interaction, in the sense that the σ and π-bonds reinforce each other, one can understand why there is correlation between the variation in $\nu(CO)$ or K(CO) and the basicity of the ligands L. However, the fact that these correlations are linear seems to imply that the π-acceptor properties of the free ligands vary inversely to their σ-donor capacities. If we realize that we are trying to differentiate small electronic effects, with indirect crude experimental parameters such as the variations on $\nu(CO)$, we at once understand why the same phenomenon can have such different explanations. It appears in any case that σ-bonding alone cannot explain the different values of K(CO) according to whether CO is *cis* or *trans* to a ligand L like phosphine [43]. It should be also pointed out that such linear correlations are generally not found using Hammett σ_H or resonance σ_R constants of the substituents R of the ligand L, as has been also recognized in metal–metal bonded species of the type $Mn(CO)_5L$ ($L = SnR_3$) [44]. However, while it seems correct to extend the use of Taft polar constants to organometallic compounds, since these constants are only related to the polarizabilities of the electronic densities of the bonds, the same may not be true for parameters such as σ_R, which involve a more complex mechanism.

In trifluorophosphine–metal complexes the d_π–d_π contributions to the bonds could become significantly more important, since the presence of the electronegative fluorine atoms lower the energy of the phosphorus $3d$ orbitals. It has been observed that in zerovalent metal derivatives when only trifluorophosphine is present, ν(PF) is always at higher frequencies than in the free ligand but that, in contrast, the corresponding values in trifluorophosphine complexes containing stronger bases (e.g. $Ni(PF_3)_{4-n}(PR_3)_n$) are always much lower [45]. This has been explained by considering that in the first case the fluorine–phosphorus p_π–d_π bond in coordinated PF_3 is strengthened by the σ phosphorus–metal bond. In the presence of a stronger base as in $Ni(PF_3)_{4-n}(PR_3)_n$, the increased electron density on the metal makes the remaining metal–phosphorus π-bonds stronger, and this corresponds to a decrease in the p_π–d_π component of the P–F bond [46]. However, in this case also the nature of the metal–phosphorus bond is not fully understood, and the i.r. evidence based on ν(PF) is at least ambiguous. In fact force constant calculations for $Ni(PF_3)_4$ have shown that K(PF) is similar in the free ligand and in the complex [47, 48].

*1.3.1.2. The Metal–Unsaturated Hydrocarbon Bond**

Infrared spectra have been also used to support the existence of the $d_\pi \rightarrow \pi^*_\mu$ back-bonding in these compounds, usually called π-complexes, which following the new IUPAC rules are now named η-complexes. The explanation of the stability of the silver(I)–alkene and platinum(II)–alkene complexes, which was difficult to understand in terms of a pure σ bond, was first proposed by Dewar [49] and Chatt and Duncanson [50] in terms of an additional π-back donation from the metal into the π^* antibonding orbital of the olefin. This view was supported by the decrease of the ν(C=C) absorptions in platinum and palladium(II) complexes by *ca.* 150–160 cm^{-1} [50]. With not very "soft" centres such as the d^{10} closed shell Ag^I ion, the σ bond between the metal and the alkene seems to be the most important, since the stability of these complexes increases with the number of alkyl substituents on the olefin, that is with the basicity of the ligand [51]. In these cases the ν(C=C) absorption is decreased by only 50–60 cm^{-1} [52]. With "softer" centres, such as the d^8 open-shell Pt^{II}, it seems that the π-bond is somewhat more important than the σ-bond, as has been shown from the determination of some stability constants [53]. In contrast, with very "soft" centres such as low oxidation state complexes (Ni^0, Pd^0, Pt^0, Rh^{+1}, Ir^{+1}) the stability seems to increase when the alkene

* For a recent review on the molecular orbital and valence bond descriptions of the bonding in metal–alkene and –alkyne complexes, see: F. R. Hartley (1972). *Angew. Chem. Internat. Edn.*, **11**, 596.

carries electron-withdrawing substituents; in these cases the ν(C=C) absorption decreases by *ca.* 200 cm^{-1} [54, 55, 56].

These effects indicate an enhancement of the π-bond component on increasing the π acidity of the alkene, as has been discussed recently for some platinum(0) olefin complexes [57]. Analogously, while the average decrease of ν(C≡C) in platinum(II) alkyne complexes is about 200 cm^{-1} [58], a greater shift was found in platinum(0) derivatives (300–400 cm^{-1}) [57, 59, 60], in accordance with a greater back-donation from the metal to the alkyne. Some semi-empirical MO calculations have shown that the Dewar–Chatt–Duncanson model for alkene complexes can be extended to the alkyne derivatives [61].

Moreover, in another theoretical study the same authors have found that for platinum(0) alkene and alkyne complexes of the type $(PH_3)_2PtL$ (L = alkene, alkyne), energetically the most favourable configuration is square planar, in agreement with known X-ray structures [62]. The calculated energies for the rotation of the ligands L have also shown that $\sigma + \pi$ interactions in the alkyne complexes are greater than in the alkene complexes, which have a lower barrier to rotation. A dp^2 hybridization of the central metal atom has been preferred to the usual sp^2 hybridization for these zerovalent complexes [62]. A subsequent theoretical study on the geometries of ethene and ethyne in these complexes has shown that the population of the π-antibonding orbital of the organic moiety definitely affects the equilibrium geometry for free or complexed alkynes and alkenes [62 *bis*].

The i.r. spectral evidence shows that electron withdrawing substituents on the alkene, such as C=O or C≡N, are not directly involved in π-coordination. The shifts to lower wave numbers of the ν(C=O) and ν(C≡N) frequencies have been attributed to the high perturbation of the alkene double bond upon coordination, and they have been related to the actual electronic density on the metal [57]. For instance, in nickel(0) compounds the shift of the C≡N stretching mode increases with the number of phosphine ligands, so that for $Ni(CH_2{=}CH{-}CN)_2$, $Ni(CH_2{=}CH{-}CN)_2(PPh_3)$ and $Ni(CH_2{=}CH{-}CN)_2(PPh_3)_2$, Δν(C≡N) is 20, 65 and 78 cm^{-1} respectively [63].

An alternative explanation of the great reduction in the C–C bond order in the alkene and alkyne complexes of metals in low oxidation states is the formation of cyclopropane and cyclopropene rings of the type shown in Fig. 1.3.

These models assume that in a V.B. formalism, a compound such as $Pt(PPh_3)_2L$ (L = alkene, alkyne) should be considered to be a derivative of platinum(II), rather than of platinum(0), with a change in the hybridization of the carbon atoms from sp^2 to sp^3 in case (a) and from sp to sp^2 in case (b). This should explain the reduced C–C bond order, and the distortion of the unsaturated organic ligand on coordination to the metal. However, the rigid V.B. formalism takes into account only the extreme situation of the electronic interaction between the metal and the alkene or alkyne. This would be a fairly reliable description of the bond, of the derivatives of very strong π-acids like tetracyanoethylene [64] or fluoroalkenes [65]. In other cases, as has been

Fig. 1.3. (a) Cyclopropane ring formed on the coordination of an alkene on the metal; (b) Cyclopropene ring formed on the coordination of an alkyne on the metal.

suggested by Mason [24, 66] the type of bonding is better represented by considering the participation in the bond of the first excited state of the hydrocarbon, which corresponds to a modern description of the Dewar–Chatt–Duncanson model. The stronger the π-back donation from the metal or the greater the participation of the first excited state of the ligand in the bond, the more the electronic description of the bond approaches the limiting case (a) or (b) in Fig. 1.3. The first excited state of an alkene corresponds in fact to a notable distortion and to a great reduction of the C–C bond order. This type of description is more convenient for describing the intermediate cases that we can have with the different substituents on the olefin, and the effects due to the other ligands bound to the metal, than the rigid V.B. formalism. Moreover, it is not limited to ligands like alkenes, alkynes, or dienes, but can be extended to other very interesting ligands such as O_2 and CS_2, in compounds of the type $Pt(PPh_3)_2(O_2)$ and $Pt(PPh_3)_2(CS_2)$ [66], where they can be considered as strong π-acids attached to the metal through the O=O or C=S double bonds, behaving as an alkene.

With this approach, not only can the distortions and the deformations in geometry of the coordinated alkene, or alkyne, be easily explained (either in terms of electronically excited states [66] or of molecular orbitals [67]), but also the planar coordination of the metal atom as in $Pt(PPh_3)_2$(alkene) complexes. Indeed the fact that the alkene is nearly coplanar with the metal–phosphine plane is probably due, as pointed out by Mason and Robertson [68], to the better π-back donation which can be achieved with this geometry. This is because the *d–p*-mixing, necessary with an alkene perpendicular to the metal–phosphine plane, is not easy owing to the rather high *d–p* energy separation in the platinum(0) atom.

As an example to clarify this theory, let us consider butadiene complexes [24]. The four p_π orbitals of a *cis*-butadiene residue form the basis for molecular orbitals of symmetries and energies listed in Table 1.VII. In simple

TABLE 1.VII
Molecular orbitals in a *cis*-butadiene structure A–B–C–D.
Data taken from ref. 24

Orbital	Energy
$B_2(1) = \frac{1}{\sqrt{1+\lambda^2}} \{A + D + \lambda (B + C)\}$	$\alpha + 1.6\,\beta$
$B_2(2) = \frac{1}{\sqrt{1+\lambda^2}} \{\lambda (A + D) - (B + C)\}$	$\alpha - 0.6\,\beta$
$A_2(1) = \frac{1}{\sqrt{1+\mu^2}} \{(A - D) + \mu (B - C)\}$	$\alpha + 0.6\,\beta$
$A_2(2) = \frac{1}{\sqrt{1+\mu^2}} \{\mu (A - D) - (B - C)\}$	$\alpha - 1.6\,\beta$

valence bond terms, we can postulate the formation of localized metal–ligand bonds in a butadiene complex, such as those shown in Fig. 1.4.

The ligand contribution to the bonding molecular orbital (I) arises from $B_2(1)$ and $A_2(1)$, whereas the σ–π bonded structure (II) has bonding molecular orbitals which have symmetries corresponding to $B_2(1)$, $A_2(1)$ and $B_2(2)$. The

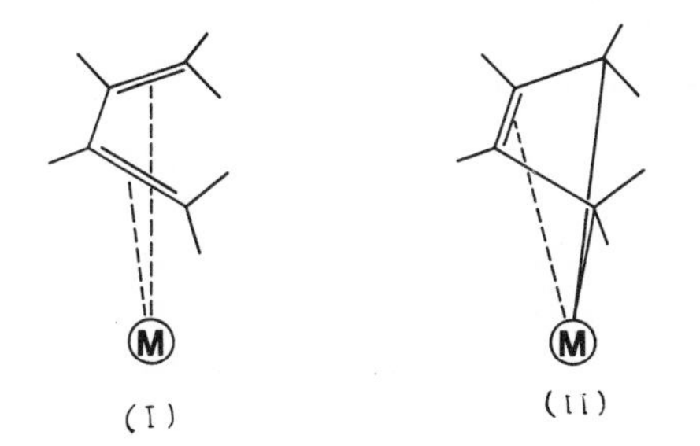

Fig. 1.4. π-bonded(I) and σ–π bonded(II) structures in a butadiene complex.

$B_2(2)$ molecular orbital corresponds to the first excited state of butadiene, and there is a gradual transition from (I) to (II), as the $B_2(2)$ orbital increasingly contributes to the bonding [69].

Different substituents on the butadiene ligand can affect the energy of the $B_2(2)$ molecular orbital with the consequence that structure (II) will be favoured by electron-withdrawing substituents, while structure (I) will be favoured by substituents such as alkyls. The amount of participation of the $B_2(2)$ molecular orbital in the bonding will also depend on the nature of the other ligands bound to the metal in the complex, as has been discussed in detail by Churchill and Mason for a series of butadiene–metal complexes of known structures [24]. A rather similar problem to that discussed above has been presented by the recent discovery of tetraazadiene complexes like $Fe(CO)_3(Me_2N_4)$ [70]. The X-ray structure of this compound [71] cannot be interpreted by either of the two structures reported in Fig. 1.5.

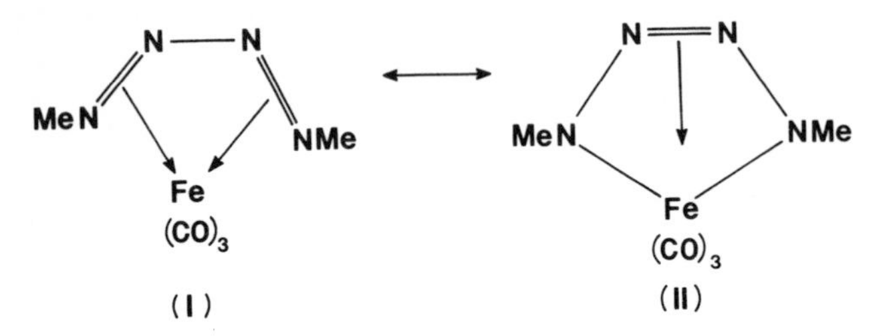

Fig. 1.5. π-bonded (I) and σ–π bonded (II) structures in a tricarbonyliron–tetraazadiene complex.

In the i.r. spectrum no bands assignable to ν(N=N) have been detected, while the ν(CO) pattern is similar to that of the analogous iron–butadiene complex [70]. It has been proposed that π-back donation from the iron d_π orbitals to the π-antibonding orbitals of the tetraazadiene ligand is probably important, in relieving the excess of negative charge from the metal.

1.3.1.3. The Metal–NO, –SO_2, and –O_2 Bonds

In complexes of the series of isoelectronic ligands CO, CN^-, NO^+, N_2 and of the related molecules R–NC, $R{-}C{\equiv}C^-$ and $R{-}N{\equiv}N^+$, the bonding appears to be quite similar in most cases. The ligands usually form an angle of about 180° with the metal, that is they are linearly coordinated through their carbon (CO, CN^-, CNR, $[C{\equiv}CR]^-$) or nitrogen atoms (NO, N_2, $[RN{\equiv}N]^+$)*. This allows extensive overlap between the filled d metal orbitals and the π^* antibonding orbitals of the ligands. It is sometimes said that CO, NO^+, N_2 and CN^- coordinate in a similar manner because they have the same external electronic configuration, $(\sigma_s)^2\,(\sigma_z^*)^2\,(\pi_{y,z})^4\,(\sigma_x)_2$.

The highest occupied bonding orbital has σ symmetry and metal-ligand bond formation using this orbital is energetically preferred [72, 73,

* For arylazo derivatives the M–N–N– unit is linear, but the R–N–N moiety forms an angle of about 120° (see for example: W. E. Carroll, and F. J. Lalor (1973). *J. Organometallic Chem.* **54**, C37).

74]. However for nitric oxide the type of bonding in metal complexes is not so clearly defined as, for instance, for carbon monoxide and the cyanide anion. Nitric oxide is one of the simplest molecules with an odd number of electrons, and it has considerable stability (bond dissociation energy, 151 kcal mol^{-1}). In terms of simple MO theory, the external electronic structure of its ground state is $(\sigma_s)^2$ $(\sigma_s^*)^2$ $(\pi_{y,z})^4$ $(\sigma_x)^2$ $(\pi_{y,z}^*)^1$, where the highest singly occupied MO is an antibonding π orbital. Loss of this electron with formation of NO^+ should thus reinforce the N—O bond, as experimentally observed. In fact the nitrosyl cation has a shorter bond length, a greater dissociation energy (251 kcal mol^{-1}) and a higher vibration frequency (*ca.* 2220 cm^{-1} compared with 1876 cm^{-1}) than does NO. In many complexes where nitric oxide acts as a ligand, we can assume that an electron transfer from NO to the metal preceeds the coordination of NO^+ through the nitrogen lone pair [75]. π-Back donation from the metal into the π^* antibonding orbital of NO^+ might explain why the ν(NO) stretching frequencies in nitrosyl complexes, are markedly lower than in the free nitrosyl cation and in many cases than in nitric oxide itself [75, 76].

The rigid orbital approximation above, where it is assumed that the MO scheme of free NO is preserved during bond formation in a nitrosyl complex, has been recently questioned [77]. It has in fact been pointed out that in the process $NO \rightarrow NO^+ + e^-$ there is a significant reorganization of the charge density in orbitals other than the one involved in the ionization process. Moreover, it was found that the presence of an electron in the π^* orbital of NO does not have a pronounced weakening effect upon the N—O bond, and that only the redistribution of electronic charge following the ionization process can account for the stronger bond in the nitrosyl cation. If so, π-back bonding to the π^* orbital of NO^+ could not explain the lowering of the ν(NO) frequencies in the complexes, although an alternative explanation has not yet been proposed [77].

A discussion of the type of bonding in nitrosyl complexes must also consider another interesting feature of this ligand. Nitric oxide can in fact coordinate to transition metal ions in either a linear (as above discussed) or bent manner. Earlier interpretations of the ν(NO) in the region 1700–1500 cm^{-1} in nitrosyl complexes, were based on the assumption that there was a transfer of an electron from the metal to NO, the NO^- then acting as a normal ligand with donation of two electrons to the metal [78]. Even before, for a long time the only derivative of NO^- was considered to be the red isomer of $[Co(NH_3)_5(NO)]^{++}$ which was later found to have a quite different structure. This compound is in fact the dimeric cation, $\{[Co(NH_3)_5]_2N_2O_2\}^{4+}$, where the hyponitrite ion is asymmetrically bonded in a *trans* conformation to the two cobalt moieties [79], and the i.r. absorptions assignable to the $N_2O_2^{2-}$ ligand are at 1136, 1046 and 932 cm^{-1}. In an accurate determination of the structure of

$[IrCl(CO)(NO)(PPh_3)_2]BF_4$ it has been shown that a bent M—N—O linkages in a metal nitrosyl complex is possible [80]. In this case NO^+ is believed to be coordinated as a Lewis acid accepting an electron pair from the weak base iridium(I). The nitrogen atom must thus have an sp^2 hybridization, with a lone pair occupying one hybrid orbital. In this situation, an Ir—N—O angle of 120° is, expected, in good agreement with the observed value of 124°. Moreover, a long Ir–N bond length of 1.97 Å was clearly found, notably longer than the metal-nitrogen bond distance in nitrosyl complexes wherein NO is bonded in a linear manner.

The ν(NO) at 1680 cm^{-1} found in this complex is in the region where the compounds of the type R–N=O, which are bent, absorb [80]. Although the oxidation state of the metal has little meaning, particularly in this type of compound, nitrosyl complexes where a bent M–N–O linkage is present can be considered to be derivatives of NO^-. It has been in fact pointed out that the following formulations:

are equivalent and may represent two of the canonical forms which may be used to describe the bent M–N–O bonding [81]. This point emphasizes how unrealistic it can be to speculate on a rigid formalism in the actual oxidation state of the metal in nitrosyl complexes. In any case, ν(NO) can be of very little help in discriminating between linear or bent coordination of NO in the metal complex. The iridium–nitrosyl derivative $Ir(NO)(PPh_3)_3$, provides a good example for demonstrating this assumption. In a recent X-ray determination of its structure, a linear Ir–N–O linkage with approximately tetrahedral co-ordination around the metal was suggested as the most likely [82]. Assuming that the nitrosyl ligand is present as NO^+, the metal assumes a formal –1 oxidation state in a d^{10} configuration. This is in agreement with the reported tetrahedral structure. However, ν(NO) in this complex is very low (1600 cm^{-1}) and definitely in a region which was generally accepted as evidence for the presence of a bent M–N–O linkage. We should consider that the high electron density on the metal allows extensive π-back donation to the ligand, with a great reduction in the N–O bond order. In this case a very short metal–nitrogen bond is expected. The Ir–N distance of 1.67(2) Å is in fact among the shortest metal–nitrogen bond distances determined in metal–nitrosyl complexes. At the same time, a corresponding lengthening of the N–O bond occurs, and the value of 1.24 Å for the N–O bond distance is one of the highest yet reported. However, it has been pointed out that direct correlation between ν(NO) and the metal–nitrogen bond length is not always possible. The known NO^+ complexes have ν(NO) in the range 1600–1845 cm^{-1}, while for the NO^- complexes the range is 1525–1720 cm^{-1}, with a large area of overlap (1600–1720 cm^{-1}). In conclusion, a definite distinction between NO^+ and NO^- in metal complexes is not possible using ν(NO) values alone [83].

The bonding description in nitrosyl complexes in terms of NO^+ and NO^- in the linear and bent arrangements respectively, has been more recently considered too formal and simple a view [84]. On considering an orbital correlation between a square pyramid with C_{4v} symmetry (NO linearly bound) and C_s symmetry (NO bent) an important correlation was found. The doubly-degenerate π-bonding level delocalized over the metal d_{xz}, d_{yz} orbitals and the nitrosyl π^* (NO) functions in C_{4v} symmetry, splits in the reduced C_s symmetry into a metal d_{yz} level and a nonbonding orbital localized on the nitrosyl nitrogen atom (84). This corresponds to a change from an orbital having essentially metal character, to one localized on the nitrosyl group.

Such a correlation is directly related to the process involving the transition from a nitrosyl ligand formally coordinated as NO^+ to one which can be considered as NO^-. The energy level ordering in C_{4v} symmetry for a 20 electron system like $M(NO)L_4$ (10 electrons in the ligand σ functions, 4 electrons in the π (NO) set and d^6 metal ion) shows that a significantly σ^* antibonding orbital is left empty. In a 22 electron system like $[Ir(NO)Cl(CO)(PPh_3)_2]^+$ a C_s symmetry is preferred. In this case the σ^* level, which is antibonding with respect to the metal–nitrosyl bond in C_{4v} symmetry, correlates with the largely nonbonding d_{xz} level in C_s symmetry. Interestingly, in a 21 electron system like $[Fe(NO)\text{-}(S_2C_2(CN)_2)_2]^{2-}$, an intermediate structure with a slightly bent nitrosyl group was found [85].

Recently Ibers explained the existence of the NO^+ and NO^- complexes by other considerations [81, 83]. He suggested that for a bent NO^- complex of a d^6 metal ion with square pyramidal geometry, two electrons are mainly localized in a π^*(NO) orbital with the resulting electronic configuration $(d_{xz})^2(d_{yz})^2(d_{xy})^2(\pi^*(NO))^2(d_{z^2})^0(d_{x^2-y^2})^0$. The two electrons are in a doubly degenerate π^* orbital, which results in a triplet ground state. According to the Walsh rules [86], an angular distortion which removes the degeneracy leads to a net gain in energy. The NO^+ species will be favoured when the metal d orbitals are more stable than the π^* (NO) orbitals, as in the case of a metal in a high oxidation state. These two situations correspond to a weak metal–nitrosyl π-bonding. For metals in low oxidation states the metal d and π^* (NO) orbitals have comparable energies. Their strong overlap should raise the π^* (NO) energy, leading to the electronic configuration $(d_{xz})^2(d_{yz})^2(d_{xy})^2(d_{z^2})^2(d_{x^2-y^2})$ $(\pi^*(NO))$ for an $M(d^8)$–NO^+ complex.

A loss of degeneracy of the doubly degenerate π^*(NO) orbitals, with a consequent non-linearity of the M–N–O moiety, has also been attributed to the symmetry about the metal in a complex like $Ni(N_3)(NO)(PPh_3)_2$ [87]. For the highly symmetric π-$C_5H_5Ni(NO)$ complex, the two d_{metal}–π^*(NO) interactions are equivalent, leaving the degeneracy of the π^*(NO) orbitals in the complex. This is not the case for $Ni(N_3)(NO)(PPh_3)_2$, in which the two π^*(NO) orbitals do not belong to the same representation and need not interact equally with the metal *d* orbitals. The last point to discuss about the infrared spectra of nitrosyl derivatives is concerned with compounds having ν(NO) in a region lower than 1500 cm^{-1}. Earlier interpretations for the few complexes which absorb in this region were based on the assumption that the type of bonding was involving or a NO^- or a bridging NO ligand [75]. However, in some cases coordination by the hyponitrite ion, $N_2O_2^{2-}$, was demonstrated by X-ray diffraction studies. The reaction between nitric oxide and $Pt(PPh_3)_3$ in a completely deoxygenated atmosphere affords a compound analysing as $Pt(PPh_3)_2(NO)_2$ [88]. This complex possesses strong absorptions at 1285, 1240 and 1062 cm^{-1}, which are consistent with the presence of a hyponitrite ligand. Moreover, the heterogeneous reaction between $PtCl_2(PPh_3)_2$ and sodium hyponitrite affords a product mixture which clearly shows infrared absorptions at the same frequencies as those observed in the complex $Pt(PPh_3)_2(NO)_2$ [88].

The observed stretching frequencies for hyponitrite complexes are thus very far from the region where nitrosyl complexes absorb. On the contrary a bridging NO ligand may absorb in a region which can have some overlap with the absorptions of derivatives of the formally NO^- group. In fact for $Ru_3(CO)_{10}(NO)_2$, which contains double nitrosyl bridges, ν(NO) are at 1517 and 1500 cm^{-1} (KBr) [88 *bis*].

The amphoteric character of NO is also shown by the SO_2 ligand, which can act either as a Lewis acid or a Lewis base [80*a*]. If the SO_2 were acting as a Lewis acid, one would expect a long metal-sulphur bond and an essentially pyramidal geometry about the sulphur atom. This type of coordination was found in the iridium-SO_2 complex, $IrCl(CO)(SO_2)(PPh_3)_2$ [89]. In contrast, for a complex like $[Ru(NH_3)_4Cl(SO_2)]Cl$ a much shorter ruthenium-sulphur bond and a planar Ru–SO_2 fragment was found [90]. In the i.r. spectra the iridium complex shows ν(SO) at 1048, 1198, 1185, and 559 cm^{-1} and the ruthenium one at 1100, 1301, 1278, and 552 cm^{-1} respectively.

These absorptions can be assigned as the ν_1 sym. stretch., ν_3 asym. stretch., and ν_2 bending of the coordinated SO_2 molecule [91]. Sulphur dioxide shows ν(SO) at 1147, 1330, 1308, and 521 cm^{-1} in the solid state, and the stretching frequencies are lowered upon coordination on a transition metal, as expected. They are lower in the iridium than in the ruthenium complex, and this may be due to the different type of coordination, although any conclusions based on the i.r. spectra should have the same limitations as those for the related nitrosyl derivatives. In the zerovalent platinum-sulphur dioxide complex, $Pt(PPh_3)_2(SO_2)$, $\nu(SO)$ are at 1045 and 1195 cm^{-1} [92], much lower than the values reported for ruthenium, and this could be due to a bent coordination as for the iridium–SO_2 adduct.

It is interesting to note that the NO^- fragment is isoelectronic with O_2. On this basis, one would expect dioxygen complexes, with a bent M—O—O linkage. This type of bonding has been in fact suggested for monomeric cobalt complexes of the type:

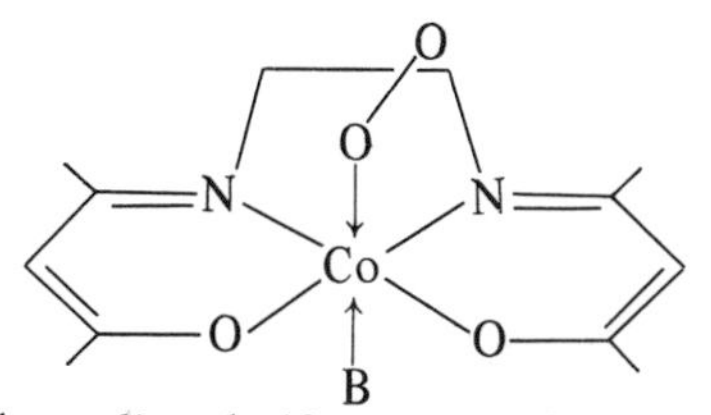

where B is a base such as dimethylformamide [93]. However, on the basis of e.p.r. studies [94], these complexes are better viewed as derivatives of cobalt(III) and the superoxide O_2^- ion, instead of cobalt(II)—O_2 adducts. When dioxygen is bound to very soft centres like $Pt(PPh_3)_2$ in a platinum derivative such as $Pt(PPh_3)_2(O_2)$, the type of bonding is completely different, and they can be considered derivatives of the peroxo O_2 group with structures like O—O over M (a triangle, with both O atoms bonded to M),

or as derivatives of O_2 coordinated in a manner similar to an olefin [66] (see 1.3.1.2). In this type of complex $\nu(OO)$ occurs as a strong band at about 800–900 cm^{-1}. This assignment has been recently confirmed by an i.r. study of the isocyanide and phosphine dioxygen complexes $M(RNC)_2(O_2)$ (M = Ni, Pd) and $Pt(PPh_3)_2(O_2)$, using $^{18}O_2$ [95].

In a cobalt-oxygen complex like the one considered above, $\nu(OO)$ appears as a strong band at 1120–1240 cm^{-1}. These absorptions should be compared with those of gaseous oxygen ($\nu(O_2) = 1555$ cm^{-1}, Raman) and of solid H_2O_2 (1380 cm^{-1}). The very low $\nu(OO)$ in the peroxo type complexes seems to imply a certain mixing of the normal $\nu(OO)$ and $\nu(MO)$ vibrations.

The π coordination of the dioxygen molecule has been attributed to the fact that the bonding π-orbitals, $\pi_{2pz,2py}$, are of higher energy than the σ_{2px} orbital in the dioxygen molecule orbital scheme [96]. The opposite should be true for the NO^- ligand (and perhaps for O_2^-) and therefore NO^- is expected to utilize its σ orbitals in coordinating to a metal. It has also been shown [96] that the σ orbitals for oxygen in its valence state, and presumably also for NO^-, have the same form as filled sp^2 hybrid orbitals on the oxygen atom.

These considerations should exclude the possibility of a bent M–O–O linkage, but in an X-ray structure determination of a monomeric oxygen complex of the type reported above [93], such a bond has been found [93 *bis*]. In this case a superoxo O_2^- ion seems to be present, and we can imagine that in a non-rigid orbital approximation a significant reorganization of the molecular orbitals is involved in the process $O_2 \rightarrow O_2^-$, resulting in a different electronic configuration for the ion.

While CO, N_2 and CN^- do not form complexes where they are present in a reduced form, NO and O_2 are able to accommodate an electron in a ligand antibonding orbital. This is probably due to the much lower energy of the $\pi^*_{2py,2pz}$ orbitals for NO and O_2 than for CO, CN^- and N_2 [72, 73, 74]. For CO, CN^- and N_2 the $\pi^*_{2py,2pz}$ orbitals are of higher energy than metal *d* orbitals [30b, 97, 98]. In NO and O_2 these orbitals are of approximately the same energy as metal *d* orbitals, and in some cases are easier occupied than the higher energy metal *d* orbitals.

1.3.1.4. The Metal–Dinitrogen Bond

The dinitrogen molecule has a very high dissociation energy (225.8 kcal mol^{-1}), which is reflected in a Raman-active vibrational frequency at 2330.7 cm^{-1}. On mixing the 2*s* and 2*p* orbitals on nitrogen because of their relatively low energy separation, we have the configuration

$$(\sigma_s)^2 (\sigma_s^*)^2 (\pi_{y,z})^4 (\sigma_x)^2$$

for the ground state of the molecule (see also 1.3.1.3).

This is confirmed by spectroscopic evidence, which indicates that the odd electron in N_2^+ is in a σ orbital [99]. The highest occupied orbital is thus a σ orbital, and not the degenerate π orbital, as in acetylene, and an end-on coordination, M ← N≡N, is expected, as found for carbon monoxide. This type of coordination is so far the only one observed in the crystal structures of metal–dinitrogen complexes. The first empty orbital is the degenerate antibonding orbital $\pi^*_{2py,pz}$.

The dinitrogen molecule has been considered as a potential ligand for many years, but dinitrogen complexes were first prepared only in 1965. Orgel had suggested that a simultaneous transfer of electrons from the σ orbital of the ligand to the metal, and from the metal to a π orbital of the ligand, should correspond to bring the molecule, whether carbon monoxide or dinitrogen, in an excited state in which a n–π transition has taken place; the energy of n–π upper state of CO being of about 6 eV in comparison with 7.3 eV for for N_2. This suggestion was used for explaining the inability of nitrogen to act as ligand analogously to carbon monoxide [13]. The recent isolation of dinitrogen complexes demonstrated that this hypothesis is not valid. However, dinitrogen complexes are generally less stable than the corresponding carbonyl derivatives, and up to now a pure metal–dinitrogen complex such as $M(N_2)_x$ has not yet been isolated. In fact other ligands must be present in order that N_2 may coordinate to a metal atom.

As expected nitrogen is a less effective σ donor than carbon monoxide. Its σ_x orbital is in fact of lower energy than the corresponding orbital of CO by *ca*. 1.7 eV [74]. Moreover, since the π^* orbital of N_2 is 0.3 eV higher than the antibonding orbital of CO, it is to be expected that a metal requires a higher charge density to interact satisfactorily with N_2 than with CO. Furthermore, the π^* orbitals of CO overlap more efficiently with metal *d* orbitals, since they are derived from 68% C_{2p} and 32% O_{2p}, and the latter are more diffuse towards the metal than the corresponding nitrogen orbital [100].

On the other hand the low reactivity of N_2 has been attributed by Chatt essentially to the energetically unfavourable state of the highest occupied orbital, which has an energy of −15.5 eV and it is much more stable than the corresponding orbitals of CO and the degenerate pair of π orbitals or acetylene (−11.4 eV) [101].

Infrared studies of dinitrogen chemisorbed on nickel, palladium and platinum show absorptions at around 2220–2000 cm^{-1} [102]. This suggests an electronic interaction with the absorbed nitrogen, and "coordination" to the bulk metal, similar to that found in dinitrogen complexes. The relatively high extinction coefficient of the observed i.r. band is an indication of a considerable polarization of the adsorbed N_2 molecule. The first dinitrogen complex isolated, $[Ru(NH_3)_5N_2]^{2+}$, has a strong i.r. absorption at 2118–2167 cm^{-1}, depending on the anion [103]. All other complexes having N_2 as a non-bridging ligand show absorptions ranging from *ca*. 2150 to 2000 cm^{-1}. It follows that molecular nitrogen coordinated in this type of complex has a weak electronic interaction with the metal or at least an interaction comparable to that found in the dinitrogen chemisorbed on a metal surface. This situation is not dissimilar to that found in terminal metal–carbonyl complexes. Free carbon monoxide absorbs at 2143.3 cm^{-1}, while carbon monoxide chemisorbed on nickel shows a band at 2035 cm^{-1} which has been attributed to the linearly absorbed M–CO molecules [104]. In metal complexes, terminal carbon monoxide has stretching frequencies ranging from *ca*. 2100 to 1900 cm^{-1}. The i.r. evidence thus seems to suggest a greater electronic interaction for metal–nitrogen than for metal–carbon monoxide, since the variation $\Delta\nu(CO) = \nu(CO)_{free}$-$\nu(CO)_{terminal}$ is less than the corresponding $\Delta\nu(N_2)$ values. This seems to contradict what we have discussed above, on the basis of the relative MO energies for these ligands, and the inertness of nitrogen to reduction, even when it is coordinated to a transition metal in a low oxidation state. However, as we have seen for all the types of bonding previously discussed, i.r. data do not appear to be conclusive regarding the strength and the type of metal–ligands interaction. Particularly significant seems to be the fact that in some cases, stretching frequencies and force constants do not lead to the same conclusions. Moreover, it has been found by means of overlap population calculations, that the π^* orbitals on N_2 are more destabilizing to the N≡N bond than those of CO are to the C≡O bond [105].

Therefore an equal amount of π-back-donation should result in a greater reduction in the N≡N than in the C≡O bond order. Chatt has considered that the disagreement between infrared data and stability should be mainly attributed to poor σ-donor property of nitrogen compared with CO, since the frequency shifts are dependent on both the σ and π properties of the ligand [106]. Removal of electron density from the orbitals of CO and N_2, as a consequence of the σ donation to the metal, in fact results in an increase in σ bond strength for these molecules [97].

These orbitals are slightly antibonding with respect to the σ framework of the two ligand species, and the antibonding character is larger for carbon monoxide. The balance between σ bond strengthening and π bond weakening must thus be considered in the interpretation of i.r. data, and this new point of view can now explain the apparently anomalous behaviour of dinitrogen. Interestingly the force constant for the CO^+ ion is greater than for the corresponding neutral molecule [35] (Table 1.VIII).

TABLE 1.VIII

Force constants (mdyn Å^{-1}) for neutral molecules and related ionic species. Data taken from ref. 35

	K_f(mdyn Å^{-1})
N_2	22.97
N_2^+	20.10
CO	19.02
CO^+	19.80
CN^-	16.9
HCN	18.8

For either CO and CN^-, the σ and π bonds are polarized towards the more negative oxygen or nitrogen atom. When CO loses an electron, the bonds are less polarized and therefore stronger [107]. Similarly, when the CN^- group bonds to other atoms through the lone pair on carbon, the negative charge on the cyanide group is decreased and the C≡N bond becomes more symmetrical and thus stronger [108]. This implies that a pure σ bond should raise the stretching frequencies both of CO and CN^-. This is in agreement with what has been discussed above, but this interpretation is not extensible to nitrogen, which is not polarized in the neutral molecule and for which the loss of one electron results in a *decreased* force constant.

Binuclear complexes in which the nitrogen molecule forms a bridge between two transition metals have also been isolated. In $\{Ni[P(C_6H_{11})_3]_2\}_2N_2$ for instance, one would expect a double weakening of the N≡N bond, but in this case nitrogen remains unreactive [109]. For these symmetrical complexes the ν(N≡N) is only Raman active as in the free ligand, and, at the moment, we cannot follow the decreased nitrogen bond order through its ν(N≡N) as only few data of this type are available [99b].

However, the X-ray structure of $\{[(C_6H_{11})_3P]_2Ni\}_2N_2$ has shown that in the linear Ni–N–N–Ni system, the N–N distance (1.12 Å) is only slightly longer than in free nitrogen (1.0976 Å) and compares well with other known distances of monodentate dinitrogen complexes (1.12–1.16 Å), without any additional lengthening due to the bis-coordination [109b]. In this particular compound the nitrogen molecule is enclosed in a cage formed by four of the cyclohexyl rings, and this seems to be the main reason for the stability of this complex.

Although the number of dinitrogen complexes is continuously growing, many more data are still needed in order to allow more significant comparisons with other related complexes. In particular, it seems that the formation of dinitrogen complexes is related to a very delicate balance between dinitrogen and the other ligands bound to the same metal atom. For instance, while $IrCl(N_2)(PPh_3)_2$ is very easy to prepare, the corresponding iodide or similar complexes with other phosphines appear to be too unstable to be characterized [99, 101].

Finally, we must conclude that rough parameters such as stretching frequencies should be regarded as giving merely a qualitative approach to the understanding of the metal–ligand electronic interaction. In any case force constants should be a more realistic parameter, although we must remember that they are related to the curvature of an energy potential function and not only to bond dissociation energies. Moreover, the i.r. parameters are the final result of the actual bonds in a given molecule and cannot, for instance, discriminate between σ and π contributions which in some cases work in opposite directions with respect to the strengthening or lengthening of a bond within the ligand.

Meaningful discussions are reliable only for i.r. data in a strictly related series of compounds, having the same geometries and the same types of ligands, while a comparison between the i.r. data of two different ligands such as dinitrogen and carbon monoxide, even when these are bound to the same complex moiety, can be very misleading.

In recent communications it has been proposed that the integrated infrared intensities (from which the dipole moment derivatives can be calculated) be used as a measure of π-electron delocalization in the ligand [109, *bis*]. Using this parameter, it has been confirmed that N_2 is a less effective σ-donor and π-acceptor than CO, as expected from theoretical considerations.

1.3.2. n.m.r. Spectroscopy

1.3.2.1. Chemical Shifts

Proton chemical shifts have been tentatively used as an indirect measure of π-back donation from the metal to ligands such as alkenes [57] and quinones [110] in compounds of formula $Pt(PPh_3)_2L$ (L = alkene, quinone). A large shift to high field of the signals of the alkene protons has been observed ($\Delta\tau$ *ca.* 3, L = alkene; $\Delta\tau = 1.7$, L = *para*-benzoquinone). Such a large shift, typical of alkene complexes of low-valent transition metals, is practically absent in platinum(II)-alkene derivatives, while in silver(I)–alkene complexes, where the σ contribution to the metal–alkene bond is rather high, a slight shift to low field is observed [111]. These effects seem to support an enhanced contribution of the π-bond for zerovalent platinum derivatives, in accord with the previous discussion of the i.r. spectra (1.3.1.2). However, the same effects should be expected if the zerovalent metal alkene derivatives form a cyclopropane ring (Fig. 1.3 (a)), where the alkene hydrogens are bound to an essentially sp^3 hybridized carbon in the complex.

N.m.r. spectroscopy can also be used with fluorine compounds, because of the spin $I = \frac{1}{2}$ of the ^{19}F nucleus. Chemical shifts are notably larger with fluorine atoms than with protons, and a substantial deshielding of the ^{19}F nucleus in fluoroolefin-metal complexes relative to that in the free ligands has been observed [112].

In a tetrafluoroethylene complex such as $IrCl(CO)(PPh_3)_2(C_2F_4)$ two multiplets are present, the lowest *ca.* 20 p.p.m. to low field of free tetrafluoroethylene. However in $Ni(PPh_3)_3(C_2F_4)$ a much lower downfield shift is present (*ca.* 2 p.p.m. to low field of free tetrafluoroethylene) and this has been taken as evidence that the metal–alkene bond in this case is better interpreted by the Dewar–Chatt–Duncanson model than by formation of a rigid cyclopropane ring.

The observed chemical shifts are in all cases at higher fields (+5 p.p.m.) than those of the α fluorines in $C_4F_8Fe(CO)_4$, prepared from $Fe(CO)_5$ and C_2F_4, which was found to contain an octafluorotetramethyleneiron ring rather than two fluoroalkenes η-bonded to the iron atom [113]. As we will discuss later for fluorophosphine complexes, fluorine chemical shifts are the result of many different effects which do not allow a simple discussion of their values in terms of the metal–alkene bond. A general feature of the ^{19}F n.m.r. spectra of fluorophosphine–metal complexes is the appearance of the resonance at much lower field than the free ligand value [45]. This low field shift on coordination is similar to that observed for perfluoroalkyl-transition metal complexes [114]. It has been found that ^{19}F chemical shifts δ_F, in the series $F_2P{-}X$, vary with the electronegativity of the atom X directly attached to the phosphorus atom [115, 116]. The observed decrease of δ_F with increasing electronegativity of the X ($X = CH_3$, $\delta_F = +92.9$ p.p.m.; X = F, $\delta_F = +34.0$ p.p.m.; $CFCl_3$ as internal reference) can be explained mainly by electronegativity considerations, analogous to those employed for the interpretation of ^{1}H chemical shifts in series such as CH_3–X. Beside the electronegativity of X, the paramagnetic term, due to the presence of low-lying excited states, may contribute to fluorine chemical shifts [117]. For fluorophosphines coordinated to nickel in a series of complexes of the type NiL_4 (L = PF_2X) [116], a downfield shift $\Delta\delta_F$ of *ca.* 20–30 p.p.m. is observed. The higher the chemical shift of fluorine in the free ligand, the larger the low-field shift of fluorine on complexation. There is a linear correlation between $\Delta\delta_F$ and the electronegativity of the substituent X, the greatest effects being observed for the least electronegative substituents. Since the slope of the correlation ($\Delta\delta_F$ in NiL_4 versus electronegativity of X) was found to be identical to that for the related molybdenum compounds, $Mo(CO)_3L_3$, it was concluded that the π-bond contribution to δ_F is negligible, while the anisotropy of the metal atom, the paramagnetic term, and the changes in bond angles (see later), are expected to be responsible for the observed changes in δ_F. However, effects due to the metal and to its formal oxidation state on the values of δ_F have not been studied to any great extent.

Proton n.m.r. spectra of alkylphosphines coordinated to a transition metal in series of compounds such as $M(CO)_5L$ [M = Cr, Mo, W; L = PPh_2Me, PPh_2Et, $PPh_2(Bu^t)$] showed that the proton chemical shifts move downfield upon coordination as is expected from the greater electronegativity of phosphorus in the coordination compound compared with that in the free phosphine [118].

As with fluorine, the phosphorus nucleus ^{31}P (100% natural abundance) has a nuclear spin I = $\frac{1}{2}$ and ^{31}P nuclear magnetic resonance studies now occupy a very important place in the development of coordination and organometallic compounds with trivalent phosphorus ligands. The first ^{31}P n.m.r. study of metal complexes having tricovalent phosphorus as ligands was reported many years ago for a series of nickel(0) derivatives [121]. The trivalent phosphorus compounds where phosphorus is bound to an electronegative atom such as O, N, S, or halogen usually have δ_P between –95 and –250 p.p.m. (85% aqueous H_3PO_4 as reference) [119]. For PH_3, δ_P = +241, while for the trialkyl- or triaryl-phosphines, δ_P are slightly above or below zero (Table 1.IX) [116, 119, 120, 121, 128 *bis*, 152]. In general no convenient quantitative correlation has yet been observed between δ_P and the nature of the atoms or groups bonded to phosphorus. As a matter of fact no single parameter appears to dominate δ_P although for a series of fluorophosphines PF_2X, δ_P increases (shifts to higher field) with increasing electronegativity of the atom X directly attached to the phosphorus atom, this behaviour being unrelated to any single hypothesis [115, 116].

When a similar correlation was attempted using the Taft σ^* parameters for X, irregular variations were observed even with fairly closely related compounds [45].

TABLE 1.IX

Chemical shifts $\delta_P{}^a$ for some trivalent phosphorus compounds.

Phosphine	δ_P (p.p.m.)
PH_3	+ 241
$P(CH_3)_3$	+ 61.6
$P(C_6H_5)(CH_3)_2$	+ 47.55
$P(n\text{-}C_3H_7)_3$	+ 33.0
$P(n\text{-}C_4H_9)_3$	+ 32.6
$P(C_6H_5)(C_4H_7)_2$	+ 30.24
$P(CH_3)_2(t\text{-}C_4H_9)$	+ 28.66
$P(C_6H_5)_2(CH_3)$	+ 27.68
$P(C_6H_5)(n\text{-}C_3H_7)_2$	+ 27.34
$P(C_6H_5)(n\text{-}C_4H_9)_2$	+ 26.02
$P(C_2H_5)_3$	+ 20.4
$P(C_6H_5)_2(n\text{-}C_3H_7)$	+ 17.6
$P(C_6H_5)_2(n\text{-}C_4H_9)$	+ 17.1
$P(C_6H_5)(C_2H_5)_2$	+ 16.2
$P(C_6H_5)_2(C_2H_5)$	+ 12.0
$P(n\text{-}C_3H_7)_2(t\text{-}C_4H_9)$	+ 8.74
$P(C_6H_5)_3$	+ 5.9
$P(n\text{-}C_4H_9)_2(t\text{-}C_4H_9)$	+ 4.38
H_3PO_4	0.00
$P(C_2H_5)_2(t\text{-}C_4H_9)$	− 6.86
$P(C_6H_5)_2(t\text{-}C_4H_9)$	− 17.32
$P(n\text{-}C_3H_7)(t\text{-}C_4H_9)_2$	− 26.33
$P(C_6H_5)(t\text{-}C_4H_9)_2$	− 37.99
PF_3	− 97.4
P_4O_6	− 113
$P(SC_2H_5)_3$	− 115.6
$P(SC_3H_7)_3$	− 118.2
$P\{N(C_2H_5)_2\}_3$	− 118.2
$P\{N(CH_3)_2\}_3$	− 122.0
$P(OC_6H_5)_3$	− 126.8
$P(O\text{-}p\text{-}C_6H_4CH_3)$	− 127.6
$P(OC_2H_5)_3$	− 136.9
$P(i\text{-}OC_3H_7)_3$	− 136.9
$P(OCH_3)_3$	− 140.0
$PCl_2(C_6H_5)$	− 161.6
PI_3	− 178.4
$PCl_2(OCH_3)$	− 180.5
$PCl_2(CH_3)$	− 191.2
$PF_2(C_6H_5)$	− 208.3
PCl_3	− 219.4
PBr_3	− 227.4
$PF_2(CH_3)$	− 250.7

[a] These δ_P values could be slightly different from similar data reported in the literature because different solvents were used. Data taken from ref. 116, 119, 120, 121, 128 *bis* and 152.

The temperature independent paramagnetic contribution, which arises from a field-induced mixing of the ground state with low-lying excited states, seems to be dominant in determining ^{31}P chemical shifts.

Moreover, when the substituent groups either have π-electrons (e.g. phenyl) or unshared electron pairs (e.g. F, N, O) there is the possibility of formation of a partial double bond with the phosphorus atom. The effect of π-bonding will be to further lower δ_P. Another factor which may contribute to changes in δ_P, for example in PF_2–X compounds, is related to the polarization of the lone pairs on X, which lowers δ_P.

On considering the series of phosphines reported on Table 1.X, it is evident that any consideration based solely on the inductive effects of the substituents cannot explain the behaviour of δ_P for all the ligands considered. In particular, for chlorophosphines changes in δ_P are scattered, and this may reflect the relative importance of Cl–P π-bonding in such compounds.

TABLE 1.X

δ_P (p.p.m.) for different series of phosphines. Data taken from Table 1.IX

Phosphine	δ_P (p.p.m.)
$P(OC_6H_5)_3$	− 126.8
$P(C_6H_5)_3$	+ 5.9
$P(CH_3)_3$	+ 61.6
PF_3	− 97.4
$PF_2(C_6H_5)$	− 208.3
$PF_2(CH_3)$	− 250.7
PCl_3	− 219.4
$PCl_2(C_6H_5)$	− 161.6
$PCl_2(CH_3)$	− 191.2
PF_3	− 97.4
PCl_3	− 219.4
$P(CH_3)_3$	+ 61.6

There are several features which can influence the ^{31}P n.m.r. parameters in complexes [116, 121]. A pure σ bond from phosphorus to the metal should result in a decrease in shielding of the phosphorus nucleus due to the decreased electron density on the atom. The effect of d_π–d_π bonding on δ_P from the metal to phosphorus has been considered from two opposing viewpoints. While Meriwether and Leto [121] assume that a π-bond should increase the phosphorus shielding because of a drift of electrons from the metal to phosphorus, Reddy and Schmutzler [116] attribute a negative contribution to δ_P to this type of bonding, synergically reinforced by the σ bond, because of the increase in the paramagnetic term. Among the other factors it has been considered that the anisotropy of the metal atom and the polarization of the lone electron pairs should have only minor importance [116]. The increase in the coordination number of phosphorus corresponding to a change in hybridization, and the changes in the bond angles of the ligand on complexation, could affect δ_P in a critical way [116, 119]. It may be interesting to compare the shift in δ_P by increasing the coordination of phosphorus, both by complexation to low-valent metals and by the formation of dative bonds with electron deficient molecules.

An increased coordination number for phosphorus generally results in an increased δ_P, both for phosphines with organic substituents and for fluorophosphines [45, 116]. Analogously, δ_P in dialkylaminofluorophosphine complexes with boron σ-acceptors occur at higher field than in the free ligands [122]. The same happens in the series $P_4O_6 \cdot (BH_3)_x$ (x = 1, 2 or 3), where positive shifts of 22.6, 14.3, and 9.4 p.p.m. respectively for the resonance of the coordinated phosphorus atoms were observed [123], the shifts decreasing from $x = 1$ to $x = 3$.

These shieldings can be attributed to the decreased inductive effects of oxygen or fluorine on δ_P, as a consequence of the pure σ-donation to the BR_3 acceptors. In contrast, when P_4O_{10} is bound to a transition metal as in the series of complexes $P_4O_{10}\{Ni(CO)_3\}_x$ (x = 1, 2, 3, 4) a net downfield shift of about 13–20 p.p.m. is observed, which slightly increases from $x = 1$ to $x = 4$ [124]. This may be due to a decreased σ-bond strength in the nickel derivatives in comparison with the boron compounds, and/or to the presence of a significant amount of $d\pi$–$p\pi$ bonding between phosphorus and nickel.

However, it appears very difficult to discriminate clearly between the effect on δ_P due to a d_π–d_π metal phosphorus bond and the other factors which can contribute to δ_P.

Very good correlations were found between ^{31}P chemical shifts in $M(CO)_5(PPh_2R)$ compounds (M = Cr, Mo; R = alkyl) and the group contribution of R to the chemical shift of the free tertiary phosphines [118]. This is a rather surprising result, as it implies that the metal–phosphorus interaction is practically insensitive to the nature of R through the whole series of compounds for a given metal.

Except for the chlorophosphines, δ_P is generally shifted to lower field in the complexes of metals in zero oxidation states in comparison with the free ligand. [125]. In a series of nickel(0) complexes, $Ni(CO)_2(PR_3)_2$ (R = alkyl, aryl), a nearly constant deshielding of δ_P in comparison with the free ligand ($\Delta\delta_P = -42 \pm 3$ p.p.m.) was observed [121]. It was assumed that the major effect on δ_P is the strong σ-donor bond from P to Ni, while any contribution from d_π–d_π back donation from Ni to P must be either very small or constant throughout the series. However, it appears rather unrealistic to assume that two ligands such as

$P(C_6H_5)_3$ and $P(C_2H_5)_3$, for instance, are equivalent in their σ donor properties towards the same acceptor. Considerable anomalies were observed in series of complexes such as $Ni(CO)_{4-x}(PR_3)_x$ ($x = 1-4$) [121]. Only for R = Cl does δ_P regularly increase from $x = 1$ to $x = 4$, but with chlorophosphines δ_P is higher in the complexes than in the free ligands. For R = OC_2H_5 there is a decrease in δ_P, but the compound with $x = 4$ does not fit this trend. For R = C_6H_5 or C_2H_5 while δ_P is lower than in the free ligands, it increases on going from $x = 1$ to $x = 2$. Finally, for $x = 4$, when R = Cl, F, or OC_2H_5, and for the mixed $PCl_2(C_6H_5)$ ligand, δ_P values are scattered. This behaviour is difficult to rationalize and it may also be determined by different amounts of bond rehybridization of the phosphine ligands on coordination, particularly for the chlorophosphines [116, 121].

However, it appears that with ligands usually considered as weak σ donors and strong π acceptors [PCl_3, $P(OC_2H_5)_3$], there is little variation in δ_P with increasing substitution, while stronger effects are found with better σ donors and weaker π acceptors as PPh_3 and PEt_3. This possibly reflects the minor importance of their π bond in respect to the σ bond in the metal–phosphorus electronic interaction.

It has also been observed that by changing the stereochemistry of the complex and the oxidation state of the metal the effects on δ_P are not very relevant [121]. For instance, for the square planar nickel(II) NiL_2X_2 complexes (L = $P(n-C_4H_9)_3$; X = Cl, $C{\equiv}C{-}C_6H_5$), δ_P is not greatly different from the δ_P value in the tetrahedral nickel(0) complexes $NiL_2(CO)_2$. Some other results obtained with L = $P(C_6H_5)_3$ were not rationalized at all, as the great differences on δ_P observed by changing the metal, its oxidation state and the stereochemistry of the complexes do not allow any apparently simple explanation.

In a more recent work where 60 derivatives of the type $M(CO)_{m-n}(PR_3)_n$ (M = Ni, $m = 4$; M = Cr, Mo, W, $m = 6$; $1 < n < 4$; R = F, Cl, OCH_3, SCH_3, $N(CH_3)_2$, CH_3, C_2H_5) have been studied, the following general conclusions were reached [126]: (i) the chemical shift of coordinated phosphorus decreases with increasing substitution, but this tendency is weaker or even reverses when $n > 3$; (ii) for metal carbonyl derivatives of Group VI, δ_P for phosphorus *trans* to CO occurs at higher field than for phosphorus *trans* to another phosphorus; (iii) the nature of the central atom seems to be the major factor on determining δ_P values of the coordinated phosphines. However, it was concluded that from so large a number of data it is impossible to distinguish between σ and π-bonding effects upon the δ_P values, since both types of bond should cause deshielding of phosphorus once coordinated to a metal.

By studying the phosphorus chemical shifts in a series of complexes having phosphites as ligands with different metals in a d^{10} configuration (Ni^{0}, Cu^{I}, Ag^{I}, Hg^{II}), a qualitative correlation of δ_P with the metal oxidation number was observed [127]. For Cu^{I}, Ag^{I} and Hg^{II} a net shielding was observed. Such shielding becomes progressively greater from the +1 to the +2 state. The same is true for nickel(II) derivatives. These shieldings were attributed to the decreased inductive effects of oxygen in the phosphites as a consequence of a pure σ-donation to a metal in a relatively high oxidation state. In contrast, the high electron density in $Ni^{(0)}$ allows d_π–d_π back bonding to become important, and this should explain the more negative δ_P values in the nickel(0) complexes than in the free ligand.

More recently it was shown that the correlation between the oxidation state of the metal and the ^{31}P resonances in complexes is strictly valid only for systems which are both isoelectronic and isostructural [128]. In such systems in fact all other factors which can influence δ_P are minimized. Discrepancies were observed in complexes having the same ligand but different stereochemistries, and with metals in different oxidation states. In fact some phosphite complexes were shown to have δ_P lower than in the free ligand, even when the metals are in a +1 or +2 oxidation state. Moreover, in some cases a reverse correlation of δ_P with the oxidation state of the metal was found. For instance δ_P is greater in NiL_4 than in $[CoL_5]ClO_4$ ($L = P(OCH_2)_3CR$), and also in $[AgL_4]NO_3$ compared with $[NiL_5](ClO_4)_2$.

It is interesting to note that for two palladium(II) complexes of the type *cis* and *trans*-PdL_2Cl_2 (L = phosphine), both upfield and downfield shifts in respects to the free ligand were observed with different L [152]. However, in spite of the different factors which can affect δ_P, it has been shown that it is possible to predict the coordination shift, $\Delta\delta_P$, of a tertiary phosphine from the equation $\Delta\delta_P = A\ \delta_P + B$, where A and B are constants when homologous series of complexes are considered [128 *bis*]. This equation holds for complexes such as $M(CO)_5L$ (M = Cr, Mo, V), *trans*-$MCl(CO)L_2$ (M = Rh, Ir), $Ni(CO)_2L_2$, etc. (L = phosphine), and the deviations from the observed $\Delta\delta_P$ are within a few p.p.m. The downfield shift upon complex formation is greater the higher the δ_P of the free phosphine.

As found for ^{19}F chemical shifts (see above), once coordinated to nickel in a series of complexes like NiL_4 [116], a ^{31}P downfield shift $\Delta\delta_P$ of *ca.* 20–35 p.p.m. was observed for the fluorophosphine ligands F_2PX.

The higher the chemical shift of phosphorus in the free ligand, the greater its low-field shift on complexation. In this case, there is a linear correlation between $\Delta\delta_P$ and the electronegativities of the substituent X, the greatest effect being shown by the most electronegative substituent X; this is opposite to the behaviour of $\Delta\delta_F$. This seems to be due, at least in part, to a d_π–d_π interaction between the metal and the phosphorus atom, which increases the paramagnetic contribution, and which is favoured by the most electronegative substituents.

This effect is more evident for NiL_4 complexes than for related $Mo(CO)_3L_3$ derivatives, that is the slope of the correlation between $\Delta\delta_P$ and the electronegativities of X, for the former is more negative.

From the above discussion it can be seen that there should be a direct correlation between $\Delta\delta_P$ and the carbonyl stretching frequencies (cm^{-1}), since both should increase in compounds such as $Ni(CO)_2(F_2PX)_2$ and $Mo(CO)_3(F_2PX)_3$ on increasing the electronegativity of X. This was in fact observed [116]. The opposite effect is noted as fluorine is successively substituted by less electronegative atoms or groups, as expected.

The steric effects of bulky substituents R such as the *t*-butyl group on the ^{1}H, ^{19}F, ^{31}P n.m.r. spectra for ligands of the type $PF_{3-n}R_n$ (n = 0–3) have been investigated and also for their nickel, chromium, molybdenum and tungsten carbonyl derivatives [129]. This work summarized very well how experimental δ_P values can be used to predict bond angles within the free phosphine ligand. The effect of the bulky substituent R on δ_P, which have much lower negative values when n = 1 or 2, have also been rationalized. The most important results obtained with metal carbonyl complexes of such ligands are mainly related to coupling constants, and this will be discussed in the next section.

In conclusion, chemical shifts, particularly for nuclei like ^{1}H, ^{19}F and ^{31}P, can give some information about the metal–ligand bond. However, these shifts are the result of many different factors operating together, and at present they cannot be interpreted unambiguously, least of all for solving the problem of detection of π-back donation. In some cases they seem strongly to support the existence of such a bond, but at the moment these results are not conclusive.

1.3.2.2. Coupling Constants

Parameters such as coupling constants should give direct information about the σ interaction between the metal and the ligand, when nuclei directly involved in the metal ligand bond such as ^{31}P, ^{13}C, ^{183}W, or ^{195}Pt are considered. These coupling constants can be considered to be a function of the *s* orbital participation to the bond. Since only the *s*-electron wave functions have a finite value at the nucleus it is generally supposed that a coupling between two atoms, involved in a bond, can be transmitted only through the σ-electronic interaction, in which the *s* orbitals can participate. Coupling constants between nuclei of different ligands bound to the same metal will be also examined and we will consider how coupling constants between nuclei within a ligand change upon complexing of the ligand to a transition metal.

Tungsten-183 (nuclear spin 1/2, natural abundance 14.3%) causes splitting of the phosphorus signals in its phosphine derivatives, and this type of interaction has been widely studied. In terms of the synergic interaction between σ and π bonds, the behaviour of J_{W-P} coupling constants in a wide series of $W(CO)_5PR_3$ compounds has been interpreted as an indication of the strengthening of σ bonds by π interactions [118, 130]. The less basic ligands have the larger P–W coupling constants (for R = Bu^tn, J_{W-P} = 200 Hz, while for R = OPh J_{W-P} = 411 Hz). If only the σ interaction was involved in the metal-ligand bond one would expect the best σ-donors to have the largest coupling constants, since the coupling should be transmitted through orbitals having *s* character. At the same time, a relationship was found between J_{W-P} and ν(CO) (E mode) and it was suggested that J_{W-P} is a better measure of π-acceptor ability (or inversely as σ-donor ability) than ν(CO). As an alternative explanation it was also suggested that back-donation by tungsten 5*d* electrons to phosphorus could render the Fermi contact mechanism more effective by deshielding the 6*s* bonding electrons of tungsten. The shielding of the 3*s* bonding electrons on phosphorus as a consequence of the increased electronic charge in the 3*d* orbitals should in fact be less important, because the same quantum number is involved in the case of phosphorus. In a subsequent study, Grim and Wheatland considered J_{W-P} for a series of disubstituted $W(CO)_4(PR_3)_2$ compounds, with a *cis* and *trans* stereochemistry [131]. After widely reviewing the i.r. evidence reported for and against π-bonding between phosphorus and transition metals, they considered that phosphorus–tungsten coupling constants are a more reliable parameter for detecting the existence of a π-bond, as they arise directly from the bonds in question. Values of J_{W-P} for *trans* compounds were always found to be larger than J_{W-P} for *cis* derivatives, although they do not change appreciably from ligand to ligand for compounds of the same stereochemistry (for R = Bu^tn, $J_{(W-P)cis}$ = 225 and $J_{(W-P)trans}$ = 265 Hz). This was explained assuming that phosphorus competes more effectively for tungsten $d\pi$ electrons *trans* to itself in the *trans* isomers, compared to the carbon monoxide *trans* to it in the *cis* isomer. It was also observed that the ^{31}P chemical shift of the *trans* isomer is considerably downfield from the *cis* isomers, and this could be in agreement with the preceding discussion of δ_P values (1.3.2.1). However a greater J_{W-P} should then be expected for the disubstituted than for the monosubstituted tungsten derivatives. This is always the case for the *trans*-derivatives and for the *cis* and *trans* PBu^{tn}_3 derivatives (it must be noted however that for the monosubstituted butyl derivative a value for J_{W-P} of 227 Hz was reported [134], 27 Hz greater than the previous value) but the opposite is true for the other *cis*-derivatives. This may reflect the relatively greater importance of the ligand *trans* to phosphorus in the determination of the amount of metal-phosphorus π-back donation compared with the degree of substitution in the complexes. Although this technique appears to be very promising for solving the problem of the existence of a metal–phosphorus π-back-donation, more recent results were found to be in net contrast to that reported above, and this reminds us of the situation reached by using only i.r. data in order to solve this problem (1.3.1.1). By studying the series $W(CO)_5PR_3$, where phosphines of constrained structure were used, Keiter and Verkade have found that J_{W-P} increases with the electronegativity of the atoms directly bound to phosphorus in the ligand, and a very good quantitative correlation was observed [132]. This can be simply explained by considering that phosphorus *s* character concentrates in orbitals which are directly towards the most electropositive substituents [133], and this explanation does not require any π-bonding as a coupling mechanism. Moreover, no significant correlation was found between J_{W-P} and ν(CO), force constants, or the $\Delta\sigma$ and $\Delta\pi$ parameters derived by Graham as a measure of the σ and π-bonding capacities of the phosphine ligands [38]. Similar results were obtained from a study of J_{W-P} as a function of the degree of substitution in the complexes in analogous tungsten derivatives [134]. The coupling constants appear to be determined mainly by the nature of the group *trans* to phosphorus, and by the electronegativity of the substituent groups on phosphorus which directly affect the value of $|Sp(O)|^2$ (the magnitude of the valence state *s*-orbital of phosphorus evaluated at the nucleus) and α^2 (the *s* character of the phosphorus lone-pair hybrid) terms of the equation which gives the phosphorus-metal coupling constant. Both terms increase with increase in the electronegativity of the substituents.

In support of the view that the correlation between $J_{(^{183}W-^{31}P)}$ and the CO stretching frequency in the tungsten complexes, when observed, can be interpreted by a mechanism involving σ-bonds only, a very good linear relation between $J_{(^{183}W-^{31}P)}$ in $W(CO)_5PR_3$ compounds and $J_{(^{31}P-H)}$has been found, as determined for the corresponding phosphine ligands dissolved in 100% sulphuric acid [151]. In fact the majority of the variations in $J_{(^{31}P-H)}$ appears to arise from differences in the effective nuclear charge (Z_P) on phosphorus, determined by the different inductive effects of the groups attached to phosphorus. Decrease of electron density at phosphorus increases Z_P and $J_{(^{31}P-H)}$.

Measurements of the coupling constants $J_{(^{57}Fe-^{31}P)}$ (iron contains 2.19% of ^{57}Fe with I = 1/2) for $Fe(CO)_4(PEt_nPh_{3-n})$ (n = 1-3) has only been briefly reported [135]. These coupling constants increase from 25.9 (n = 3) to 27.4 Hz (n = 1) in agreement with the data reported for the analogous tungsten derivatives (see above).

^{195}Pt magnetic resonance spectra have not been extensively studied, and the majority of the work is concerned mainly with platinum(II) derivatives [136]; the same is true for the examination of Pt–P coupling constants in ^{31}P n.m.r. studies [137].

The applications of ^{13}C magnetic resonance to organometallic chemistry have so far been rare and difficult [138], but the use of pulsed Fourier and noise decoupling techniques in the case of compounds containing C–H bonds provide a very powerful method for studying an enormous number of these compounds in the near future.

The carbon-13 nuclear magnetic resonance spectra of carbonyl metals of the type $M(CO)_6$ (M = Cr, Mo, W) and of their monosubstituted derivatives (M = W) have been reported [139]. Besides some observations on the ^{13}C chemical shifts and correlations with $\nu(CO)$, several $^{183}W-^{13}C$ coupling constants were measured, and a comparative study with $J_{^{183}W-^{31}P}$ for the same molecules is in progress.

By using the labelled complexes bis(perdeuteriotriphenylphosphine)-platinum(ethyne-1,2-^{13}C) and (ethene-1-^{13}C), Cook and Wan have measured the coupling constants $J_{^{13}C-H}$ for the unsaturated ligands coordinated to platinum [140]. The use of a deuterated triphenylphosphine was necessary in order to clear the region where the protons of the coordinated ethyne absorb. These protons show a 5.30 p.p.m. downfield shift upon coordination (free ligand $\tau = 8.20$, $CDCl_3$ as solvent), opposite to the behaviour of ethene, for which an upfield shift of 2.73 p.p.m. is observed (free ligand $\tau = 4.69$, $CDCl_3$ as solvent) (see 1.3.2.1). The large downfield shift of the protons of the coordinated alkyne ligand is expected, on considering for instance the resonance of free cyclopropene ($\tau = 2.99$).

Unlike the $J_{^{13}C-H}$ value for a σ-bonded methyl group for *trans*-$Pt(PPh_3)_2(^{13}CH_3)Cl$, $J_{^{13}C-H} = 131.0$ Hz the "normal" $J_{^{13}C-H}$ value being 125 Hz, the $J_{^{13}C-H}$ coupling constants obtained for coordinated ethene and ethyne are considerably different (Table 1.XI) [140]. The magnitude of the coupling constants agrees with a reduction in carbon–carbon bond order of the unsaturated molecules upon coordination to metals. Assuming that a co-

TABLE 1.XI

Proton resonance data for L_2PtX and X in C_6D_6. Data taken from ref. 140

Compound	J_{Pt-H} (Hz)	$J_{^{13}C-H}$ (Hz)	$J_{trans P-H}$ (Hz)	$J_{cis P-H}$ (Hz)
$^{13}C_2H_4$	–	156	–	–
$^{13}C_2H_2$	–	250	–	–
$L_2Pt(^{12}CH_2{=}^{12}CH_2)^a$	60.0	–	unresolved	unresolved
$L_2Pt(^{13}CH_2{=}^{12}CH_2)$	60.0	146.5	unresolved	unresolved
$L_2Pt(^{12}CH{\equiv}^{12}CH)$	58.0	–	17.4	16.6
$L_2Pt(^{13}CH{\equiv}^{13}CH)$	59.0	210.0	19.5	14.0

L = perdeuteriotriphenylphosphine; X = ethene or ethyne.
[a] See ref. 141; L = non-deuterated triphenylphosphine.

ordinated methyl group can be regarded as being an sp^3 hybrid, the percentage of *s* character of the hybrid orbital at carbon (ρ) and the ^{13}C–H coupling constant (131 Hz) can be related by the expression: $\rho = 0.19J$. This, assuming linearity, leads to ρ values for the ethene and ethyne complexes of 28 and 40%, respectively [140]. The valence bond approach to the bonding in such complexes again seems to be inadequate, being too rigid a formalism, and unable to account for the intermediate situations which can be determined by the nature of the metal, its oxidation state, the presence of other ligands bound in the complex, and the effects due to the substituents present in the unsaturated ligands (see 1.3.1.2). The ^{13}C n.m.r. parameters in platinum(0)- and platinum(II)- alkene and -alkyne derivatives indicate that the bonding in zero and divalent complexes differs only in magnitude and not in mode [140 *bis*]. For the alkene derivatives J_{Pt-C} does not correlate with the increased shielding of the olefinic carbons, which on the other hand can be related to the metal-to-olefin π-back donation. J_{Pt-C} thus seems to be dominated by the Pt 6*s* orbital contribution to the metal-unsaturated hydrocarbon σ-bond. The magnitude of the $J_{^{195}Pt-H}$ coupling constants in platinum alkene complexes of the type described above (Table 1.XI) has been used as an indication of the amount of π-back donation from the metal to the alkene [57, 110]. The larger the π-back donation, the greater should be the distortion of the ligand from planarity. This allows mixing of the σ and π bonds within the unsaturated ligands where, in order to have some coupling with the metal, the carbon $2_{p\pi}$ orbitals must have some *s*-character. However, in a recent investigation it has been suggested that the ^{195}Pt–H coupling constants in the alkene ligand in a series of platinum(II)-alkene complexes are more complex in origin than would be expected if only the Fermi contact term is considered [142]. The transmission of spin coupling appears to occur by the same mechanism in both σ- and π-bonded situations, and therefore this coupling is not a reliable indication of the amount of *s*-character of the metal–alkene bond.

Internal olefin proton–proton coupling has also been carefully analysed [142]. Upon coordination there is a decrease in magnitude of all coupling constants involving the protons originally bound to a sp^2 hybridized carbon, and they approach the values found for the fully saturated systems, $J_{HH\,trans}$ changing by about 50% while $J_{HH\,gem}$ becomes zero. The observed changes are consistent with some contribution from a structure in which the bonding to the metal is via two metal–carbon σ-bonds, but the complexes are probably still best thought of in terms of π-bonding, at least for the platinum(II)–olefin derivatives [142].

The ^{31}P–H coupling constants in this type of compound (Table 1.XI) invariably show that $J_{trans} > J_{cis}$. This means that in a rigid planar arrangement in which the alkene occupies two vicinal coordination positions, the phosphorus *trans* to a carbon atom couples more strongly than the phosphorus in *cis* position. This follows the general behaviour of coupling constants between nuclei in different stereochemical positions, and as a limiting case J_{cis} was not detected when using alkenes bearing electron-withdrawing substituents, such as maleic anhydride [57]. This highly directional effect shows that the platinum–alkene bond has a rather directional distribution of the electron density. Moreover in some cases J_{trans} could not be detected even at low temperatures, which excludes a rapid dissociation or rotation of the alkene around the metal ligand axis, [57] and this is difficult to explain. Such a rotation has never been observed for low-valent alkene complexes, in contrast to complexes with metals in higher oxidation state which show a facile intramolecular rotation of the alkene about the metal–olefin axis [142]. This seems also to support the view that a stronger π-back donation, as present in zerovalent alkene complexes, does not allow a rotation around the metal–olefin double bond. However, an interpretation in terms of a rigid three-membered ring in formally non-zerovalent complexes could also explain such behaviour.

The ^{19}F spectra of the fluoroalkene complexes $(PPh_3)_2Pt(C_2F_3Z)$ generally have a very complex nature because of the non-equivalence of the phosphorus and fluorine nuclei [65]. However the "AB" resonance pattern of the CF_2 group of the perfluoropropene complex ($Z = CF_3$) enabled the coupling constant J_{AB} to be evaluated (188 Hz). This value is similar to the value observed for J_{FF} in substituted fluorocyclopropane systems, and is very different from the value for a vinylic coupling constant. It suggests that the perfluoropropene moiety is essentially σ-bonded to the metal, as has been previously discussed (1.3.1.2).

By changing the other ligands bound to platinum in a series of tetrafluoroethene complexes of the type described above, it has been found that J_{Pt-F} has the highest values when nitrogen–donor ligands are present. For instance with L = $NH_2CH_2CH_2NH_2$, J_{Pt-F} = 501 Hz; L = $Ph_2PCH_2CH_2PPh_2$ J_{Pt-F} = 316 Hz; L = 2 PPh_3, J_{Pt-F} = 278 Hz. [143]. This may be taken as an indication of increased σ-character in the metal–alkene bond or, in other words, of a stronger platinum–alkene bond, as expected on the basis of an increased charge transfer from the metal to the fluoro-alkene ligand. Similar studies on hexafluorobut-2-yne-platinum complexes came to analogous conclusions [144], but we have already pointed out how platinum–ligand coupling constants are complex in origin.

Phosphorus–phosphorus coupling constants have also been widely studied. By a qualitative estimation of P–P coupling constants in the series of complexes $Ni(CO)_{4-x}L_x$ (x = 2, 3, 4) (L = 4-methyl-2,6,7-trioxa-1-phosphabicyclo-[2.2.2]octane), it has been found, in one of the first works in the field, that J_{PP} increases with increasing substitution (x = 3, J = 1, x = 4, J = 15 Hz) [145]. It has also been shown that generally J_{PP} is easily detectable for *trans*-disubstituted derivatives [for instance in $Mo(CO)_4L_2$ complexes (L = 2,8,9-trioxa-1-phosphaadamantane) J_{PP} = 201 ± 50 Hz] [146], while J_{PP} appears to follow the order $Fe(CO)_3L_2 > Mo(CO)_4L_2 > W(CO)_4L_2 > Cr(CO)_4L_2 > Ni(CO)_2L_2$, which seems to have some correlation with the geometry of the complexes and of course with the nature of the metal in compounds of the same type. For a series of mixed ligands complexes, $M(CO)_4LL'$ (M = Cr, Mo, W), studied in order to measure P–P couplings directly from the ^{31}P spectra in compounds where the phosphorus atoms are chemically non-equivalent, similar results were obtained. It has also been observed that for the two *cis* and *trans* isomers (L =

PPh_3, L′ = $PPh(C_4H_9)_2$; M = W) $J_{PP\,trans}$ (49 Hz) is notably larger than $J_{PP\,cis}$ (21 Hz) [147].

An explanation in terms of an increased $d_\pi(P)d_\pi(W)d_\pi(P)$ bonding in the *trans* derivatives was considered to be irrelevant, as *trans* P–Ir–H couplings are larger than *cis* P–Ir–H couplings, while hydrogen cannot be involved in any π-bonding [148]. These results have been confirmed by other authors, who studied similar mixed ligand complexes [149]. They observed that larger values of J_{PP} in the *trans* than in the *cis*-molybdenum carbonyl disubstituted complexes may be the consequence of the fact that the σ-bonding electrons of *trans*-phosphorus nuclei share the same P_σ-metal molecular orbital whereas those *cis* do not. However, they also pointed out that the opposite behaviour for coupling constants can be observed in chromium complexes [149, 150].

The effects of the electronegativity of the substituents on phosphorus on magnitude of J_{PP} have been studied by comparing a large number of disubstituted PF_3-carbonyl complexes with similar derivatives of other trivalent phosphorus ligands [150]. It has been generally found that J_{PP} fall in the general order $PR_3 < P(OR)_3 < PF_3$. This has been attributed to the increasing *s* character of the metal-phosphorus bond rather than to a different π-bonding character, according to Bent arguments, that is to say that during the rehybridization the *s* character is concentrated in orbitals directed toward electropositive substituents [133]. However, it has also been considered that an additional (and perhaps dominating) effect may cause J_{PP} to rise with the electronegativity of the phosphorus substituents, that is with the increase in phosphorus nuclear charge from PR_3 to PF_3. In fact the nuclear charge could appear, as for the ^{13}C–H interaction, as a cubed term in the coupling equation, while the percent *s* character is only squared [150, 151]. The signs of phosphorus–phosphorus coupling constants in complexes of the type *cis* and *trans*-$M(CO)_4L_2$ (M = Cr, Mo, W), *trans*-$Fe(CO)_3L_2$ (L = $P(CH_3)_3$, $P\{N(CH_3)_2\}_3$, $P(OCH_3)_3$) were determined [152, 153]. J_{PP}'s in the *trans* complexes are generally positive (with the exception of some *trans* chromium derivatives), while the *cis* zerovalent group VI complexes have negative values in all cases. The most significant factor affecting ^{31}P–^{31}P coupling appears to be the electronegativity of the substituents on phosphorus [152, 154]. J_{PP} appears to increase with electronegativity in the *trans* complexes, while in the *cis* systems it seems to decrease. An explanation of such behaviour without invoking any π bond in the metal–phosphorus interaction has been shown to be possible, as discussed above [154].

We will now briefly discuss how coupling constants between nuclei within a ligand like a fluorophosphine change upon complexing of the ligand to a transition metal. For a series of phosphines PF_2–X, the P–F coupling constants, J_{P-F}, were found to increase with increasing electronegativity of X [116]. This may be due mostly to the increased *s* character in the P–F bond on increasing the electronegativity of X, as discussed above for J_{PP} coupling constants. It is also known that J_{P-F}s, in complexes with different coordination numbers, decrease with increasing number of atoms bound to phosphorus. For a wide variety of coordination compounds of zerovalent nickel and molybdenum of the types $Ni(CO)_2L_2$, NiL_4 and $Mo(CO)_3L_3$ (L = fluorophosphine), the coupling constants between phosphorus and fluorine were always found to be smaller than those in the free ligands, except in a few cases, where the phosphorus is bonded to two dialkylamino groups [116]. It was assumed that as the *s* electrons of phosphorus, are tied up on coordination to form a σ bond, the hybridization of the phosphorus orbitals involved in forming bonds with fluorine atoms, as well as the bond angles, would change and that these changes were responsible for the variations in J_{P-F} on complex formation.

In a series of phosphines $PF_{3-n}R_n$ (n = 0–3) (R = Bu^t), the coupling constants J_{P-H}, J_{F-H} and J_{P-F} decrease on increasing n [129]. Upon coordination to a metal, in complexes such as $Ni(CO)_3L$ or $M(CO)_5L$ (M = Cr, Mo, W) (L = $PF_{3-n}R_n$), while J_{P-F} and J_{F-H} decrease, J_{P-H} increases with respect to the free ligands. It has been proposed that the quantity $(\Delta J)/J_L$ (where J_L is the P–F or P–H coupling constant in the free ligand, ΔJ is the change of these values on complex formation) [129] be used in place of the relative "π-donor capacity" of the metal carbonyl fragments, as well as of the "π-acceptor strength" of the phosphine ligand L, as obtained from i.r. spectra data [155]. This parameter decreases with increasing n, while for a given phosphine L, it decreases in the order $W(CO)_5L > Mo(CO)_5L > Cr(CO)_5L >$ $Ni(CO)_3L$. This series does not agree with the one obtained from i.r. data. However, since these new parameters are related to changes within the ligand L, they should provide immediate information about the overall complexing properties of L and the effects due to the transition metal.

As for the chemical shifts previously discussed, the coupling constants can also give us general information about the nature of the metal-ligand bonds. Moreover, general rules can be derived in order to predict the effects on coupling constants by complex formation for a given ligand. However, all the effects that we have discussed above can be interpreted in terms of σ- and/or π-bonds and again we cannot discriminate between them.

1.3.2.3. Steric Effects

Among the factors that are important in determining the stability of the complexes, we have not considered up to now the role of steric effects, having taken into account only the electronic properties of the ligands. However, this problem has been studied in detail by ^1H and ^{31}P n.m.r. spectroscopy, particularly for zerovalent nickel phosphine complexes. It has been shown that the size of the ligands, rather than (or together with) their electronic character, primarily determines the stability of such compounds [156].

Ligand-competition experiments were carried out in which a toluene solution of a NiL_4 complex (0.2 M) was treated with a competing ligand L′ (0.8 M). The total concentration of each ligand being the same, it was possible to determine the extent of exchange from the relative intensities of free and bound ligand resonances in the ^{31}P n.m.r. spectra:

$$NiL_4 + 4\,L' \rightleftharpoons NiL_{4-n}L'_n + n\,L$$

These experiments were carried out using 24 ligands having phosphorus as donor atom, and 12 NiL_4 complexes. From product distributions, the ligands could be ordered in a series according to the relative stability of the complexes. Semiquantitative relative ligand bonding abilities were also determined.

From these data the 24 ligands were ordered in the following stability series:

$$P(C_6H_5)(OC_2H_5)_2 \gtrsim P(OCH_2)_3CCH_3 = P(OCH_2)_3CC_2H_5 = (CH_3)_2PCH_2CH_2P(CH_3)_2 =$$
$$= P(OCH_3)_3 = P(OC_2H_5)_3 = P(OCH_2CH_2Cl)_3 \gtrsim P(OCH_2CCl_3)_3 \gg P(O\text{-}p\text{-}C_6H_4CN)_3 \gtrsim$$
$$\gtrsim P(OC_6H_5)_3 = P(O\text{-}p\text{-}C_6H_4CH_3)_3 = P(O\text{-}p\text{-}C_6H_4OCH_3)_3 = P(CH_3)_3 = P(CH_3)_2CF_3 \gtrsim$$
$$\gtrsim P(C_6H_5)_2(OC_2H_5) \gg P(C_2H_5)_3 = P(C_4H_9)_3 \sim P(C_6H_5)_2(C_2H_5) \sim P(C_6H_5)_3 \sim$$
$$\sim P(O\text{-}o\text{-}C_6H_4CH_3)_3 \gg P(i\text{-}C_3H_7)_3 > P(t\text{-}C_4H_9)_3,\ P(o\text{-}C_6H_4CH_3)_3,\ P(C_6F_5)_3.$$

The order of ligands in this series does not correlate with the electronic properties of the phosphorus ligands, as determined by their influence on the symmetric carbonyl stretching frequency in the corresponding $Ni(CO)_3L$ compounds [42] (see 1.3.1.1). It was generally expected that ligands which are strongly electron-withdrawing (strong "π-acids") should give the most stable complexes, but this was not the case. In fact in the series reported above, one finds for instance that $P(CH_3)_3$ (a good electron donor) forms a remarkably stable nickel(0) complex, while $P(C_6H_5)_3$ (one of the best π-acceptors) fails to bond to nickel(0) when in competition with $P(CH_3)_3$.

Moreover, there is no general correlation between the ^{31}P chemical shift of the free ligand with its electronic properties, and the same is true also for the coordination chemical shift (see also 1.3.2.1).

In order to have some measure of the steric hindrance for each ligand, Tolman has constructed models of the $Ni–PR_3$ stereochemical arrangements, as shown in Fig. 1.6.

A Ni–P distance of 2.28 Å was chosen. The apex angle of the subtending cone, illustrated in Fig. 1.6, was assumed to be a quantitative measure of the steric effects for the various ligands. This is because it encloses the van der Waals radii of the ligand freely rotating around the Ni–P axes, and folded back while maintaining C_3 symmetry. The results are reported on Table 1.XII. In the case

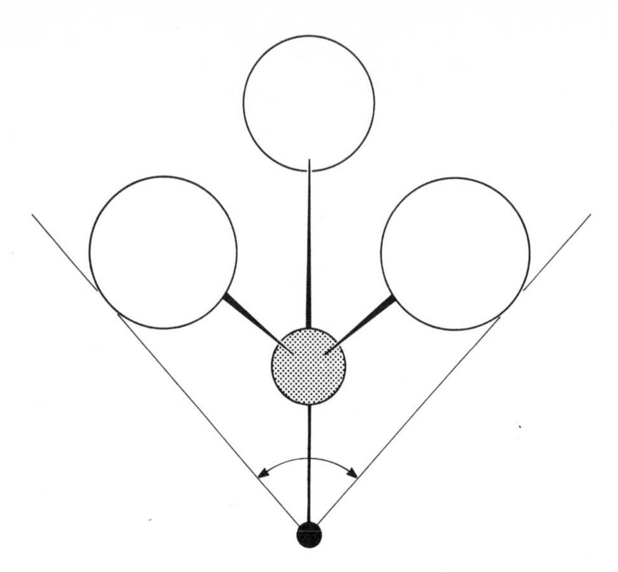

Fig. 1.6. Apex angle of the subtending cone in a Ni–PR_3 fragment. ● = nickel; ◍ = phosphorus; ○ = R.

where ligands approach cylindrical symmetry around the metal–phosphorus bond, coordination numbers of six, four, three or two should not be possible for ligands having respectively cone angles larger than 90°, 109.5°, 120° or 180°. Since the ligands have symmetries less than cylindrical, somewhat larger ligands can be accommodated by constraining one ligand into the other [156].

It is interesting to note that among the most common ligands, some have very large cone angles, for example $P(C_6H_5)_3$ (145°), $P(C_6H_{11})_3$ (179°), $P(t\text{-}C_4H_9)_3$ (182°). A fair semiquantitative correlation was found between ligand binding ability (as above defined) and the ligand cone angles. Once the tetrahedral angle is exceeded, ligands form complexes of decreasing stability as their size increases. From this point of view we can understand why $P(C_6F_5)_3$ (cone angle 184°) forms a much less stable zerovalent nickel complex than $P(CH_3)_3$ (cone angle 118°).

It follows that the size of the ligands rather than their electronic character is important in determining the stability of these zerovalent nickel complexes. These results were confirmed by reacting the same ligands with $Ni(CO)_4$. The degree of carbonyl substitution by the various phosphines and phosphites was shown to depend on the size of the ligands, that is the largest ligands gave less-substituted nickel carbonyl derivatives. The degree of substitution was estimated semiquantitatively from the relative intensities of the carbonyl i.r. absorptions of the various reaction products. The most striking exception is given by PCl_3, a ligand which replaces CO on $Ni(CO)_4$ only with difficulty in

TABLE 1.XII

Ordering of various phosphorus ligands PR_3 and $P(OR)_3$ according to size. Data taken from ref. 156

PR_3	$P(OR)_3$	Minimum cone angle (deg.°)
PH_3		87 ± 2
	$P(OCH_2)_3CCH_3$	101 ± 2
PF_3		104 ± 2
	$P(OCH)_3(CH_2)_3$	106 ± 2
	$P(OCH_3)_3$	107 ± 2
	$P(OC_2H_5)_3$	109 ± 2
	$P(OCH_2CH_2Cl)_3$	110 ± 2
	$P(O\text{-}i\text{-}C_3H_7)_3$	114 ± 2
$P(CH_3)_3$		118 ± 4
	$P(OC_6H_5)_3$	121 ± 10
PCl_3		125 ± 2
PBr_3		131 ± 2
	$P(OCH_2CCl_3)_3$	131 ± 10
$P(C_2H_5)_3$		132 ± 4
$P(C_4H_9)_3$		130 ± 4
$P(CF_3)_3$		137 ± 2
$P(C_6H_5)_3$		145 ± 2
$P(i\text{-}C_3H_7)_3$		160 ± 10
	$P(O\text{-}o\text{-}C_6H_4CH_3)_3$	165 ± 10
$P(C_6H_{11})_3$		179 ± 10
$P(t\text{-}C_4H_9)_3$		182 ± 2
$P(C_6F_5)_3$		184 ± 2
$P(o\text{-}C_6H_4CH_3)_3$		194 ± 6

spite of its rather small size. For this phosphine, ligand competition experiments were not possible owing to side reactions. Some other discrepancies were noted [156], and they were attributed to electronic and chelation effects. Although the approach to the measurement of steric effects of the phosphine ligands leaves some doubts about the absolute value of parameters such as the minimum cone angles (in particular a fixed Ni–PR_3 bond distance was used for all phosphines, while it is expected that bond distances vary from ligand to ligand), they are still meaningful for comparisons between ligands. In fact if a ligand gives a longer Ni–P bond, a smaller cone angle should result. However, a longer Ni–P bond means a less stable complex, or a weaker metal-phosphorus electronic interaction, and this is indicated by a cone angle greater than the one actually present in the relative scale.

From these considerations, we conclude that the electronic effects have been overestimated in the past, at least in some cases, and it is to be expected that the problem of the influence of steric effects on the stability of complexes will be considered more carefully in the future. Moreover, not only will the stability towards substitution or exchange reactions be studied, but also the general chemical behaviour of these complexes will be reconsidered in the light of this important effect. At the same time, kinetic investigations could greatly contribute to the understanding of this important problem.

During this study it has also been observed that doublet resonances for L in NiL_3L' complexes, and triplets for L in $NiL_2L'_2$ (corresponding to quartet and triplet resonances for L', respectively), are consistent with the expected tetrahedral coordination in these complexes [156]. Phosphorus–phosphorus coupling constants were found to be in the range 10–45 Hz. While no simple correlation was found between the changes in δ_P of L in $NiL_{4-n}L'_n$ as n increases and the nature of L', δ_P values were found in most cases to decrease regularly (shift to lower field) as n increases.

The isolation of a tris-phosphite nickel(0) complex in the case of $P(O{-}o{-}C_6H_4CH_3)_3$, but not with $P(O{-}p{-}C_6H_4OCH_3)_3$ or $P(OC_2H_5)_3$, was attributed to steric effects [157]. The first two ligands are in fact electronically equivalent, while the cone angles are 165, 121 and 109° respectively. The large steric bulk of trio-*o*-tolyl phosphite destabilizes the tetrakis complex so much that the tris complex can be isolated. From this point of view the existence of a bis-phosphinenickel(0) complex like $Ni\{P(C_6H_{11})_3\}_2$ [109] can also be explained (cone angle for this ligand 179°).

Similar results have been obtained from n.m.r. studies on PtL_4 complexes [L = $PCH_3(C_6H_5)_2$, $P(CH_3)_2C_6H_5$ and $P(CH_3)_2C_6F_5$] [158]. A rapid exchange has been observed for L = $PCH_3(C_6H_5)_2$ at room temperature of the phosphines (no Pt–H or P–H coupling in the ^1H n.m.r. spectrum). Lowering the temperature first resulted in the appearance of platinum satellites ($J_{^{195}Pt-CH_3}$ = 20 Hz), and at −10° a broad singlet due to free phosphine was observed. Finally below −30° the exchange of the phosphine was frozen and the methyl proton signal of the free phosphine became a doublet, with one-third the intensity of the complexed phosphine. Further lowering the temperature did not result in any splitting of the broad methyl signal of the complexed ligand due to ^{31}P. Although the tricoordinated species was the most stable below −30°, it was not possible to isolate this compound. For L = $P(CH_3)_2C_6H_5$ and $P(CH_3)_2C_6F_5$, free phosphine was never observed in the n.m.r. spectrum. However in the former case, J_{Pt-CH_3} appeared at 25°, while in the latter case platinum satellites were observed only below 10°. It was concluded that the relative stabilities of the PtL_4 species decrease in the order: $P(CH_3)_2C_6H_5 > P(CH_3)_2C_6F_5 \gg PCH_3(C_6H_5)_2 > P(C_6H_5)_3$. The last member of the series was included because it is well known that tris-triphenylphosphineplatinum(0) can be easily isolated

[159a] as well as the tetrakis complex [159b]. This order of stability does not appear to correlate with the electronic properties of the ligands, whilst steric effects again seem to be more relevant.

Another interesting example which emphasizes this problem is related to a report on the kinetic investigation of addition of oxygen to Vaska's compounds [160]:

$$[IrCl(CO)(PR_3)_2] + O_2 \underset{k_{-1}}{\overset{k_2}{\rightleftharpoons}} [O_2IrCl(CO)(PR_3)_2]$$

$$K_2 = \frac{k_2}{k_{-1}} = \frac{[O_2IrCl(CO)(PR_3)_2]}{[IrCl(CO)(PR_3)_2]\,[O_2]}$$

The rates of oxygen addition and the stability of the resulting dioxygen adducts increase with increasing basicity of the substituents R, provided the substituents have comparable steric hindrance. In fact log k_2 and $\Delta H_2^{\ddagger(-1)}$, the enthalpy of activation, are inversely proportional to the Hammett σ constants for the *para* substituents in a series of *para*-substituted arylphosphines, or directly proportional to the basicity of the tertiary phosphines. These basicities are inferred from the ν(CO) in the initial complexes, which decrease as $\Delta H_2^{\ddagger}$ increases. With $P(C_6H_{11})_3$, k_2 is the lowest observed, whilst this phosphine was the most basic used in the work. Moreover when R = C_6F_5 or *o*-$CH_3C_6H_4$, no reaction with oxygen was observed. While in the former case this effect could be also attributed to a lack of the required electron density on the metal, the inertness in the latter case must be largely due to the steric effect of the *ortho*-CH_3 groups, which overcomes the favourable electronic disposition. These observations strongly support the results obtained by Tolman for the NiL_4 complexes [156], since the less reactive iridium complexes, and the least stable oxygen adduct, are found with the phosphines with the greatest steric hindrance.

Similar results were obtained by studying the kinetic and thermodynamic parameters of the same reaction, but with hydrogen or alkyl halides as the entering molecule instead of oxygen [161].

1.3.3. Photoelectron Spectroscopy

X-ray photoelectron spectroscopy (also called ESCA) studies the energy distribution of the electrons emitted from X-ray-irradiated compounds [162]. With this technique both the core and valence electrons can be studied in principle, while with ultraviolet photoelectron spectroscopy (PS) only the valence electrons can be studied [163, 164]. These new types of spectroscopy are finding wide application, and X-ray electron spectroscopy appears to be particularly appropriate for qualitative and quantitative studies on large molecules. Together with some work on the measurement of inner-orbital binding energies of phosphorus compounds [165], two ESCA studies are strictly related to the problems that we are now discussing [166, 167]. Several related complexes of platinum in low oxidation states have been investigated [166]. The binding energies of Pt $4f_{7/2}$ electrons were assumed to depend only on the electronic charge transferred from the metal to the ligand, and the positive charge on platinum in the complexes was determined from the binding energy data. The variations in binding energies within a closely related series of compounds such as $Pt(PPh_3)_2L$ (L = $2PPh_3$, C_2H_4, PhC≡CPh, CS_2, O_2, 2Cl), can be considered to be solely related to changes in the localized charge on the atom. It was found that the binding energy of the platinum $4f_{7/2}$ level increases in the order $2PPh_3 < PhC_2Ph \leqslant C_2H_4 < CS_2 < O_2 < Cl_2$, suggesting that the positive charge on platinum increases in the same order. The phosphorus–platinum interaction does not change significantly throughout the series, as shown from the essentially constant phosphorus $2p$ binding energies for the complexes studied. This fact supports the premise that $4f_{7/2}$ binding energies are mainly determined by the electronic charge transferred from platinum to the ligand L. For the two compounds at the opposite side of the series, i.e. $Pt(PPh_3)_4$ and $Pt(PPh_3)_2Cl_2$, it was assumed that the metal transfers 0 and 2 electrons respectively to the L ligand, although it is evident that such electron transfers do not correspond to the actual electronic situation in the complexes. However the Pt $4f_{7/2}$ binding energy in $Pt(PPh_3)_4$ (71.7 eV) is very near to that found in metallic platinum (71.2 eV), while for $Pt(PPh_3)_2Cl_2$ (73.2 eV) the binding energy is not dissimilar to that in K_2PtCl_4 (73.4 eV) [166, 167].

A linear dependence of the binding energies on the metal to ligand charge transfer in the intermediate cases was also assumed. The estimated metal–ligand charge transfers were: 0.7 ± 0.2, 0.8 ± 0.2, 1.3 ± 0.2, 1.8 ± 0.2 electrons for L = PhC_2Ph, C_2H_4, CS_2 and O_2 respectively. Although these values are probably high by about 0.4 electron [166], they seem to support the idea that an increased π-back-donation takes place on going from L = $2PPh_3$ to L = O_2, up to the limiting case where a nearly complete charge transfer is present (L = 2Cl). In a more recent investigation this problem has been reconsidered for a wide series of compounds, and the results are somewhat different from those reported earlier [167]. For the series of platinum olefin complexes $Pt(PPh_3)_2L$ (L = C_2H_4, C_2Cl_4, C_2F_4, $C_2(CN)_4$), a nearly constant binding energy of 73.1 ± 1 (eV) was found for the Pt $4f_{7/2}$ electrons. Such invariance was also found for Pt $4f_{5/2}$ binding energies. In particular for L = C_2H_4 a value 0.7 eV higher than that previously reported [166] was observed. This is an important result, because it implies that even with an olefin which does not bear electron withdrawing substituents, such as ethylene, the charge transfer from platinum to the ligand is as strong as that found for the limiting case $Pt(PPh_3)_2Cl_2$. The authors concluded that the Lewis basicity of the bis(triphenylphosphine)platinum(0) moiety is so strong that the substituent groups on the olefin play but a minor role in determining the extent to which charge transfer from metal to ligand antibonding orbitals takes place. Similar results were obtained for rhodium derivatives such as $Rh(PPh_3)_2ClL$.

In contrast, for $Ir(CO)(PPh_3)_2XY$ complexes (X = Cl, I; Y = O_2, CO, C_2F_4, $C_2(CN)_4$, $(NO^+)\ BF_4^-$), marked differences were observed for the Ir 4*f* binding energies as a function of the nature of Y. These results are in accordance with the previously proposed greater π-basicity of $Pt(PPh_3)_2$ with respect to $IrCl(CO)(PPh_3)_2$, for instance on the basis of i.r. and n.m.r. data [57]. However, it is rather disappointing that even in complexes with olefins such as ethylene, which show an enhanced reactivity and lability in comparison to similar compounds with olefins of stronger π-acidity [57, 141], the binding energies do not differ from normal values. This might be a consequence of the strict validity of the Pauling Electronegativity Principle for low oxidation state complexes, but it might also be a consequence of the fact that this new technique is not yet refined enough to discriminate small electronic effects.

Further data on other phosphine–olefin derivatives will be of great value to investigate this problem in more detail.

Molecular core binding energies for some zerovalent transition metal carbonyls have also been recently measured, and it has been found that C_{1s} and O_{1s} binding energies of carbon monoxide are lower in the complexes, in comparison to free ligand [168]. This suggests an overall electron transfer from metal to ligand, and supports the idea that on complex formation the stabilizing effect is mainly the π-back-donation from metal to carbon monoxide, which results in an overall electron drift from metal to ligand.

The photoelectron spectra of $Ni(PF_3)_4$ and $Pt(PF_3)_4$ have also been analysed [169]. On complex formation the change in energy of the non-bonding fluorine orbitals should reflect the net result of σ and π-bonding between metal and phosphorus. Such orbitals show very little differences in energy between the free ligand and both complexes.

Thus in each complex, σ-donation (from the lone pair) must be compensated for by π-back-bonding to produce only little, if any, net charge migration [169]. Moreover, there is a greater lowering of the energies of the orbital localized on the phosphorus atom (lone pair) in the platinum compared with the nickel complex, indicating a greater degree of σ-donation in the former. It follows that the extent of electron delocalization by σ and π mechanisms is greater in the platinum than in the nickel complex although the effects compensate each other. This point will be reconsidered later (see 1.4). The phosphorus–metal σ-donation will tend to shorten the P–F bond lengths, while the π-back-donation will have the opposite effect. Although little net charge migration takes place in these complexes, as above discussed, the lengthening due to σ-donation seems to dominate. In fact the P–F bond lengths decrease from 1.569 Å in PF_3 to 1.561 in $Ni(PF_3)_4$ and to 1.546 Å in $Pt(PF_3)_4$ [170], the greatest shortening being observed for the complex where a greater degree of σ-donation is indicated by the photoelectron spectroscopy data.

The binding energy of the 1*s* core electrons of the nitrogen atoms in transition metal–nitrosyl, oxides, nitrides and dinitrogen complexes has been examined [170 *bis*]. Bent nitrosyls have low core binding energies, implying a relatively high negative charge on the nitrogen atom. From the above discussion it appears that photoelectron spectroscopy is an appropriate technique for studying metal–ligand electronic interactions in complexes. The first results in the field favour the existence of metal-to-ligand π-back donation, and it is hoped that more data that will help to solve this problem will be collected in the near future.

1.3.4. Mössbauer Spectroscopy

The Mössbauer effect, or gamma-resonance spectroscopy, consisting in the recoilless resonance absorption and emission of gamma quanta by nuclei may usefully be applied to the study of some organometallic compounds [171].

The isomeric (or chemical) shift of resonance line (IS or δ) is proportional to the difference between the *s*-electron densities at the nuclei of the absorber and emitter of gamma quanta. The effect of the *p* and *d* electrons consists only of a different screening of the *s*-electrons. The quadrupole splitting of the resonance line (QS or ΔE) appears when nuclei with spin $I \geqslant 1$, in consequence of their quadrupole moment Q, interact with the electric field gradient created by atomic electron shells and neighbouring ions. The quadrupole splitting depends on the presence of the *p* and *d* electrons in the atoms. Mössbauer spectroscopy has very useful applications in the study of molecular structure and symmetry, of kinetics and reaction mechanisms, and of properties of solids, but in particular we are interested on its applications to the study of the nature of chemical bonds or the electronic structure of molecules [171]. Unfortunately, although in principle this technique can be applied to a large number of elements most of the research has so far been dedicated to nuclei such as ^{57}Fe and ^{119}Sn, while iridium was employed only in the first of Mössbauer's experiments.

Mössbauer studies on compounds containing a transition metal group IV B metal bond will not be considered here.

Complexes of the type $Fe(CO)_3$(diene) and the corresponding cations have been studied by Mössbauer spectroscopy [172]. The IS was found to increase (indicating decreased *s* electron density at the metal) on going from the neutral diene derivatives to their salts, but the differences are not very large. This trend was attributed to the fact that the positive charge in the cation is mainly localized on the diene ligand. This produces a shift of the electronic density from the central iron atom to the ligand, through both the σ and π bonds. The σ contribution to the metal–olefin bond is of major importance and it is responsible for the increase in value of the IS in the cations. In fact the shift of electronic density along the π-bond causes a decrease in the population of the $3d$ iron orbitals, and consequently a decrease of the IS, owing to the weakening of the screening of the iron s electrons.

The observed increase on IS in the series: $Fe(CO)_4\{P(OPh)_3\} \leqslant Fe(CO)_4(PPh_3) < Fe(CO)_4$(*trans*-cinnamaldehyde) $< Fe(CO)_4$(maleic anhydride) $< [Fe(CO)_4(\pi\text{-allyl})]^+(BF_4^-)$ was also attributed to the diminishing Lewis basicity of the ligands through the series [173]. Again the σ contribution to the bond appears to be of major importance. In fact we would expect the π-acidity of the ligands to increase from triphenylphosphine to π-allyl. If π-back donation was predominant, then the IS order should be reversed. The minor importance of π-bond in these derivatives is not unexpected, since the other ligands bound to iron are four CO molecules and they presumably delocalize the majority of the π-electron density on the metal.

The weakening of the metal–ligand bond was assumed to be responsible for the observed decrease of the quadrupole splitting (QS becomes smaller with increasing IS). The ^{57}Fe and ^{121}Sb Mössbauer parameters have been determined for the complexes $Fe(CO)_{5-n}(MPh_3)_n$ ($n = 1, 2$; M = P, As, Sb), and it was concluded that the Fe–M bond has primarily σ-character [173 *bis*].

For a series of iron carbonyl derivatives of fluorocarbon ligands of the type $DC{=}CD(CF_2)_x$ (D = PPh_2, $AsMe_2$; x = 2,3), the isomer shifts span only a relatively narrow range of 0.20 ± 0.04 mm sec^{-1} and are fairly insensitive to the nature of the ligands which donate electrons (DCH_2CH_2D derivatives also fall in the same range) [174]. Analogously, $[Fe(CN)_6]^{4-}$ and $Fe(CO)_5$ show practically identical δ values [171a]. In these cases we must conclude that the reduced σ-bonding ability of CO with respect to CN^- for instance, is compensated by an enhanced π-acidity of carbon monoxide.

Mössbauer spectroscopy can thus give us very important information concerning the nature of chemical bonds. IS values can be used directly for the classification of chemical valence states. Still it is difficult to discriminate in a quantitative way between the contributions of the screening electrons (d and p) and the direct contributions of the s electrons to the IS. In fact both a strong σ-ligand–metal bond, and a strong π-metal–ligand bond, affect the IS in the same way. However, some of the results considered above again favour the existence of π-bonds.

1.3.5. Kinetic and Thermodynamic Data

The substitution reactions of metal carbonyl complexes have been widely explored. The vast majority of such reactions proceed according to a rate law which is either first order in the complex and zero order in the entering ligand L (dissociative mechanism), or first order in both the complex and L (associative mechanism) [175]. Although the reaction mechanisms are now well established, the kinetic parameters do not indicate unambiguously the nature of the metal-ligand bond in these complexes. The kinetics and mechanisms of substitution reactions of tetrakis(trifluorophosphine)-nickel(0) and -platinum(0) complexes have been studied [176]. In this interesting work the nature of the metal-PF_3 bond is discussed and compared with those of the metal-CO and metal-$P(OC_2H_5)_3$ bonds, previously studied for the related tetrakis derivatives of the nickel triad [177, 178].

The rate laws for the exchange and substitution reactions:

$$ML_4 + L' \longrightarrow ML_3L' + L$$

(M = Ni: L = CO: $L' = C^{18}O$, $P(C_6H_5)_3$; L = $P(OEt)_3$: $L' = C_6H_{11}NC$; L = PF_3: $L' = C_6H_{11}NC$. M = Pd: L = $P(OEt)_3$: $L' = P(OEt)_3$. M = Pt: L = $P(OEt)_3$, $L' = P(OEt)_3$; L = PF_3: $L' = C_6H_{11}NC$) were found to be first-order in complex concentration, and zero order in ligand L′ concentration. On these bases the most plausible mechanism consists of a rate-determining dissociation of a ligand L, followed by rapid addition of the entering ligand L′:

$$ML_4 \longrightarrow ML_3 + L \quad \text{(slow)}$$

$$ML_3 + L' \longrightarrow ML_3L' \quad \text{(fast)}$$

The rate constants and activation parameters for the exchange and substitution reactions are reported on Table 1.XIII.

The positive entropies of activation are consistent with the dissociative mechanism. Moreover, since the mean Ni–C bond dissociation energy measured by calorimetry is 35 kcal mol^{-1} [179], the enthalpy of activation for $Ni(CO)_4$ is consistent with the rupture of a Ni–C bond [177]. The relative rates of reaction with L = $P(OC_2H_5)_3$ follow the order Ni $<$ Pd $>$ Pt, while the enthalpies of activation for M–P bond rupture vary in the reverse order Ni $>$ Pd $<$ Pt.

Assuming that the reverse reactions of the dissociation process have a small activation energy, the enthalpies of activation represent a direct measure of the bond dissociation energies. This order of reactivity does not correlate with that of the divalent complexes of the nickel triad for which the rates of substitution of low-spin d^8 complexes of the type $M^{II}L_4$ decrease in the order Ni $>$ Pd $\gg$ Pt [180]. In the latter case, since the M–L bond is to be considered to have essentially σ character, in consequence of the positive oxidation state of the metals, the electronegativity of the central metal atom primarily determines the M–L bond strength [178]. On the other hand it would not be easy to explain the order of reactivities for the zerovalent compounds in terms of σ-bonding along without invoking π-bonding arguments. It has been pointed out by Nyholm [12] that on the basis of the ionization potential of the metal atoms in the spin-paired states ($d^{10} \rightarrow d^9$), the abilities to form d_π bonds decrease in the

TABLE 1.XIII

Rate constants and activation parameters for exchange and substitution reactions of ML_4 complexes in toluene.

	L		
	CO^a	$PF_3{}^b$	$P(OC_2H_5)_3{}^c$
		M = Ni	
$K(25°)$ (s^{-1})	$2.03 \cdot 10^{-2}$	$2.06 \cdot 10^{-6}$	$9.94 \cdot 10^{-7}$
$\Delta H^\ddagger$ (kcal mol^{-1})	22.3 ± 0.2	28.4 ± 1.1	26.2 ± 1.0
$\Delta S^\ddagger$ (eu)	8.4 ± 1	10.7 ± 1.9	1.8 ± 3.0
		M= Pd	
$K(25°)$ (s^{-1})	–	–	$2.07 \cdot 10^3$
$\Delta H^\ddagger$ (kcal mol^{-1})	–	–	22.0 ± 2.3
$\Delta S^\ddagger$ (eu)	–	–	30.4 ± 8.6
		M = Pt	
$K(25°)$ (s^{-1})	–	$1.05 \cdot 10^{-1}$	$2.60 \cdot 10^{-2}$
$\Delta H^\ddagger$ (kcal mol^{-1})	–	22.5 ± 0.9	27.5 ± 1.6
$\Delta S^\ddagger$ (eu)	–	[illegible]2.3 ± 1.4	26.8 ± 4.4

[a] see ref. 177; [b] see ref. 176; [c] see ref. 178

order Ni(5.81) ≫ Pt(8.20) ⩾ Pd(8.33) (see also 1.4). Thus the stability of the nickel(0) compounds seems to be essentially due to π-bonding, while that of the platinum(0) derivatives is due to σ-bonding. Since palladium has a small tendency to form π-bonds and an intermediate tendency to form σ-bonds, it forms the least stable compounds [178]. For L = PF_3, the Ni–P bond strength is greater than the Pt–P bond, whereas for L = $P(OC_2H_5)_3$ the values are similar. This seems to indicate that for L = PF_3, π-bonding is more important than σ-bonding [176], which is in agreement with the general behaviour of PF_3 as ligand. On the other hand from the activation enthalpies (Table 1.XIII), it appears that the bond strength for the Ni–PF_3 bond is closer to the Ni–$P(OEt)_3$ bond than to the Ni–CO bond. In particular, the Ni–PF_3 bond appears to be considerably stronger than the Ni–CO bond [176], while it is generally assumed that these two ligands have very similar bonding properties.

It has also been observed qualitatively that $Ni\{P(OC_6H_5)_3\}_4$ reacts with $C_6H_{11}NC$ much faster than does $Ni\{P(OC_2H_5)_3\}_4$, while the nickel(0) complex with L = $P(OCH_2)_3C(CH_2)_2Me$ reacts neither with $C_6H_{11}NC$ nor with CO [178]. These results must be compared with the high reactivity generally exhibited by the PPh_3 derivatives of the nickel triad, and they emphasize once more the importance of steric effects in the case of bulky ligands such as PPh_3, and to a lesser extent $P(OC_6H_5)_3$, as has been postulated to explain some kinetic investigations (1.3.2.3). Ligand-exchange reactions in platinum–alkyne complexes have been recently reinvestigated [181]. For substitution reactions of the type:

$$(PPh_3)_2Pt(alkyne) + alkyne' \rightleftharpoons (PPh_3)_2Pt(alkyne)' + alkyne$$

(alkyne = $C_6H_4(X)C{\equiv}CH$ [phenyl ring bearing C≡CH and X]; alkyne′ = $p\text{–}NO_2C_6H_4C{\equiv}CH$),

an alternative mechanism to the dissociative mechanism previously proposed has been suggested [182]. The rate-determining step was assumed to be a stereochemical change in the starting complex, possibly involving rotation of the coordinated alkyne through 90°, to give a pseudotetrahedral complex. This internal rearrangement is followed by coordination of either alkyne or alkyne′. The loss of the originally coordinated alkyne completes the sequence [181]. The lowest activation energies were found for the arylalkyne having X with the lowest Hammett σ values. This is in agreement for both mechanism proposed (dissociative or through an internal stereochemical change) if it assumes that the π-back-donation from the metal to the alkyne plays a relevant role in the metal-alkyne bond. In this case the strongest π-bonds are expected for X having the highest Hammett σ-values. A stronger π-bond means a higher activation energy for the rotation or for the dissociation of the alkyne.

Similar results were obtained by studying the stability constants of olefin–nickel(O) complexes of the type Ni(bipyridyl)(olefin) [183]. The following equilibrium was studied spectroscopically:

$$\text{Ni(bipy)(solvent)} + \text{olefin} \overset{K}{\rightleftharpoons} \text{Ni(bipy)(olefin)} + \text{solvent}$$

The stability of these olefin complexes is greater for olefins having strongly electron-withdrawing substituents, i.e. for olefins with larger Alfrey–Price e values [184], which are considered to reflect the polarity or electron density of the vinyl groups. Such values in fact linearly correlate with the Hammett σ constants [185]. A linear relation between $\log K$ at 30° and e values was found:

$$\log K = 2.3\,e + 0.3$$

These results imply that for the formation of a stable complex the π-bond between the olefin and the nickel atom is more important than the σ-bond. The energy levels of the lowest antibonding orbitals of olefins can be estimated from the ionization potentials [20, 186] and from $\pi \rightarrow \pi^*$ excitation energies [187] of the olefins. The energies of the lowest vacant orbitals of various olefins have also been calculated [188]. There is a roughly linear relation between $\log K$ of the olefin-nickel complexes considered above and the energies of the π^* orbitals in the related olefins, that is log K increases as the energy of the π^* orbital decreases. No similar relationship was found by using the energy of the highest occupied orbitals of olefins [183]. These results strongly suggest that π-back-bonding between the filled d orbitals of the metal and the π^* orbital of the olefin is essential for the formation of a stable complex between olefin and nickel. In one of the most recent works in the field [189], the equilibrium constants for ethylene complex formation via the following reaction:

$$ML_3 + C_2H_4 \rightleftharpoons ML_2(C_2H_4) + L$$

$$(M = Ni, Pd, Pt; L = \text{phosphine})$$

were determined. The results are reported on Table 1.XIV.

In contrast to previous results, neither the ML_3 complexes nor the $ML_2(C_2H_4)$ complexes dissociate in solution [189], and the reaction with ethylene does not involve any ML_2 species. For L = PPh_3 the equilibrium constant K varies in the order Ni $\gg$ Pt $>$ Pd. This is in complete accordance with our previous discussion if we consider that the changes in K for the different metals depend more on changes in metal–olefin bond strength (that is the π-back donation from metal to olefin) than on changes in metal–phosphorus bond strength (see the rate constants for exchange reactions of ML_4 complexes). For a given metal with different ligands, K varies in the order $P(p\text{-}C_6H_4CH_3)_3 > PPh_3 > P(m\text{-}C_6H_4CH_3)_3$.

TABLE 1.XIV

Equilibrium constants for reaction in benzene at 25°:

$$ML_3 + C_2H_4 \overset{K}{\rightleftharpoons} ML_2(C_2H_4) + L$$

Data taken from ref. 189

ML_3	K
$Ni(PPh_3)_3$	300 ± 40
$Pt\{P(p\text{-}C_6H_4CH_3)_3\}_3$	0.21 ± 0.002
$Pt(PPh_3)_3$	0.122 ± 0.003
$Pt\{P(m\text{-}C_6H_4CH_3)_3\}_3$	0.07 ± 0.02
$Pd\{P(p\text{-}C_6H_4CH_3)_3\}_3$	0.016 ± 0.002
$Pd(PPh_3)_3$	0.013 ± 0.002
$Pd\{P(m\text{-}C_6H_4CH_3)_3\}_3$	0.004 ± 0.001

The largest value, with the more basic $P(p\text{-}C_6H_4CH_3)_3$ phosphine, is consistent with a stronger metal–olefin bond because of greater π-back-donation from metal to olefin, while for $P(m\text{-}C_6H_4CH_3)_3$ the low K value may be due to steric effects [189].

1.4. THE ELECTRON CONFIGURATION AND THE ROLE OF THE METAL

In coordination compounds with essentially covalent bonds, the formation and stability of the complexes is often controlled by the so-called inert gas or nine-orbital rule. In these cases the nd, $(n + 1)s$, and $(n + 1)p$ orbitals of the metals are in general occupied either by non-bonding lone pairs on the metal, or by bonding pairs shared with the ligands. However the zerovalent derivatives of metals having a d^{10} configuration may be stable in a coordinatively unsaturated state even in the solid state. As discussed above (see 1.3.2.3) this may be due to steric effects, although an interpretation in terms of electronic effects has also been given. It has been shown that when the metal has a very high d–p energy separation, as in the case of platinum(0) derivatives [68], it can achieve a larger amount of back donation in a planar pseudo-trigonal geometry than with tetrahedral coordination.

If there is no strong d–p mixing the number of possible back-donation π-bonds can be lowered from five to two in a tetrahedral configuration [190]. However, such an explanation is not entirely satisfactory, because the tendency to form coordinatively unsaturated species is sometimes exhibited by ligands with poor "π-acidity", and sometimes by ligands with strong π-acidity. In any case it could be interesting to have an idea of how much electron density can be removed from a metal in a coordination compound through π-back-donation. Either the ionization potentials of the gaseous metals in their spin-paired valency state, as has been discussed by Nyholm [12], or the $(n - 1)d \rightarrow np$ promotion energies which roughly parallel the ionization potentials (Table 1.XV), can be taken as a rough guide to this.

In the nickel triad, the aptitude to form d_π-bonds should follow the order: $Ni^{(0)} \gg Pt^{(0)} \geqslant Pd^{(0)}$. This is in agreement with the fact that nickel gives a well known and stable metal carbonyl, whilst the corresponding palladium and platinum derivatives are unstable and have not been synthesized. In contrast the values of the energies of atomization of the metals and those necessary for spin-pairing suggest that the preparation of $Pd(CO)_4$ at least should provide no difficulty [12]. Further support to the π-back-donation hypothesis is given by the fact that substituted carbonyls of the type $Pt(CO)_2(PR_3)_2$ and $Pt(CO)(PR_3)_3$ can be easily prepared [159a, 191]. The presence of the basic phosphine ligands apparently increases the donor capacity of the platinum atom; at the same time there are fewer CO groups competing for double bonding. Since on the basis of promotion energies palladium has the poorest donor capacity, it is not surprising that while $Pd(CO)(PPh_3)_3$ can be prepared, the bis-carbonyl derivative could not be obtained [192]. Finally the longest wavelength maxima of the three $M(PPh_3)_3$ complexes, which are considered to correspond to the metal-to-ligand electron transfer, were found at 393, 332 and 322 nm for Ni, Pt and Pd respectively [189] in agreement with what has been discussed above.

TABLE 1.XV

Ionization potentials of spin-paired states (eV), $(n-1)\ d \rightarrow np$ promotion energies (eV) and related data for some transition elements. Data taken from ref. 12 and 179

Metal atom	Ionization potential of spin-paired state (eV)	$(n-1)d \rightarrow np$ promotion energy (eV)	Heat of atomization + spin pairing energy (eV)	Heat of reaction (kcal mol^{-1}) $1/n\ M^*(g) + CO^*(g) \rightarrow 1/n\ M(CO)_n(g)$
Cr	4.01	—	11.0	87
Fe	3.90	—	10.8	89
Co	6.67	0.85	—	—
Rh	7.31	1.60	—	—
Ir	7.95	2.4	—	—
Ni	5.81	1.72	6.2	77
Pd	8.33	4.23	3.8	—
Pt	8.20	3.28	6.4	—

This approach also agrees with kinetic and thermodynamic data (1.3.5). These considerations do not fit with the results of photoelectron spectroscopy of $M(PF_3)_4$ (M = Ni, Pt) complexes (1.3.3) since in these cases it can be concluded that the extent of electron delocalization by σ and π mechanisms is greater in the platinum than in the nickel complex where the effects of the σ and π bond compensate each other.

On the basis of the heats of vaporization and the energies of spin pairing only, the carbonyls of chromium and iron should be much less stable than $Ni(CO)_4$ (Table 1.XV) [12]. Conversely, π-back-donation from the metal to CO should take place much more readily for chromium and iron than for nickel. The overall heats of formation of these carbonyls from the metals in their spin-paired state have been estimated by Cotton, Fischer and Wilkinson [179] (Table 1.XV) and support the view that π-bonding is very important for the stability of such complexes.

For the cobalt triad, both the ionization potentials and the promotion energies (Table 1.XV) suggest that iridium(0) should have the poorest π-basicity. This is in agreement with the instability of $Ir_2(CO)_8$ [195], whilst the $[Ir(CO)_4]^-$ ion, in which an excess of negative charge on the metal is present, can be readily obtained [193]. Moreover, the introduction in the molecule of a basic phosphine ligand allows the isolation of $\{Ir(CO)_3(PPh_3)\}_2$ [194].

This poor π-basicity of iridium(0) is in contrast with the strong increase in coordinative reactivity (still related to promotion energies) on going from rhodium to iridium in their +1 oxidative state [196]. However, the formal positive charge on the metal may even reverse the order of promotion energies [12].

Although no promotion energy data are available for the iron and chromium triad, the very low ionization potentials of the first two members of the series (Table 1.XV) suggest that these elements should exhibit a very high tendency to π-back-bond to the ligands. This may explain, for instance, why the $M(CO)_5$ carbonyls (M = Fe, Ru, Os) can be easily prepared, as can the corresponding derivatives with PF_3 as ligands, whilst, up to now, only mixed $M(CO)_x(PPh_3)_{5-x}$ derivatives have been obtained, although in these cases steric effect must be kept in mind. Moreover from the rates of substitution of $M(CO)_6$ carbonyls (M = Cr, Mo, W) which vary in the order Cr $<$ Mo $>$ W [175], it is expected that the promotion energies follow the same order.

At this point, we should recall that the ease by which electron density can be removed from the metal in a complex can be greatly affected by the nature of the ligands. This means that one can modify the reactivity of a metal complex by varying, for instance, the basicity of the ligands, and this represents a very important point for the understanding of the coordinative reactivity of complexes with the metal in a low oxidation state.

In particular when a vacant site is present (or it has been generated by dissociation in solution) in a complex of a metal in a low oxidation state (which means in general a low promotion energy) one should expect a high reactivity toward dissociative or coordinative addition [197]:

$$L_nM + XY \longrightarrow L_nM(X)(Y) \qquad \text{or } L_{n-1}M(X)(Y) + L$$

$$L_nM + Z \longrightarrow L_nM(Z) \qquad \text{or } L_{n-1}M(Z) + L$$

($XY = H_2$, HCl, RCl, RCOCl; $Z = CH_2{=}CH_2$, $CH{\equiv}CH$, O_2, CO, SO_2, etc.). In the first case there is a dissociation of the entering molecule XY, with the oxidation of the metal and the energy variation ΔE of the reaction can be expressed approximately as [197]:

$$\Delta E = E_{MX} + E_{MY} - (E_{XY} + P)$$

In this equation E_{XY} is the energy of the X–Y bond and P is the promotion energy corresponding to the process: $d^n \rightarrow d^{n-2}$. By changing the ligands L in the L_nM complex it is possible to modify P. The more basic ligand L, the lower should be P; in this case ΔE is greater, and thus the reaction with the same entering molecule XY, becomes easier. This discussion emphasizes how in a chemical reaction, the nature of the metal and the properties both of L and XY are strictly interdependent.

1.5. CONCLUDING REMARKS

More than ten years ago, Nyholm considered it surprising that only a few unambiguous experimental results were available to support the hypothesis of the existence of d_π-bonding between metals and certain ligands [12]. In spite of the large amount of research carried out since then to solve this problem, we must now conclude that a precise answer still cannot be given. However, in our opinion the results achieved by new physicochemical techniques such as photoelectron and Mössbauer spectroscopy seem to favour the existence of such a bond. In any case, the hypothesis of the existence of the π-back-donation from metals to ligands has been useful both from a practical and a theoretical point of view.

In fact, since 1950 this has been the leading hypothesis for all subsequent research in the field of low oxidation state complexes, and it has given numerous important and often completely unexpected results. Though there are still only a few experimental results that demonstrate the existence of back donation it is also true there are even fewer experimental results definitely contradicting this idea. The difficulties met with in reaching a conclusion are related to the synergic nature of this bond, which tends to be reinforced by the concomitant σ-donation from ligands to metals, but at present we do not have the necessary means to discriminate clearly between them.

Much of the interpretation of the experimental data in order to support or contradict the existence and/or the importance of the π-bond appears to be doubtful. In particular the results from infrared spectroscopy (1.3.1) seem to be uncertain, and susceptible to significant revisions. A correct assignment of metal–phosphorus stretching frequencies, necessitates the use of the metal isotope technique (1.3.1.1); moreover it must be considered on discussing metal–ligand bond orders in terms of vibrational data the possibility of a mixing of various symmetry modes in the normal modes (1.3.1.1). Besides, we have seen how i.r. parameters generally depend upon the joint contribution of σ and π interactions to the metal–ligand bond, so that the modification of a bond within the ligand cannot be only ascribed to the π component of the metal–ligand electronic interaction (1.3.1). Chemical shifts (1.3.2.1) and coupling constants (1.3.2.2) appear to be too complex in their origin for clarifying this problem at present. On the other hand kinetic and thermodynamic data (1.3.5) provided convincing explanations for low oxidation state complexes when the role of the metal as a π-donor base was considered (1.4). Particularly significant is the agreement between these experimental results and the predictions based on the promotion energies for the metals of the nickel triad. We must also recall that in many cases only electronic effects have been considered, whilst we have seen that steric effects can be very important in determining the stereochemistry and the reactivity of complexes (1.3.2.3). Theoretical calculations have shown that metals in low oxidation states and certain ligands have orbitals of suitable symmetry and energy for large overlaps in the π-bond. This has been demonstrated particularly for a ligand such as carbon monoxide (1.3.1.1). However, in the case of bonds between metal and unsaturated hydrocarbons some semi-empirical MO calculations have also shown how the π-bond is entirely acceptable from a theoretical point of view (1.3.1.2). On the other hand, the metal d orbitals are closer in energy to the π^* antibonding orbitals in ligands such as NO and O_2 than in ligands such as CO, N_2 and CN^-, and this explains why the former are able to accept an electron in an antibonding orbital (1.3.1.3). It is expected that more refined calculations will allow in the future more exact determination of the extent of the electron delocalization from the metal to the ligand, and vice versa, in this type of covalent bond.

REFERENCES

1. L. Mond, C. Langer and F. Quinke, *J. Chem. Soc.*, **57**, 749 (1890).
2. L. Mond and F. Quinke, *J. Chem. Soc.*, 604, (1891).
3. M. Berthelot, *Compt. Rend.*, **112,** 1343 (1891).
4. J. W. Eastes and W. M. Burgess, *J. Amer. Chem. Soc.*, **64,** 1187 (1942).
5. J. J. Burbage and W. C. Fernelius, *J. Amer. Chem. Soc.*, **65,** 1484 (1943).
6. L. Malatesta, *Endeavour,* **28,** 30 (1969).

7. L. Pauling, "The Nature of the Chemical Bond", Cornell University Press, Ithaca, New York (1939).
8. C. L. Deasy, *J. Amer. Chem. Soc.*, **67**, 152 (1945).
9. L. Pauling, *J. Chem. Soc.*, 1461 (1948).
10. M. F. Amr El Sayed and R. K. Sheline, *J. Amer. Chem. Soc.*, **80**, 2047 (1958).
11. G. Giacometti, *J. Chem. Phys.*, **23**, 2068 (1955).
12. (a) R. S. Nyholm and M. L. Tobe, *Adv. Inorg. Chem. Radiochem.*, **5**, 1 (1963).
    (b) R. S. Nyholm, *Proc. Chem. Soc.*, 273 (1961).
13. L. E. Orgel, "An Introduction to Transition Metal Chemistry", Methuen and Co., London (1960).
14. L. M. Venanzi, Accad. Nazl. Lincei, IV Corso Estivo di Chimica Varenna 415 (1959).
15. J. J. Kaufman and W. S. Koski, *J. Amer. Chem. Soc.*, **82**, 3262 (1960).
16. H. C. Brown and Y. Okamoto, *J. Amer. Chem. Soc.*, **80**, 4979 (1958).
17. G. Distefano, G. Innorta, S. Pignataro and A. Foffani, *J. Organometallic Chem.*, **14**, 165 (1968) and references therein.
18. W. A. Henderson, Jr. and C. A. Streuli, *J. Amer. Chem. Soc.*, **82**, 5791 (1960).
19. R. W. Taft, Jr., "Steric Effects in Organic Chemistry", M. S. Newmann, ed., John Wiley and Sons Inc., New York, N.Y. (1956).
20. K. Watanabe, *J. Chem. Phys.*, **26**, 542 (1957).
21. A. Foffani, S. Pignataro, G. Distefano and G. Innorta, *J. Organometallic Chem.*, **7**, 473 (1967) and references therein.
22. L. D. Pettit, *Quart. Rev.*, **25**, 1 (1971) and references therein.
23. R. G. Wilkins, *Quart. Rev.*, **16**, 316 (1962).
24. M. R. Churchill and R. Mason, *Adv. Organometallic Chem.*, **5**, 93 (1967).
25. J. Chatt and F. A. Hart, *J. Chem. Soc.*, 1378 (1960).
26. T. L. Brown and D. J. Darensbourg, *Inorg. Chem.*, **6**, 971 (1967).
27. (a) D. M. Adams, "Metal-Ligand and Related Vibrations", E. Arnold (Publishers) (1967).
    (b) D. W. James and M. J. Nolan, *Progr. Inorg. Chem.*, **9**, 195 (1968).
    (c) E. Maslowsky, Jr., *Chem. Rev.*, **71**, 507 (1971) and references therein.
    (d) D. M. Adams and P. J. Chandler, *Chem. Comm.*, 69 (1966).
    (e) K. Nakamoto, "Infrared Spectra of Inorganic and Coordination Compounds" (Second Edition), Wiley–Interscience, John Wiley and Sons (1970).
28. See for instance: (a) V. G. Myers, F. Basolo and K. Nakamoto, *Inorg. Chem.*, **8**, 1204 (1969); (b) R. L. Keiter and J. G. Verkade, *Inorg. Chem.*, **9**, 404 (1970).
29. K. Shobatake and K. Nakamoto, *J. Amer. Chem. Soc.*, **92**, 3332 (1970).
30. (a) S. F. A. Kettle, *Spectrochim. Acta*, **22**, 1388 (1965).
    (b) K. G. Caulton and R. F. Fenske, *Inorg. Chem.*, **7**, 1273 (1968).
31. N. A. Beach and H. B. Gray, *J. Amer. Chem. Soc.*, **90**, 5713 (1968).
32. E. W. Abel, R. A. N. McLean, S. P. Tyfield, P. S. Braterman, A. P. Walker, and P. J. Hendra, *J. Mol. Spectroscopy*, **30**, 29 (1969).
33. (a) D. E. Clegg and J. R. Hall, *Spectrochim. Acta.*, **23A**, 263 (1967).
    (b) G. T. Behnke and K. Nakamoto, *Inorg. Chem.*, **6**, 440 (1967).
34. L. H. Jones, *Spectrochim. Acta*, **19**, 329 (1963).

35. L. H. Jones, "Inorganic Vibrational Spectroscopy", Vol. 1, Marcel Dekker, Inc., (1971).
36. F. A. Cotton and C. S. Kraihanzel, *J. Amer. Chem. Soc.*, **84**, 4432 (1962).
37. F. A. Cotton, *Inorg. Chem.*, **3**, 702 (1964).
38. W. A. Graham, *Inorg. Chem.*, **7**, 315 (1968).
39. M. Bigorgne, *J. Inorg. Nucl. Chem.*, **26**, 107 (1964).
40. R. J. Angelici and M. D. Malone, *Inorg. Chem.*, **6**, 1731 (1967).
41. M. Bigorgne, *J. Organometallic Chem.*, **2**, 68 (1964).
42. C. A. Tolman, *J. Amer. Chem. Soc.*, **92**, 2953 (1970) and references therein.
43. R. P. Stewart and P. M. Treichel, *Inorg. Chem.*, **7**, 1942 (1968).
44. R. Ugo, S. Cenini and F. Bonati, *Inorg. Chim. Acta*, **1**, 451 (1967).
45. J. F. Nixon, *Adv. Inorg. Chem. Radiochem.*, **13**, 364 (1970) and references therein.
46. T. Kruck, *Angew. Chem. Internat. Edn.*, **6**, 53 (1967).
47. A. Loutelier and M. Bigorgne, *Bull. Soc. Chim. France*, **11**, 3186 (1965).
48. L. A. Woodward and J. R. Hall, *Spectrochim. Acta*, **16**, 654 (1960).
49. M. J. S. Dewar, *Bull. Soc. Chim. France*, **18**, C79 (1951).
50. J. Chatt and L. A. Duncanson, *J. Chem. Soc.*, 2939 (1953).
51. M. A. Muhs and F. T. Weiss, *J. Amer. Chem. Soc.*, **84**, 4697 (1962).
52. G. Bressan, R. Broggi, M. P. Lachi and A. L. Segre, *J. Organometallic Chem.*, **9**, 355 (1967) and references therein.
53. R. G. Denning, F. R. Hartley and L. M. Venanzi, *J. Chem. Soc.* (*A*), 324 (1967).
54. W. J. Bland and R. D. W. Kemmit, *J. Chem. Soc.* (*A*), 1278 (1968).
55. W. H. Baddley, *J. Amer. Chem. Soc.*, **90**, 3705 (1968).
56. W. H. Baddley, *J. Amer. Chem. Soc.*, **88**, 4545 (1966).
57. S. Cenini, R. Ugo and G. La Monica, *J. Chem. Soc.* (A), 409 (1971).
58. J. Chatt, L. A. Duncanson and R. G. Guy, *Chem. and Ind.*, 430 (1959).
59. E. O. Greaves, C. J. L. Lock and P. M. Maitlis, *Canad. J. Chem.*, **46**, 3879 (1968).
60. J. Chatt, G. A. Rowe and A. A. Williams, *Proc. Chem. Soc.*, 208 (1957).
61. J. H. Nelson, K. S. Wheelock, L. C. Cusachs and H. B. Jonassen, *J. Amer. Chem. Soc.*, **91**, 7005 (1969).
62. K. S. Wheelock, J. H. Nelson, L. C. Cusachs and H. B. Jonassen, *J. Amer. Chem. Soc.*, **92**, 5110 (1970).
62*bis*. J. H. Nelson, K. S. Wheelock, L. C. Cusachs, and H. B. Jonassen, *Inorg. Chem.* **11**, 422 (1972).
63. G. N. Schrauzer, *J. Amer. Chem. Soc.*, **82**, 1009 (1960).
64. C. Panattoni, G. Bombieri, U. Belluco and W. H. Baddley, *J. Amer. Chem. Soc.*, **90**, 799 (1968).
65. (a) M. Green, R. B. L. Osborne, A. J. Rest, and F. G. A. Stone, *Chem. Comm.*, 502 (1966).
(b) *ibidem, J. Chem. Soc.* (*A*), 2525 (1968).
66. R. Mason, *Nature*, **217**, 543 (1968); *Chem. Soc. Rev.*, **1**, 431 (1972).
67. A. C. Blizzard and D. P. Santry, *J. Amer. Chem. Soc.*, **90**, 5749 (1968).
68. R. Mason and G. Robertson, *J. Chem. Soc.* (*A*), 492 (1969).
69. S. F. A. Kettle and R. Mason, *J. Organometallic Chem.*, **5**, 97 (1966).
70. M. Dekker and G. R. Knox, *Chem. Comm.*, 1243 (1967).
71. R. J. Doedens, *Chem. Comm.*, 1271 (1968).

72. R. C. Sahni and E. J. Lorenzo, *J. Chem. Phys.*, **42**, 3619 (1965).
73. H. Brion and M. Yamazaki, *J. Chem. Phys.*, **30**, 673 (1959).
74. B. J. Ransil, *Rev. Mod. Phys.*, **32**, 245 (1960).
75. (a) B. F. G. Johnson and J. A. McCleverty, *Progr. Inorg. Chem.*, **7**, 277 (1966).
(b) W. P. Griffith, *Adv. Organometallic Chem.*, **7**, 211 (1968).
(c) N. G. Connelly, *Inorg. Chim. Acta Rev.*, **6**, 47 (1972).
76. J. Lewis, R. J. Irving and G. Wilkinson, *J. Inorg. Nucl. Chem.*, **7**, 32 (1958).
77. P. Politzer and R. H. Harris, *J. Amer. Chem. Soc.*, **92**, 1834 (1970).
78. P. Gans, *Chem. Comm.*, 144 (1965).
79. B. F. Hoskins, F. D. Whillans, D. H. Dale and D. C. Hodgkin, *Chem. Comm.*, 69 (1969).
80. (a) D. J. Hodgson, N. C. Payne, J. A. McGinnety, R. G. Pearson and J. A. Ibers, *J. Amer. Chem. Soc.*, **90**, 4486 (1968).
(b) D. J. Hodgson and J. A. Ibers, *Inorg. Chem.*, **7**, 2345 (1968).
81. D. M. P. Mingos and J. A. Ibers, *Inorg. Chem.*, **10**, 1035 (1971).
82. V. G. Albano, P. Bellon and M. Sansoni, *J. Chem. Soc.* (*A*), 2420 (1971).
83. D. M. P. Mingos and J. A. Ibers, *Inorg. Chem.*, **10**, 1479 (1971).
84. C. G. Pierpont and R. Eisenberg, *J. Amer. Chem. Soc.*, **93**, 4905 (1971).
85. A. I. M. Rae, *Chem. Comm.* 1245 (1967).
86. A. D. Walsh, *J. Chem. Soc.*, 2266 (1953).
87. J. H. Enemark, *Inorg. Chem.*, **10**, 1952 (1971).
88. S. Cenini, R. Ugo, G. La Monica and S. D. Robinson, *Inorg. Chim. Acta*, **6**, 182 (1972).
88*bis*. J. R. Norton, J. P. Collman, G. Dolcetti and W. T. Robinson, *Inorg. Chem.*, **11**, 382 (1972).
89. S. J. La Placa and J. A. Ibers, *Inorg. Chem.*, **5**, 405 (1966).
90. L. H. Vogt, Jr., J. L. Katz and S. E. Wiberley, *Inorg. Chem.*, **4**, 1157 (1965).
91. L. Vaska and S. S. Bath, *J. Amer. Chem. Soc.*, **88**, 1333 (1966).
92. R. Ugo, *Co-ordination Chem. Rev.*, **3**, 319 (1968) and references therein.
93. A. R. Crumbliss and F. Basolo, *J. Amer. Chem. Soc.*, **92**, 55 (1970).
93*bis*. G. A. Rodley and W. T. Robinson, *Nature*, **235**, 438 (1972).
94. B. M. Hoffman, D. L. Diemente and F. Basolo, *J. Amer. Chem. Soc.*, **92**, 1961 (1970).
95. (a) K. Hirota, M. Yamamoto, S. Otsuka, A. Nakamura and Y. Tatsuno, *Chem. Comm.*, 533 (1968).
(b) A. Nakamura, Y. Tatsuno, M. Yamamoto and S. Otsuka, *J. Amer. Chem. Soc.*, **93**, 6052 (1971).
96. J. S. Griffith, *Proc. Roy. Soc.*, Ser. A, **235**, 23 (1956).
97. K. G. Caulton, R. L. De Kock and R. F. Fenske, *J. Amer. Chem. Soc.*, **92**, 515 (1970).
98. P. T. Monaharan and H. B. Gray, *J. Amer. Chem. Soc.*, **87**, 3340 (1965).
99. (a) G. Henrici-Olivé and S. Olivé, *Angew, Chem. Internat. Edn.*, **8**, 650 (1969).
(b) J. Chatt and G. J. Leigh, *Quart. Rev.*, **26**, 121 (1972) and references therein.
100. J. A. Pople and G. A. Segal, *J. Chem. Phys.*, **43**, 5136 (1965).
101. J. Chatt, *Platinum Metal Rev.*, **13**, 9 (1969).
102. Y. G. Borod'ko and E. A. Shilov, *Russ. Chem. Rev.*, **38**, 355 (1969).

103. A. D. Allen and C. V. Senoff, *Chem. Comm.*, 621 (1965).
104. G. C. Bond, *Disc. Faraday Soc.*, **41**, 200 (1966).
105. K. F. Purcell, *Inorg. Chim. Acta*, **3**, 540 (1969).
106. J. Chatt, D. P. Melville and R. L. Richards, *J. Chem. Soc.* (*A*), 2841 (1969).
107. W. Moffitt, *Proc. Chem. Soc.*, **A196**, 524 (1949).
108. (a) K. F. Purcell, *J. Amer. Chem. Soc.*, **89**, 247 (1967).
(b) K. F. Purcell and R. S. Drago, *J. Amer. Chem. Soc.*, **88**, 919 (1966).
109. (a) P. W. Jolly and K. Jonas, *Angew Chem. Internat. Edn.*, **7**, 731 (1968).
(b) P. W. Jolly, K. Jonas, C. Krüger and J. H. Tsay, *J. Organometallic Chem.*, **33**, 109 (1971).
109*bis*. (a) D. J. Darensbourg and C. L. Hyde, *Inorg. Chem.*, **10**, 431 (1971).
(b) D. J. Darensbourg, *Inorg. Chem.*, **11**, 1436 (1972).
110. S. Cenini, R. Ugo and G. La Monica, *J. Chem. Soc.* (*A*), 416 (1971).
111. M. L. Maddox, S. L. Stafford and H. D. Kaesz, *Adv., Organometallic Chem.*, **3**, 1 (1965).
112. G. W. Parshall and F. N. Jones, *J. Amer. Chem. Soc.*, **87**, 5356 (1965).
113. H. H. Hoehn, L. Pratt, K. F. Watterson and G. Wilkinson, *J. Chem. Soc.*, 2738 (1961).
114. E. Pitcher, A. D. Buckingham and F. G. A. Stone, *J. Chem. Phys.*, **36**, 124 (1962).
115. G. S. Reddy and R. Schmutzler, *Z. Naturforsch.*, **20b**, 104 (1965).
116. G. S. Reddy and R. Schmutzler, *Inorg. Chem.*, **6**, 823 (1967).
117. A. Saika and C. P. Slichter, *J. Chem. Phys.*, **22**, 26 (1954).
118. S. O. Grim, D. A. Wheatland and W. McFarlane, *J. Amer. Chem. Soc.*, **89**, 5573 (1967).
119. N. Muller, P. C. Lauterbur and J. Goldenson, *J. Amer. Chem. Soc.*, **78**, 3557 (1956).
120. H. S. Gutowsky and J. Larmann, *J. Amer. Chem. Soc.*, **87**, 3815 (1965).
121. L. S. Meriwether and J. R. Leto, *J. Amer. Chem. Soc.*, **83**, 3192 (1961).
122. J. F. Nixon and A. Pidcock, *Ann. Rev. N.M.R. Spectroscopy*, **2**, 345 (1969).
123. J. G. Riess and J. R. Van Wazer, *J. Amer. Chem. Soc.*, **88**, 2339 (1966).
124. J. G. Riess and J. R. Van Wazer, *J. Amer. Chem. Soc.*, **88**, 2166 (1966).
125. V. Mark, C. H. Dungam, M. M. Crutchfield and J. R. Van Wazer, "Topics in Phosphorus Chemistry", Vol. 5, M. Grayson and E. J. Griffith, Ed., John Wiley and Sons, Inc., New York, N.Y. (1967), p. 227.
126. R. Mathieu, M. Lenzi and R. Poilblanc, *Inorg. Chem.*, **9**, 2030 (1970).
127. S. I. Shupack and B. Wagner, *Chem. Comm.*, 547 (1966).
128. K. J. Coskran, R. D. Bertrand and J. G. Verkade, *J. Amer. Chem. Soc.*, **89**, 4535 (1967).
128*bis*. B. E. Mann, C. Masters, B. L. Shaw, R. M. Slade and R. E. Stainbank, *Inorg. Nucl. Chem. Letters*, **7**, 881 (1971).
129. O. Stelzer and R. Schmutzler, *J. Chem. Soc.* (*A*), 2867 (1971) and references therein.
130. S. O. Grim, P. R. McAllister and R. M. Singer, *Chem. Comm.*, 38 (1969).
131. S. O. Grim and D. A. Wheatland, *Inorg. Chem.*, **8**, 1716 (1969) and references therein.
132. R. L. Keiter and J. G. Verkade, *Inorg. Chem.* **8**, 2115 (1969).
133. H. A. Bent, *Chem. Rev.*, **61**, 275 (1961).

134. G. G. Mather and A. Pidcock, *J. Chem. Soc.* (*A*), 1226 (1970).
135. B. E. Mann, *J. Chem. Soc.* (*D*) 1173 (1971).
136. (a) A. Pidcock, R. E. Richards, and L. M. Venanzi, *J. Chem. Soc.* (*A*), 1970 (1968).
(b) A. Pidcock, R. E. Richards, and L. M. Venanzi, *J. Chem. Soc.* (*A*), 1707 (1966).
137. S. O. Grim, R. L. Keiter and W. McFarlane, *Inorg. Chem.*, **6**, 1133 (1967).
138. L. F. Farnell, E. W. Randall and E. Rosenberg, *J. Chem. Soc.* (D), 1078 (1971) and references therein.
139. O. A. Gansow, B. Y. Kimura, G. R. Dobson and R. A. Brown, *J. Amer. Chem. Soc.* **93**, 5922 (1971).
140. C. D. Cook and K. Y. Wan, *J. Amer. Chem. Soc.*, **92**, 2595 (1970).
140*bis*. M. H. Chilsholm, H. C. Clark, L. E. Manzer and J. B. Stothers, *J. Amer. Chem. Soc.*, **94**, 5087 (1972).
141. R. Ugo, G. La Monica, F. Cariati, S. Cenini and F. Conti, *Inorg. Chim. Acta*, **4**, 390 (1970).
142. C. E. Holloway, G. Hulley, B. F. G. Johnson and J. Lewis, *J. Chem. Soc.* (*A*), 1653 (1970) and references therein.
143. R. D. W. Kemmitt and R. D. Moore, *J. Chem. Soc.* (*A*), 2472 (1971).
144. H. C. Clark and R. J. Puddephatt, *Inorg. Chem.*, **10**, 18 (1971).
145. J. G. Verkade, R. E. McCarley, D. G. Hendricker and R. W. King, *Inorg. Chem.*, **4**, 228 (1965).
146. D. G. Hendricker, R. E. McCarley, R. W. King and J. G. Verkade, *Inorg. Chem.*, **5**, 639 (1966) and references therein.
147. S. O. Grim, D. A. Wheatland and P. R. McAllister, *Inorg. Chem.*, **7**, 161 (1968).
148. J. M. Jenkins and B. L. Shaw, *J. Chem. Soc.* (*A*), 1407 (1966).
149. F. B. Ogilvie, R. L. Keiter, G. Wulfsberg and J. G. Verkade, *Inorg. Chem.*, **8**, 2346 (1969).
150. F. B. Ogilvie, R. J. Clark and J. G. Verkade, *Inorg. Chem.*, **8**, 1904 (1969).
151. W. McFarlane and R. F. M. White, *Chem. Comm.*, 744 (1969).
152. R. D. Bertrand, F. B. Ogilvie and J. G. Verkade, *J. Amer. Chem. Soc.*, **92**, 1908 (1970).
153. R. M. Lynden-Bell, J. F. Nixon, J. Roberts and J. R. Swain, *Inorg. Nucl., Chem. Letters*, **7**, 1187 (1971).
154. F. B. Ogilvie, J. M. Jenkins and J. G. Verkade, *J. Amer. Chem. Soc.*, **92**, 1916 (1970).
155. W. Strohmeier, F. Guttenberger and F. J. Müller, *Z. Naturforsch.*, **22b**, 1091 (1967).
156. C. A. Tolman, *J. Amer. Chem. Soc.*, **92**, 2956 (1970).
157. L. W. Gosser and C. A. Tolman, *Inorg. Chem.*, **9**, 2350 (1970).
158. H. C. Clark and K. Itoh, *Inorg. Chem.*, **10**, 1707 (1971).
159. (a) L. Malatesta and C. Cariello, *J. Chem. Soc.*, 2323 (1958).
(b) R. Ugo, F. Cariati and G. La Monica, *Inorg. Synth.*, **XI**, 105 (1968).
160. L. Vaska and L. S. Chen, *J. Chem. Soc.* (*D*), 1080 (1971).
161. (a) L. Vaska and M. F. Werneke, *Transactions New York Academy of Sciences*, **33**, 70 (1971) and references therein.
(b) R. Ugo, A. Pasini, A. Fusi and S. Cenini, *J. Amer. Chem. Soc.*, **94**, 7364 (1972).
162. J. M. Hollander and W. L. Jolly, *Accounts Chem. Res.*, **3**, 193 (1970).

163. (a) A. D. Baker, *Accounts Chem. Res.*, **3**, 17 (1970).
(b) A. D. Baker, C. R. Brundle and M. Thompson, *Chem. Soc. Rev.*, **1**, 355 (1972).
164. S. D. Worley, *Chem. Rev.*, **71**, 295 (1971) and references therein; see also: S. Pignataro, *Chimica e Industria*, **53**, 382 (1971).
165. (a) M. Barber, J. A. Connor, M. F. Guest, I. H. Hillier and V. R. Saunders, *J. Chem. Soc. (D)*, 943 (1971).
(b) W. E. Morgan, W. J. Stec, R. G. Albridge and J. R. Van Wazer, *Inorg. Chem.*, **10**, 926 (1971) and references therein.
166. C. D. Cook, K. Y. Wan, U. Gelius, K. Hamrin, G. Johannson, E. Olsson, H. Siegbahn, C. Nordling and K. Siegbahn, *J. Amer. Chem. Soc.*, **93**, 1904 (1971).
167. R. Mason, D. M. P. Mingos, G. Rucci and J. A. Connor, *J. C. S. Dalton*, 1729 (1972).
168. (a) M. Barber, J. A. Connor, I. H. Hillier and V. R. Saunders, *J. Chem. Soc. (D)*, 682 (1971).
(b) D. T. Clark and D. B. Adams, *J. Chem. Soc. (D)*, 740 (1971).
169. I. H. Hillier, V. R. Saunders, M. J. Ware, P. J. Bassett, D. R. Lloyd, and N. Lynaugh, *J. Chem. Soc. (D)*, 1316 (1970).
170. J. C. Marriott, J. A. Salthouse, M. J. Ware, and J. M. Freeman, *J. Chem. Soc. (D)*, 595 (1970).
170*bis*. P. Finn and W. L. Jolly, *Inorg. Chem.*, **11**, 893 and 1434 (1972).
171. (a) E. Fluck, *Adv. Inorg. Chem. Radiochem.*, **6**, 433 (1964).
(b) V. I. Goldanskii, *Angew. Chem. Internat. Edn.*, **6**, 830 (1967).
(c) V. I. Goldanskii, V. V. Khrapov and R. A. Stukan, *Organometallic Chem. Rev. (A)*, **4**, 225 (1969).
(d) R. L. Mössbauer, *Agnew. Chem. Internat. Edn.*, **10**, 462 (1971).
172. R. L. Collins and R. Pettit, *J. Amer. Chem. Soc.*, **85**, 2332 (1963).
173. R. L. Collins and R. Pettit, *J. Chem. Phys.*, **39**, 3433 (1963).
173*bis*. L. H. Bowen, P. E. Garrou and G. G. Long, *Inorg. Chem.*, **11**, 182 (1972).
174. W. R. Cullen, D. A. Harbourne, B. V. Liengme and J. R. Sams, *Inorg. Chem.*, **8**, 1464 (1969).
175. (a) F. Basolo and R. G. Pearson, *Adv. Inorg. Chem. Radiochem.*, **3**, 1 (1961).
(b) D. A. Brown, *Inorg. Chim. Acta Rev.*, **1**, 35 (1967).
(c) R. J. Angelici, *Organometallic Chem. Rev.*, **3**, 173 (1968).
(d) H. Werner, *Angew. Chem. Internat. Edn.*, 7, 930 (1968).
176. R. D. Johnston, F. Basolo and R. G. Pearson, *Inorg. Chem.*, **10**, 247 (1971).
177. J. P. Day, F. Basolo and R. G. Pearson, *J. Amer. Chem. Soc.*, **90**, 6927 (1968).
178. M. Meier, F. Basolo and R. G. Pearson, *Inorg. Chem.*, **8**, 795 (1969).
179. F. A. Cotton, A. K. Fischer and G. Wilkinson, *J. Amer. Chem. Soc.*, **81**, 800 (1959).
180. F. Basolo and R. G. Pearson, "Mechanism of Inorganic Reactions", John Wiley and Sons, Inc., New York, N.Y. (1967).
181. C. D. Cook and K. Y. Wan, *Inorg. Chem.* **10**, 2696 (1971).
182. A. D. Allen and C. D. Cook, *Canad. J. Chem.*, **42**, 1063 (1964).

183. T. Yamamoto, A. Yamamoto and S. Ikeda, *J. Amer. Chem. Soc.*, **93**, 3360 (1971).
184. J. Brandup and E. H. Immergut, "Polymer Handbook", Interscience, New York, N.Y. (1966).
185. J. Furukawa and T. Tsuruta, *J. Polym. Sci.*, **36**, 257 (1959).
186. (a) J. D. Morrison and A. J. C. Nicholson, *J. Chem. Phys.*, **20**, 1021 (1952).
(b) A. Streitwieser, Jr., *J. Amer. Chem. Soc.*, **82**, 4123 (1960).
187. M. J. Kamlet, "Organic Electronic Spectral Data", Interscience, New York, N.Y. (1946).
188. T. Kagiya, Y. Sumida and T. Nakata, *Bull. Chem. Soc. Jap.*, **41**, 2239 (1968).
189. C. A. Tolman, W. C. Seidel and D. H. Gerlach, *J. Amer. Chem. Soc.*, **94**, 2669 (1972).
190. F. A. Cotton, *J. Chem. Soc.*, 5269 (1960).
191. P. Chini and G. Longoni, *J. Chem. Soc. (A)*, 1542 (1970).
192. K. Kudo, M. Hidai and Y. Uchida, *J. Organometallic Chem.*, **33**, 393 (1971).
193. L. Malatesta, G. Caglio, and M. Angoletta, *J. Chem. Soc. (D)*, 532 (1970).
194. L. Malatesta, M. Angoletta and G. Caglio, *J. Organometallic Chem.*, in press.
195. R. Whyman, *J. Organometallic Chem.*, **29**, C36 (1971).
196. L. Vaska, *Inorg. Chim. Acta*, **5**, 295 (1971).
197. S. Carrà and R. Ugo, *Inorg. Chim. Acta Rev.*, **1**, 49 (1967) and references therein.

Part 2

Complexes with Tricovalent P, As and Sb Derivatives

2.1. INTRODUCTION

Zerovalent complexes having only donor atoms belonging to Group VB are for the most part confined to the elements of the nickel triad. Some years after their discovery, there was large development of the chemistry of such compounds [1]. The Pt° and Pd° complexes with phosphines and related ligands, synthesized by Malatesta and coworkers in 1957–58 [2, 3], were suspected for some time of not being true zerovalent species. Even when their real nature was proved [4], it was initially supposed that a d^{10} configuration would not be particularly prone to coordinative reactions, because of the very high ionization potentials of palladium and platinum. However by considering the $(n-1)d^{10} \rightarrow (n-1)d^9np$ promotion energies (see 1.4), which are much lower than the ionization potentials and more reliably related to the energies involved in the coordination reactions, it later appeared quite reasonable that a closed zerovalent d^{10} shell could act as a "soft" centre towards coordinating molecules [5]. Moreover the presence of "soft" ligands such as phosphines and arsines synergically increases the "softness" of the metal atom [6] and consequently the availability of electron density for charge transfer to the coordinated molecule [5].

As a consequence, during the last five years an increasing amount of work has been done on the synthesis and reactions of new complexes of this type although this research appears to be far from finished.

2.2. NICKEL, PALLADIUM AND PLATINUM

2.2.1. Methods of Synthesis

There are many different preparative methods for the zerovalent nickel, palladium and platinum derivatives, which are generally similar to those usually employed in the chemistry of other complexes in low oxidation state [7]. The more commonly used are summarized below.

(*a*) *Direct synthesis from the metals.* The first zerovalent compound prepared, namely $Ni(CO)_4$, was first obtained by this method, i.e. from active nickel and carbon monoxide at normal temperature and pressure [8].

Pauling's hypothesis of the partial double bond character of the Ni–C bond in $Ni(CO)_4$ [9] indicated a condition for the stability of zerovalent derivatives and suggested that other ligands having an electronic structure comparable with CO might be used for this purpose. Only some years later was it found that ligands having unsaturated donor atoms of low electronegativity could behave similarly to carbon monoxide.

From a historical point of view, the first nickel(0) derivative after $Ni(CO)_4$ to be prepared by this method was $Ni(PCl_2Me)_4$, isolated by Quin [10] in very good yields, by refluxing the phosphine through a column filled with nickel as turnings or wire. This reaction, however, cannot be considered peculiar to all ligands similar to PCl_2Me; while PCl_3 and PCl_2Ph do not react, PBr_2Me and $AsBr_2Me$ react with nickel under very mild conditions [11]. Even some chelating tertiary phosphines such as *o*-$C_6H_4(PEt_2)_2$ react with Raney nickel at 160°C [12]; palladium, but not platinum, reacts directly with *o*-$C_6H_4(PEt_2)_2$ at 200°C to give the bis-diphosphine derivative [13].

The direct synthesis of $Ni(PF_3)_4$ in low yields was successfully carried out in 1965, although the first attempted synthesis was unsuccessful [14, 15]. This compound was obtained from PF_3 and pyrophoric nickel, prepared by thermal decomposition of the carbonyl at 100°C [16]. The product was contaminated by some $Ni(CO)(PF_3)_3$, which could only be derived from some residual CO, bound to, or adsorbed on, the pyrophoric nickel.

The same reaction was observed using CF_3PF_2 instead of PF_3 as ligand. This fact was taken as an indication that the attack by PF_3 was probably catalysed, in this case, by the presence of Ni–CO–Ni carbonyl bridged systems on the metal obtained from the decomposition of $Ni(CO)_4$. Support for this view comes from the study of the spectrum of CO chemisorbed on nickel, which indicates that the formation of $Ni(CO)_4$ from Ni and CO depends upon an initial formation of CO bridges in the surface [17]. In the same year, Kruck and Baur also succeeded in preparing $Ni(PF_3)_4$ from metallic nickel and phosphorus trifluoride, but under drastic conditions (100°C, 350 atm) [18].

Later on, the synthesis of the zerovalent PF_3 derivatives of Ni, Pd and Pt was similarly accomplished [19]. More recently, a large number of fluorophosphine derivatives of nickel(0), NiL_4 [L = PF_3, CF_3PF_2, $(CF_3)_2PF$, CCl_3PF_2, Me_2NPF_2, $C_5H_{10}NPF_2$], has been synthesized in good yields by reacting the liquid L and nickel metal obtained from the decomposition of Ni(II)-oxalate at 60–80°C [20].

In 1969 the synthesis of $Ni(PF_3)_4$ in quantitative yields from nickel vapour and PF_3 was reported [21]. Although this method of synthesis is the simplest in principle, practically it is limited to fluorophosphines. In principle, other types of ligands can give this reaction. For instance, the phosphites should be able to react with nickel, palladium and platinum. It is not easy to rationalize the electronic properties of the ligands which allow the direct attack on the metal. The most obvious conclusion is that only ligands which are poor bases and strong π-acids can be employed in this reaction. However, it is difficult to explain why PCl_3 does not attack nickel metal, while PCl_2Me does. Moreover o-$C_6H_4(PEt_2)_2$ was shown to react with Ni and Pd but not with Pt, and this is not in accordance with the promotion energies for these metals (1.4).

The nature of the metal surface and the method of preparation of the metals greatly affect this reaction [22]. This is particularly interesting, because the attack of the metal can be related to the catalytic properties of its surface. In practice it demonstrates that an appropriate ligand can react with the surface of a metal to form a complex, when the surface has active centres for catalytic reactions.

(*b*) *By direct replacement of ligands in tetracarbonylnickel or tetracyano-nickelate.* The complete replacement of carbon monoxide in $Ni(CO)_4$ by a ligand such as a phosphine proved to be a difficult reaction. The degree of substitution reached with different ligands has been discussed mainly in terms of electronic effects [1], but it appears clear now that steric effects also play an important role.

The first example of complete substitution in $Ni(CO)_4$ was given by PCl_3 [23] and soon afterwards $PCl_2(C_6H_5)$ and $PCl_2(OC_6H_5)$ were shown to displace all the four CO molecules by reaction at 60°C [24]. Similar reactions were obtained by using $P(NCO)_3$ and $P(NCS)_3$ [25].

Using more drastic conditions (150–200°C) complete substitution of carbon monoxide can be obtained with some chelating bis-phosphines such as o-$C_6H_4(PR_2)_2$ (R = Et, Ph) and $C_2H_4(PPh_2)_2$ [12].

The direct reaction between $Ni(CO)_4$ and PF_3 has been investigated extensively. Chatt obtained a complex mixture from which he could isolate nearly pure $Ni(CO)(PF_3)_3$, but complete substitution was not achieved [14]. Bigorgne, on the other hand, studying the same reaction by IR spectroscopy, recognized that even at 0°C all four substitution compounds are present together [26]. He isolated and characterized the tetrasubstituted $Ni(PF_3)_4$ obtained in this way.

Later the same author obtained all products by carrying on the reaction at different temperatures: $Ni(CO)_3(PF_3)$ was obtained at 0°C, containing some $Ni(CO)_4$ and $Ni(CO)_2(PF_3)_2$; $Ni(CO)_2(PF_3)_2$ almost pure at 25°C; $Ni(CO)(PF_3)_3$ at 45–55° and, finally, $Ni(PF_3)_4$, still very impure with carbonylated derivatives, by heating the reagents at 70°C for many days [27]. Under more drastic conditions Kruck obtained 80% yields of tetrakis(trifluorophosphine)nickel by reaction at 100°C and 350 atm [18]. Clark and Brimm studied the reaction mixture of $Ni(CO)_4$ and PF_3 by gas chromatography (the pure compounds have been isolated by means of this technique) and NMR, determining the physical constants (density, vapour pressure) of all the substitution products [28]. They showed that to a first approximation the equilibrium composition of the mixture can be calculated as if there was a statistical equilibrium between the ligands CO and PF_3, and this seems to indicate that the metal-to-ligand bonds are of comparable strength. On reacting $Ni(CO)_4$ with PF_3 under pressure (200 p.s.i.) at 150°C, for *ca.* 3 days, the tetrasubstituted product has only 90% purity, the other product being only $Ni(CO)(PF_3)_3$. Other fluorophosphines usefully employed in the reaction with nickel carbonyl were of the type PF_2R [R = OC_6H_5, OC_3H_7-*n* [29]; $N(CH_3)_2$, NC_5H_{10} [30]].

With $PF_2(C_6H_5)$ complete substitution of carbon monoxide also occurred [31]. The same result was obtained with the trifluoromethylphosphine, $PF_2(CF_3)$, and with more difficulty, with $PF(CF_3)_2$ [32]. With a chelating fluorophosphine of the type $C_2H_4(OPF_2)_2$ and with *p*-$C_6H_4(OPF_2)_2$, derivatives were obtained without carbon monoxide, but probably having a polymeric structure [29].

Tetrachlorodiphosphine, P_2Cl_4, has also been used in this reaction [34]. By working with an excess ligand, P_2Cl_4 seems to act as a monodentate phosphine with formation of $Ni(P_2Cl_4)_4$. However, this compound was not isolated, and its existence was inferred only from the amount of CO evolved during the reaction.

The complete replacement of carbon monoxide in $Ni(CO)_4$ has also been obtained with trisubstituted phosphites. Bigorgne described all the possible substitution derivatives of tetracarbonylnickel with trimethyl- and triethylphosphites [26]. The last derivative was also reported by Leto [35], who used the procedure reported by Bigorgne. By reacting $Ni(CO)_4$ with $P(OC_6H_5)_3$, Clark and Storrs were first able to isolate the bi- and tri-substituted products [36]. With more drastic conditions (200°), they were then able to isolate $Ni\{P(OC_6H_5)_3\}_4$. Similarly the tris(*p*-tolyl)phosphite, the tris(*p*-methoxyphenyl)phosphite and the tris(2-ethylhexyl)phosphite derivatives were obtained as well as some mixed triphenylphosphite-triphenylphosphine or triethylphosphite derivatives [36]. With cyclic phosphites offering no steric hindrance such as $P(OCH_2)_3CCH_3$ [37] and $P(OCH)_3(CH_2)_3$, [38] all the substituted derivatives were isolated and characterized.

The first of these ligands gives monosubstitution at room temperature, disubstitution in boiling chloroform (4 h), trisubstitution in boiling ethylbenzene (6 h), while the tetrasubstituted product could only be obtained by refluxing for 16 h in chlorobenzene. Recently the series of complexes $Ni\{P(OR)_3\}_4$ (R = C_2H_5, n-C_3H_7 n-C_4H_9, Ph) has been obtained more easily by conducting the reactions *under vacuum* [38 *bis*]. Some mixed halophosphites such as

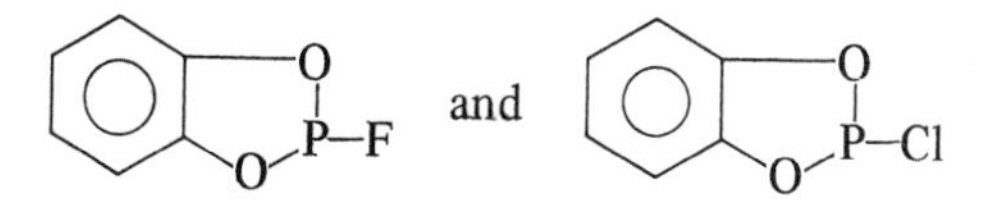

have also been successfully used in this reaction [29, 39].

From the above discussion, then, it appears that the direct replacement of carbon monoxide in tetracarbonylnickel is not a very convenient reaction for the synthesis of the tetrasubstituted nickel derivatives. The amount of Ni–C π-bonding should increase with the replacement of CO by ligands having poorer π-accepting ability than carbon monoxide [1]. This, together with steric effects, could explain why in general mono and di substitution can be readily achieved, while full substitution is much more difficult.

An interesting way for preparing zerovalent nickel derivatives uses as starting material the tetracyanonickelate(0) $K_4[Ni(CN)_4]$ [40]. Very rapid exchange reactions occur between $K_4[Ni(CN)_4]$ and ligands having P^{III}, As^{III}, Sb^{III} or heterocyclic nitrogen as donor atoms. It appears that the cyanide ion is more easily displaced from nickel than carbon monoxide. The reactions are as follows:

$$K_4[Ni(CN)_4] + 4L = NiL_4 + 4\,KCN$$

$$K_4[Ni(CN)_4] + 2D = NiD_2 + 4\,KCN$$

[L = PPh_3, $AsPh_3$, $SbPh_3$; D = $CH_2(PPh_2)_2$, $C_2H_4(PPh_2)_2$, $C_2H_4(AsPh_2)_2$]. Only with the trichelating triphosphine $CH_3C(CH_2PPh_2)_3$ (T) could the intermediate K[Ni(CN)T] be isolated [41]. Although this reaction is very neat, it was not extensively used, possibly because of the rather difficult synthesis of the starting material.

(*c*) *Exchange reactions.* We will be considering here the reactions by which the ligands (other than carbon monoxide) bound to the metal in a zerovalent complex can be replaced by a different ligand, and those by which halogen atoms, belonging to a coordinated phosphine, can be exchanged.

The trichlorophosphine can be exchanged completely with the trifluoro- and tribromo-phosphines as a result of the reaction [15]:

$$Ni(PCl_3)_4 + 4PY_3 = Ni(PY_3)_4 + 4PCl_3 \quad (Y = F, Br)$$

In an attempt to elucidate whether the reaction with PF_3 was really a ligand exchange or a halogen exchange, as happens in other cases (see later), $Ni(PCl_3)_4$ with P^{32}-labelled was used [42]. The P^{32}-activity was almost completely found in the displaced PCl_3, confirming the ligand exchange mechanism.

$Ni(PF_3)_4$ was shown to exchange with ligands such as PPh_3, $C_2H_4(PPh_2)_2$, $P(OC_6H_5)_3$ [18, 42 *bis*]. In these cases only partial substitution of PF_3 was achieved. Similarly, the exchange reaction between $Ni(CF_3PF_2)_4$ and PF_3 gives the di- and tri-substituted derivatives, but these compounds were not fully characterized [32]. Partial substitution has also been achieved between $M(PF_3)_4$ (M = Pd, Pt) and PPh_3 [115].

The related platinum derivative, $Pt(CF_3PF_2)_4$, was shown to give bis substitution in very mild conditions with ligands such as PPh_3, $PCH_3(C_6H_5)_2$ and $P(CH_3)_2C_6H_5$ [33]. A complete replacement of PF_3 in $M(PF_3)_4$ complexes, has been achieved with ligands such as PCl_3 and $PPhCl_2$ [42 *bis*] for M = Ni and $P(OPh)_3$ [115] for M = Pd, Pt. Triphenylphosphine is completely displaced from $Pt(PPh_3)_4$ by ligands such as $P(OC_6H_5)_3$ [3, 43], $P(OC_6H_4Cl\text{-}p)_3$ [3], $P(OCH)_3(CH_2)_3$ and $P(OCH_2)_3CCH_3$ [44]. The last two ligands were also used successfully with $Pd(PPh_3)_4$ [44]. The chelating diphosphine $Ph_2P{-}C{\equiv}C{-}PPh_2$ produces only the partially substituted dimeric derivatives $M_2(PPh_3)_4$-$(Ph_2P{-}C{\equiv}C{-}PPh_2)_2$ on reaction with $M(PPh_3)_4$ (M = Pd, Pt) [45]. The $M(PPh_3)_4$ complexes (M = Pd, Pt) gave similarly partially substituted stibine derivatives on treatment with $SbPh_3$ [46].

Complete replacement of isocyanide in $Pd(RNC)_2$ could easily be obtained using tertiary phosphines, arsines and phosphites [2], with formation of the corresponding PdL_4 or PdL_3 complexes.

Another useful method for the synthesis of zerovalent nickel derivatives is the replacement of olefins by suitable ligands. From bisacrylonitrilenickel(0), $Ni\{P(OCH_2CCl_3)_3\}_4$ [47], $Ni\{P(OCH_3)_3\}_4$ [47], and $Ni\{P(OC_6H_5)_3\}_3$ [48], have been obtained. By an analogous reaction $Ni\{P(CH_3)_3\}_4$ [47] and $Ni(PMePh_2)_4$ [49] have been prepared from Ni(1,5-cyclooctadiene)$_2$. Presumably the same reaction has been employed for the synthesis of $Ni\{P(OMe)_3\}_4$ [50], as it has been used for obtaining $Ni(AsMe_2Ph)_4$ and $Ni\{o\text{-}C_6H_4(AsMe_2)_2\}_2$ [206]. Similarly $Ni\{P(OMe)_3\}_4$ has been obtained by reaction of $P(OMe)_3$ with $Ni(Dq)_2$ or Ni(1,5-COD)(Dq) (Dq = duroquinone, tetramethyl-1,4-benzoquinone) [51].

Ni(1,5-cyclooctadiene)$_2$ has also been used for the synthesis of $Ni(PH_3)_4$ at $-40°C$, under PH_3 pressure. This is the first complex having only PH_3 as ligand; it decomposes above $-30°C$ and it has been characterized only by Raman and infrared spectroscopy [52]. As a consequence of the labile coordination of ethylene in complexes of the type $M(PPh_3)_2(C_2H_4)$ (M = Ni, Pd, Pt), the reaction

$$M(PPh_3)_2(C_2H_4) + 2\,L \rightarrow M(PPh_3)_2L_2 + C_2H_4$$

could have useful applications with L = PPh_3 or analogous ligands. These reactions have been recently studied in detail [53, 54, 55] and have also been applied to complexes of olefins other than ethylene [56]. However in these cases the reaction may have a different pattern [56]:

$$Pd(PPh_3)_2(\text{maleic anhydride}) + C_2H_4(PPh_2)_2 \rightarrow$$
$$\rightarrow Pd\{C_2H_4(PPh_2)_2\}_2(\text{maleic anhydride}) + 2\,PPh_3$$

This different behaviour is to be ascribed to the stronger bonding between the metal and the olefin bearing electron withdrawing substituents (see 1.3.1.2). This is also the case for the reactions of $Pt(AsPh_3)_2(C_2F_4)$ with many ligands, which lead to the displacement of triphenylarsine; however with PF_3, tetrafluoroethylene is displaced with formation of $Pt(AsPh_3)_2(PF_3)_2$ [58]. With triphenylphosphite a similar reaction takes place.

Very recently the synthesis of an interesting zerovalent palladium complex, namely $Pd(dba)_2$ (dba = dibenzylideneacetone), has been reported [59]. This complex has been shown to lose the dba ligand readily by reaction with PPh_3, giving $Pd(PPh_3)_4$. It is quite probable that $Pd(dba)_2$, as well as the analogous platinum derivative [60], will give rise to similar reactions with a vast variety of ligands. The exchange reactions of halogen atoms of a phosphine bound in a zerovalent complex have also been studied.

$Ni(PF_3)_4$ can be obtained from $Ni(PCl_3)_4$ by fluorination with reagents such as SbF_3 (with catalytic amounts of SbF_5) [15], and KSO_2F [61]. The latter reagent has also been used for preparing $Ni(PRF_2)_4$ (R = CH_3, C_6H_5, OC_6H_5) from the corresponding chloro derivatives [29, 61]. The fluorine atoms in $Ni(PF_3)_4$ exchange with ammonia giving $Ni\{P(NH_2)_3\}_4$ [61]. Solvolysis of the P–F bond is also possible with other reagents [62]:

$$M(PF_3)_4 + nM'X \rightarrow M(PF_{3-n}X_n)_4 + nM'F$$

(M = Ni, Pt; X = OR, R, NR_2; n = 1, 2, 3).

By this route the syntheses of $Ni\{P(OCH_3)_3\}_4$ [62] and $Ni(PF_3)_3(PF_2OC_2H_5)$ [63] were reported. The former compound has also been obtained from $Ni(PCl_3)_4$ and methoxide ion [63 *bis*]. The barium salt $[Ni(PF_3)_3PF_2O]_2Ba$, was first isolated from $Ni(PF_3)_4$ and $Ba(OH)_2$; a subsequent reaction with $[(C_2H_5)_3O][BF_4]$ gave the non-ionic zerovalent derivative. The use of these solvolyses render possible the synthesis of, *inter alia*, definite complexes of chlorodifluorophosphine [64]:

$$Ni\{P(NMe_2)F_2\}_4 + 8HCl \rightarrow Ni(PClF_2)_4 + 4[(CH_3)_2NH_2]Cl$$

The attempted synthesis of such complexes by the substitution reaction of $PClF_2$ on $Ni(CO)_4$ leads instead to the formation of a complex mixture [65], which can be partially separated by fractional distillation.

The starting difluoroaminophosphine complexes can be obtained according to the reaction [64]:

$$Ni(PF_3)_4 \xrightarrow[-4[R_2NH_2]F]{8\ R_2NH} Ni(PF_2NR_2)_4$$

To avoid polymerization only secondary amines [$(CH_3)_2NH$, $(C_2H_5)_2NH$, $(n\text{-}C_3H_7)_2NH$, $C_5H_{10}NH$] have been used. Such aminolyses yield monomeric, homogeneous reaction products.

The exchange reactions reported above are potentially a good method of synthesis for zerovalent compounds otherwise difficult to obtain. In particular the dibenzylideneacetone complexes of Pd and Pt seem to be very good starting materials for such reactions.

(d) Ligand elimination or addition. The stabilization of coordinatively unsaturated species such as ML_3 (and sometimes ML_2) is a common feature of the chemistry of Ni, Pd and Pt derivatives. This fact has been discussed in terms of electronic [1] and steric effects [47]. It has also been suggested that some of the ML_4 complexes may be represented as $ML_3.L$ species, with L trapped in the lattice [55].

When $L = PPh_3$ the synthesis of $Pt(PPh_3)_3$ can be achieved just by boiling the tetracoordinated complex in ethanol [66]:

$$Pt(PPh_3)_4 \xrightarrow[C_2H_5OH]{\text{boiling}} Pt(PPh_3)_3 + PPh_3$$

In many other cases such dissociation has been established in solution [55], but the tricoordinated complexes were not isolated. Such ligand elimination is however possible from $Pt\{P(C_2H_5)_3\}_4$, on heating the complex at 50–60°C *in vacuo* [132]. Strangely enough, this reaction is not possible with $Pt\{P(CH_3)_3\}_4$, which does not dissociate up to 70–80°C.

This type of ligand elimination has also been unsuccessfully attempted with $Pt(PPh_3)_2(C_2H_4)$, the volatile ethylene molecule not being eliminated [67]. In contrast, this reaction could be achieved by pumping off the ethylene from $Pd(PCy_3)_2(C_2H_4)$. In this case the white, air-sensitive, highly unsaturated $Pd(PCy_3)_2$ was obtained [68]. Similarly, nitrogen can be easily displaced from [bis(tricyclohexylphosphine)nickel] dinitrogen, $(NiL_2)_2N_2$ ($L = PCy_3$), by a stream of argon passed through a solution of the complex in toluene [69]; by adding PCy_3 to the solution, the tris complex can be obtained in quantitative yields.

The ligand addition reaction:

$$ML_3 + L' \rightarrow ML_3L'$$

has been applied to the synthesis of $Ni\{P(O\text{-}o\text{-}C_6H_4CH_3)_3\}_4$ (L = L′) [71]. With L ≠ L′ ligand exchange reactions usually occur. However the bis coordinated adduct, $Ni\{P(OC_6H_4\text{-}o\text{-}C_6H_5)_3\}_2$, easily adds tricyclohexylphosphine, forming $Ni\{P(OC_6H_4\text{-}o\text{-}C_6H_5)_3\}_2\{P(C_6H_{11})_3\}$ [105]. These methods of synthesis are not important from a preparative point of view, but they represent a very important result for the understanding of the chemistry of the compounds we are discussing here, as we shall see later.

(e) By reduction

By metals. Zerovalent nickel compounds with phosphites could be obtained in good yields by reduction of a nickel(II) halide phosphite complex, with potassium graphite (KC_8) [72]. However, this method of synthesis has not been investigated further. Reduction of phosphine complexes with sodium metal in liquid ammonia gives low yields because the reagents are insoluble in the reaction medium [73]. Related systems can be considered to be those used by Kruck for the synthesis of $Ni(PF_3)_4$ [18] and $M(PF_3)_4$ (M = Pd, Pt) complexes [74].

The nickel compound has been synthesized by reacting NiI_2, PF_3 and copper under rather mild conditions (100°C, 135 atm). Similarly, using the dry metal(II) chlorides, the palladium derivative was obrained at 100°C and 300 atm, and the platinum derivative at 100°C and 100 atm. In these cases copper was the halogen acceptor, but probably trifluorophosphine was the true reducing agent (see later).

This "reductive fluorophosphination" of metal salts is the most important method of synthesis of zerovalent fluorophosphine derivatives, particularly with respect to simplicity of the starting material and high yields [75].

The toluene-solvated $Ni(PPh_3)_3$ complex has been very recently obtained by refluxing $NiCl_2$ and PPh_3 in CH_3CN, and then reducing the resulting cooled solution with zinc dust [55]. The related red-brown, air-sensitive bis-coordinated complex, $Ni(PPh_3)_2$, has been synthesized by reduction of $Ni(PPh_3)_2Br_2$ with sodium amalgam in either CH_3CN or C_6H_6 [57].

By sodium borohydride. This method was applied to the reduction of nickel(II), palladium(II) and platinum(II) derivatives containing phosphines, tertiary diphosphines and diarsines to the zerovalent state [73, 76, 77].

As in the case of the compounds obtained by reduction with hydrazine (see later), their real nature was initially questioned but it was soon demonstrated that the reduction with $NaBH_4$ leads to true zerovalent derivatives and not to hydride species [77]. More recently sodium borohydride has been used for the preparation of $M(PPh_3)_4$ and $M\{C_2H_4(PPh_2)_2\}_2$ (M = Pd, Pt) [78], $Pd\{C_3H_6(PPh_3)_2\}_2$ [79], PtL_4 [L = $PMePh_2$, $PPhMe_2$, $PPh(OMe)_2$, $P(OPh)_3$, $AsPh_3$; L–L = $o\text{-}C_6H_4(AsMe_2)_2$] [80, 81], and $Pt(PMe_2C_6F_5)_4$ [81]. A series of zerovalent platinum complexes of arylated polytertiary phosphines and arsines has been recently synthesized by this route [82]. A biscoordinated platinum(0) complex, $Pt\{C_2(PPh_2)_2\}_2$, was also obtained; in the ditertiary phosphine $Ph_2PC{\equiv}CPPh_2$, the rigidity of the unsaturated carbon backbone forces the two phosphorus atoms to remain too far apart to chelate to a single atom [82].

Although coordinatively unsaturated, this compound has been shown to be inert toward some reagents [82]. Moreover, some nickel(0) arylphosphite complexes have been synthesized by this route in the last few years. In fact by reduction of $Ni(NO_3)_2$ with $NaBH_4$ in ethanol and excess phosphite, NiL_4 derivatives [$L = P(OC_6H_5)_3$, $P(O\text{-}m\text{-}C_6H_4CH_3)$, or $P(O\text{-}p\text{-}C_6H_4CH_3)$] have been obtained [43, 83]. Under the same conditions and with $P(O\text{-}p\text{-}C_6H_4Cl)$ as ligand, the tris derivative was isolated [83]. This is rather strange behaviour for the tris-*p*-chlorophenylphosphite, and it was found also for the corresponding chlorophosphine-platinum derivative [3].

This reduction has also been used for the synthesis of $Ni\{P(O\text{-}o\text{-}C_6H_4CH_3)_3\}_3$, using CH_3CN as solvent [71]. We briefly mention here also the reduction of nickel(II) salts with sodium dithionite, $Na_2S_2O_4$, in the presence of phosphites, which gives the corresponding NiL_4 species [84]. Sodium naphthalenide has also been used as the reducing agent in some cases [73].

By hydrazine. This is a very successful method for preparing palladium(0) and platinum(0) derivatives with tertiary phosphines, phosphites and arsines [2, 3, 85].

A zerovalent nickel(0) complex with a chelating phosphine was also obtained by reduction of $[Ni\{o\text{-}C_6H_4(PEt_2)_2\}]_2(NO_3)_2$ with NH_2NH_2 [73]. In the case of platinum reduction with hydrazine can give either zerovalent derivatives [3] or platinum(II) hydrides [86], according to the reaction conditions, the nature of the ligands, the halogen atom bound to the metal and the reaction time. The unusual stability of the presumed zerovalent complexes led to the supposition that the compounds described by Malatesta and coworkers [3] were instead hydrido complexes of platinum(II) [87].

Careful investigations, however, have confirmed that they are certainly zerovalent compounds [4, 77, 85]. This reduction is a very complicated reaction in any case [5]. Some labile intermediates have recently been isolated by reduction of *cis*-$Pt(PPh_3)_2Cl_2$ at room temperature [88]. These intermediates are cationic dehydrodiimide and amino complexes, soluble in the reaction medium (ethanol). *Trans*-$Pt(PPh_3)_2HCl$ separates from the alcoholic solution after several hours at room temperature or when the solution is warmed. This hydride can be further reduced to a zerovalent platinum complex with excess hydrazine. This seems to be the mechanism by which *cis*-$Pt(PPh_3)_2Cl_2$ can be reduced by excess hydrazine at 70–80°C to $Pt(PPh_3)_3$ [3, 5].

In these conditions some of the platinum still remains in solution and it can be recovered as a zerovalent complex by adding excess phosphine to the solution [89]. The zerovalent PdL_4 and PtL_4 complexes ($L = Ph_2PC{\equiv}CMe$) have recently been synthesized by reduction of a solution of K_2PtCl_4 (or K_2PdCl_4) and the ligand with hydrazine [45]. Zerovalent complexes of platinum having methoxodiphenylphosphine or *n*-butoxydiphenylphosphine as ligands have been similarly prepared [90]. Also $Pd(PPh_3)_4$ has been obtained from $PdCl_2$ and PPh_3 in dimethyl sulphoxide solution at *ca.* 140°C with hydrazine [91]. Dimethyl sulphoxide has also been usefully employed as solvent in the synthesis of zerovalent palladium and platinum complexes with ligands such as *meta-* or *para-*tolylphosphines from the corresponding chlorides with hydrazine as reducing agent [55].

By alcoholic potassium hydroxide. This reducing agent gives similar results to hydrazine and sodium borohydride. It was employed for the first time in the synthesis of $Pt(PPh_3)_4$ [3, 66]. Preformed hydrides such as $[PtH(PPh_3)_3]HSO_4$ can be reduced to $Pt(PPh_3)_3$ by KOH in ethanol [67]. The complexes $Pt\{P(OC_2H_5)_3\}_4$ [92, 93], $Pt(PMe_2Ph)_4$ [81] and $Pt(PMePh_2)_4$ [292] have also been obtained by this route directly from K_2PtCl_4. Reduction of aqueous solutions of nickel(II) salts at pH >7 with an excess of $P(CH_3)_3$ immediately afford the very unstable complex $Ni\{P(CH_3)_3\}_4$ [94] which can be also obtained by reaction of $Ni\{P(CH_3)_3\}_2Cl_2$ with an aqueous-alcoholic solution of KOH. A combined reduction of K_2PtCl_4 in the presence of the ligand with hydrazine and KOH has been used for the synthesis of $Pt\{P(CH_2Ph)Ph_2\}_4$ [292]. The reduction equation can be written either:

$$K_2PtCl_4 + 5L + 2\,KOH \rightarrow PtL_4 + OPPh_3 + 4KCl + H_2O$$

or

$$K_2PtCl_4 + 4L + 2\,KOH + C_2H_5OH \rightarrow PtL_4 + 4KCl + CH_3CHO + 2H_2O$$

In the first case phosphine would be the reducing agent, while in the second case ethanol directly participates in the reaction, being oxidized to acetaldehyde.

The recently reported synthesis of $Pd(PPh_3)_4$ by reduction of $Pd(PPh_3)_2Cl_2$ with sodium propoxide at room temperature in propanol in the presence of excess phosphine, [95] and the interesting study of the intermediates which can participate in the reaction [96], gave a clearer picture of this reduction. Sodium propoxide can also be used for the synthesis of $Pt(PPh_3)_4$ from $Pt(PPh_3)_2Cl_2$. Secondary and tertiary alcohols and phenols can be employed instead of propanol.

The reaction seems to proceed according to two parallel paths:

$$M(PPh_3)_2Cl_2 \xrightarrow[RCH_2O^-]{(b)} M(PPh_3)_2Cl(OCH_2R) \longrightarrow M(PPh_3)_2HCl + RCHO$$

$$M(PPh_3)_2Cl_2 \xrightarrow{(a)\ 2\,RCH_2O^-} M(PPh_3)_2(OCH_2R)_2$$

$$M(PPh_3)_2HCl \xrightarrow{+2\,PPh_3,\ -HCl} M(PPh_3)_4$$

$$M(PPh_3)_2(OCH_2R)_2 \xrightarrow[3PPh_3,\ -2RCH_2OH,\ -OPPh_3]{+H_2O} M(PPh_3)_4$$

In path (a), phosphine is the reducing agent and this is the only reaction with tertiary alcohols and phenols. Path (b) should explain the formation of oxidized products from alcohols. It has been shown that the two paths are practically equally followed [96]. Some reaction intermediates were also isolated. The phenoxy derivatives $M(PPh_3)_2(OC_6H_5)_2$ were reduced to $M(PPh_3)_4$ by PPh_3 in benzene in presence of water. Presumably in the first stage of the reaction the hydroxo complexes $M(PPh_3)_2(OH)_2$ are formed. Such derivatives were not isolated from $M(PPh_3)_2(OC_6H_5)_2$ and water. However in the case of platinum, $Pt(PPh_3)_2(OH)_2$ can be obtained from $Pt(PPh_3)_2Cl_2$ and sodium propoxide in benzene by addition of water. A subsequent reaction of $Pt(PPh_3)_2(OH)_2$ with PPh_3 in benzene gives $Pt(PPh_3)_4$ and triphenylphosphine oxide.

Related to the above discussion is the recent report on the synthesis of $Pd(PPh_3)_4$ by reductive cleavage of Pd–C bonds by action of methanol [97]. In this case formaldehyde is formed.

By metal alkyls. By reduction with aluminium alkyls, zerovalent tetra-substituted nickel(0) compounds with tertiary phosphines, arsines, stibines [98, 99] and diphosphines [100, 101] have been obtained. $Ni(SbPh_3)_4$ is one of the few reported zerovalent derivatives having a stibine as ligand [98]. In the same work [98], the synthesis of $Ni(PPh_3)_2$ was also reported, but it was shown later that the compound obtained by this way was the ethylene derivative, $Ni(PPh_3)_2(C_2H_4)$ [103].

In fact during the reduction of nickel(II) acetylacetonate with $AlEt_2(OEt)$ in presence of the phosphine, ethylene is formed from the ethyl radicals of the aluminium alkyl. This could also be the case for the reported $Ni\{P(O\text{-}o\text{-}C_6H_4CH_3)_3\}_2$, obtained in solution from nickel(II) acetylacetonate, $AlEt_3$ and excess ligand [104]. In fact the corresponding tris derivative has been shown to be undissociated in solution, and to react with ethylene giving $Ni\{P(O\text{-}o\text{-}C_6H_4CH_3)_3\}_2(C_2H_4)$ [71]. Using a stoichiometric amount of trimethyl-aluminum, the reduction of bis(2,4-pentanedionato)nickel(II) in the presence of tri-2-biphenylylphosphite leads to the orange bis-coordinated complex, $Ni\{P(OC_6H_4\text{-}o\text{-}C_6H_5)_3\}_2$ [105], which is said to be dissociated in solution, presumably because of steric factors.

The synthesis of many nickel(0) phosphites by reduction of a nickel(II) salt such as $NiBr_2$ with $AlEt_3$ is also reported in a patent [106], in which other nickel(0) derivatives with trialkylantimonite and trialkylarsenite are mentioned. As reducing agents, hydrazine, inorganic or metallorganic halides or hydrides, are also suggested. The synthesis of $Ni(PPh_3)_4$ by this route was very recently published in detail [107]. A palladium(0) derivative appears to be obtained from palladium(II) acetylacetonate, $AlEt_3$ and tricyclohexylphosphine. In this case the highly unsaturated species, $Pd(PCy_3)_2$ was obtained [108].

This reaction seems to be capable of extension to other zerovalent palladium derivatives with different phosphines, since by reduction of palladium(II) acetylacetonate in the presence of ethylene and the ligand, $PdL_2(C_2H_4)$ complexes have been obtained [L = PPh_3, PCy_3, $P(O\text{-}o\text{-}C_6H_4CH_3)_3$] [68]. It follows that the same reaction in absence of the olefin should give the corresponding PdL_3 or PdL_4 species. Another method of removing HX from hydrido compounds such as *trans*-$Pt(PPh_3)_2HX$ (X = halogen) is the reaction with *n*-butyllithium [67]. The reaction conditions are critical, but in this case the $Pt(PPh_3)_2$ complex can be isolated.

Spontaneous reduction. With ligands having reducing properties, the reduction of the metal to the zerovalent state can be brought about by the ligand, usually in ethanolic solution. This method was successfully employed for the synthesis of PdL_4 [L = PPh_3, $P(OPh)_3$], by reacting freshly precipitated palladium oxide, or palladium(II) nitrate, with an excess of the ligand [2]. The same synthesis was used for obtaining nickel(0) phosphite derivatives from nickel(II) halides [109]. In this case amines were added to the reaction in order to neutralize the acidity. The reaction works with $P(OR)_3$ when R = Me, Et, but not when R = C_4H_9, C_6H_5. Analogously $Pd\{P(OC_2H_5)_3\}_4$ [110] and nickel(0) complexes with constrained phosphite esters [44] were obtained. A similar reaction was used for the synthesis of $Ni\{P(OC_2H_5)_3\}_4$ [111] and $Ni\{P(OCH_2CH_2Cl)_3\}_4$ [47], but using CH_3CN as a solvent instead of CH_3OH.

Huttemann *et al.* [112] exploited the reducing properties of phosphites in the reaction:

$$[NiL_6]^{++}(ClO_4^-)_2 + 2NaHCO_3 \xrightarrow[\text{steam bath}]{H_2O} NiL_4 + 2NaClO_4 +$$

$$+ 2CO_2 + H_2O + \text{phosphite oxidation products}$$

[L = $P(OCH_2)_3CR$; $P(OCH)_3(CH_2)_3$] which is rather similar to the one discussed above.

Without adding amines, the reduction of a metal salt by the ligand can also be applied to the synthesis of $M\{PhP(OR)_2\}_4$ (M = Ni, Pd; R = CH_3, C_2H_5) [113].

It has been shown that the air-stable product of the reaction of $PHPh_2$ with nickel bromide, originally formulated as a square planar phosphine–phosphide complex, is in fact the zerovalent derivative $Ni(PHPh_2)_4$ [113 *bis*]. Similarly, $Pd(PHPh_2)_4$ has been obtained from $PdCl_2$ and diphenylphosphine.

The synthesis of $Pd(PF_3)_4$ has been achieved by reduction of $Pd(CO)_2Cl_2$ [114] or $Pd(C_6H_5CN)_2Cl_2$ [18] with PF_3 at room temperature, under pressure, for many days. For the synthesis of PtL_4 complexes [L = PF_3, CF_3PF_2, $(CF_3)_2PF$], the reaction can be conducted directly with platinum(II) chloride and the ligand at 60–80°C in a sealed tube [20]. However when L = PF_3, the yields were very poor and could be improved by working at higher temperature and pressure [115]. K_2PtCl_4 was also used instead of $PtCl_2$ [33]; the same reaction was applied to $PdCl_2$ [115]. In this paragraph we can also include the reactions in which an organic ligand bound to the metal can act as a reducing agent during the reaction with a ligand such as a phosphine.

Nickel [116] and palladium [117] bis π-allyl can be reduced to the zerovalent tetracoordinated species by reaction with PPh_3 or PEt_3; in contrast the corresponding platinum complex does not react in this way, but it gives the addition product, $Pt(PPh_3)_2(\sigma\text{-allyl})_2$ [117], with a rearrangement of the allyl ligand. In the case of palladium, similar reactions with PPh_3 can be obtained starting from (π-cyclopentadienyl)(π-cyclohexenyl)palladium, $C_5H_5PdC_6H_9$ [118], or from acetylacetonato-8-(acetylacetonyl)cyclo-oct-4-enyl palladium, $(C_8H_{12}acac)Pd(acac)$ [119]. In the latter case the reaction has proved possible also with $AsPh_3$ and $SbPh_3$ on the palladium complex.

Once more the platinum analogue does not give the zerovalent species by this route, probably because the metal-olefin π-bond and the metal–carbon σ-bond are stronger in the case of platinum than in the case of palladium [119]. (π-Allyl)(π-cyclopentadienyl)palladium also reacts easily with $PMePh_2$, forming the corresponding zerovalent tetracoordinated complex [204]. The same reaction has been used for the syntheses of $Pd\{P(OMe)_3\}_4$ [120, 121], $Pd\{P(OPh)_3\}_4$, $Pd\{P(OMe)_2Ph\}_4$, and $Pd\{AsMe_2(CH_2Ph)\}_4$ [121].

Another useful starting material for the synthesis of nickel(0) derivatives is bis(π-cyclopentadienyl)nickel(II), (nickelocene), $Ni(C_5H_5)_2$. This reacts with numerous aliphatic and aromatic phosphites giving the corresponding NiL_4 species [122, 123]. Dimethylnickelocene is a better reagent, being more soluble in the reaction medium. This reaction has also been used for the synthesis of $Ni\{PPh(OEt)_2\}_4$ and of $Ni\{P(OCH_2)_3CCH_3\}_4$ [47] and the detailed preparation of $Ni\{P(OC_6H_5)_3\}_4$ has been reported [93]. Similarly the π-allyl methyl complexes

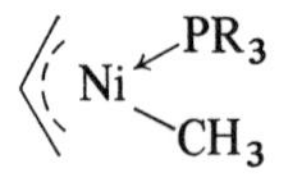

can be reduced by excess triphenylphosphite to the zerovalent derivative $Ni\{P(OPh)_3\}_4$ [124].

Fluorophosphines such as PF_3, CF_3PF_2, $(CF_3)_2PF$, CCl_3PF_2, Et_2NPF_2 and $C_5H_{10}PF_2$ react with nickelocene under mild conditions forming the corresponding NiL_4 complexes [125]. The PF_3 derivative can also be obtained at room temperature under a moderate pressure [126].

$Ni\{P(OC_6H_5)_3\}_4$ has been synthesized from nickel(II) *t*-butoxide, $Ni\{(CH_3)_3CO\}_2$, and triphenylphosphite at 120°C [127]. Bis-tritylnickel, $Ni(CPh_3)_2$, can also be reduced by triphenylphosphine to the zerovalent complex $Ni(PPh_3)_4$ [128].

An easy method of preparation of $Pd(PPh_3)_3$ and $Pd\{P(OC_6H_5)_3\}_3$ was recently discovered during study of the reaction of nucleophiles with the cationic species $[\pi\text{-}C_4H_7Pd(PR_3)_2]^+$ (R = C_6H_5, OC_6H_5; C_4H_7 = methallyl) [129]. In fact by adding benzylamine to a solution containing $(\pi\text{-}C_4H_7PdCl)_2$ and the phosphorus ligand, benzylammonium chloride, methallylbenzylamine and the PdL_3 complexes were isolated. This seems the best route to the synthesis of tricoordinated palladium(0) species, otherwise not easily accessible.

However the presence of the nucleophile does not seem to be necessary in order to obtain the zerovalent species. In fact the addition of allyl halides to $Pt(PPh_3)_4$ is a reversible reaction [130]. Addition of an excess of triphenylphosphine to a solution of $[Pt(PPh_3)_2(\pi\text{-}C_3H_3R_2)]X$ (R = H, X = Cl, Br; R = CH_3, X = Cl), obtained from $Pt(PPh_3)_4$ and allylhalides, yields $Pt(PPh_3)_3$ or $Pt(PPh_3)_4$ and the allyl halide. Similarly bis(π-allyl)nickel or palladium halides react with fluorophosphines PRF_2 in the following way [131]:

$$R^1\text{-}(\pi\text{-allyl})M(\mu\text{-}X)_2M(\pi\text{-allyl})\text{-}R^1 + 8PRF_2 \longrightarrow 2CH_2{=}CR^1{-}CH_2X + 2M(PRF_2)_4$$

(M = Ni; R^1 = H; X = Br; R = Me_2N)
(M = Pd; R^1 = H, CH_3; X = Cl; R = F, Me_2N, Et_2N)

The very reactive trialkylphosphine–platinum(0) complexes, $Pt(PR_3)_4$ (R = Et, Me), have been similarly prepared through displacement of the π-borallyl ligand from $PtB_3H_7(PR_3)_2$ by excess phosphine [132].

As a curiosity we can also mention that by reacting $Pd\{P(OPh_3)\}_2$-$\{(CF_3)_2CO\}$ with $P(MePh_2)_3$, hexafluoroacetone is displaced with formation of the mixed palladium(0) derivative, $Pd\{P(OPh)_3\}_2(PMePh_2)_2$ [121]. Reactions between bis(pentane-1,4-dionato)palladium(II) with triphenylmethyl tetrafluoroborate and ligands L such as PPh_3 and $SbPh_3$, which lead to probable cluster species of formula $(PdL_n)_x$ (n-intermediate between 2 and 3), are reported in the next section, where similar cluster derivatives are described.

Photochemical reactions. The oxalato complexes $ML_2(C_2O_4)$(M = Pt, Pd; L = PPh_3, $AsPh_3$) and $Pt\{C_2H_4(PPh_2)_2\}(C_2O_4)$, can be reduced with ultraviolet light with evolution of carbon dioxide. The reaction of oxalatobis(triphenylphosphine)platinum(II) has been studied in detail and it has been shown that in the presence of PPh_3, $Pt(PPh_3)_4$ is formed [133]. In the absence of free phosphine, a bright yellow unreactive compound of formula $Pt_2\{P(C_6H_5)_3\}_4$ was isolated which, however, in the preliminary communication [134] had been formulated as

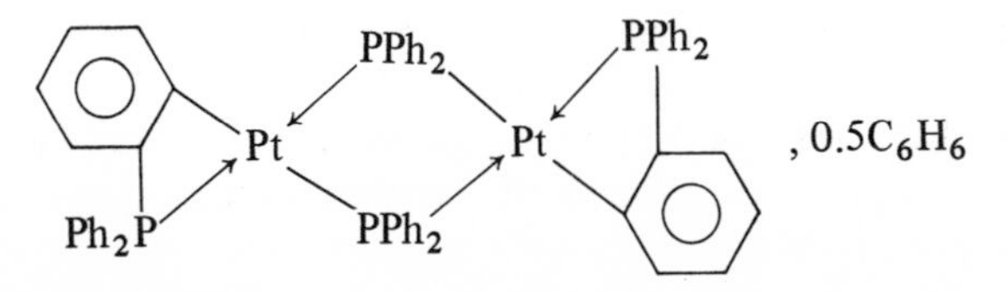

A similar photochemical reaction has been performed on carbonato complexes of the type $PtL_2(CO_3)$ (L = PPh_3, $AsPh_3$, $PMePh_2$) [135]. In the presence of phosphine, $Pt(PPh_3)_3$ is formed from $Pt(PPh_3)_2(CO_3)$. By photolysis in the absence of free phosphine, the same yellow dimer [133] obtained from the oxalato complex was obtained.

However, this reaction is complicated by the fact that by thermal decomposition in the dark of $Pt(PPh_3)_2(CO_3)$ without any addition of phosphine, an orange stable compound was isolated analysing as $Pt_2(PPh_3)_4$ but different from the one reported above. The two derivatives can be separated by thin layer chromatography. Presumably, only an X-ray structure determination will clarify this situation, which is rendered more complex by the existence of platinum-phosphine cluster compounds such as $[Pt(PPh_3)]_4$ and the unstable $[Pt(PPh_3)_2]_x$, where x is possibly 3 [136].

A related palladium complex of formula $[Pd(PPh_3)_2]_x$ has been obtained from $Pd(PPh_3)_4$ and vinyl chloride, possibly through a labile η-complex [137]. This yellow compound was shown to react with oxygen and trichloroethylene giving $Pd(PPh_3)_2 \cdot 3/2O_2$ and $Pd(PPh_3)_2Cl(CH{=}CCl_2)$ respectively.

The same complex has been obtained from $Pd(PPh_3)_4$ and allene, and it has been proposed that a labile intermediate allene complex is formed [138].

Moreover PdL_n complexes where n is intermediate between 2 and 3 (L = PPh_3, $AsPh_3$, $SbPh_3$) are the products of the reaction between bis(pentane-2,4-dionato)palladium(II) with fluoboric acid (or triphenylmethyl tetrafluoborate) and the ligand L, but their nature has not been studied in detail [139]. With dimethylphenylarsine the tetracoordinated complex was obtained.

In conclusion many reducing systems can be used for the preparation of Ni, Pd and Pt zerovalent derivatives, and only in very few cases do such reactions lead to hydrides and not to zerovalent compounds. Such reduction reactions, together with the various methods of synthesis reported above, in practice allow the isolation of a very large number of zerovalent derivatives, whose chemistry has been widely explored.

2.2.2. Survey of Reported Compounds

The nickel, palladium and platinum(0) derivatives reported in Tables 2.I, 2.II and 2.III, show that these three metals have a strong tendency to be stabilized in coordinatively unsaturated species with ligands other than halophosphines. We have already discussed how this fact has been explained in terms of electronic or steric effects (1.3.2.3 and 1.4). In view of these different interpretations, it is surprising that few X-ray structural determinations have been done on such complexes.

The molecular structure of $Pt(PPh_3)_3$ has shown that the platinum atom lies at a level of 0.1 Å out of the plane of the three phosphorus atoms and exhibits a

TABLE 2.I

Nickel(0) complexes with tricovalent P, As and Sb derivatives

Compound	Method of synthesis	References
Halophosphines and related ligands		
$Ni(PF_3)_4$	$Ni(PCl_3)_4 + SbF_3$	15
	$Ni(PCl_3)_4 + PF_3$	15, 27
	$Ni(PCl_3)_4 + KSO_2F$	61
	$Ni(CO)_4 + L$	18, 26, 27, 28
	$Ni + L$	16, 18, 19, 20
	$NiI_2 + Cu + L$	18
	$Ni(\pi\text{-}C_5H_5)_2 + L$	125, 126
$Ni(PCl_3)_4$	$Ni(CO)_4 + L$	23, 140
	$Ni(PF_3)_4 + PCl_3$	42 *bis*
$Ni(PBr_3)_4$	$Ni(PCl_3)_4 + PBr_3$	15
$Ni(PClF_2)_4$	$Ni\{P(NMe_2)F_2\}_4 + HCl$	64
$Ni(PMeF_2)_4$	$Ni(PMeCl_2)_4 + KSO_2F$	61
$Ni(PPhF_2)_4$	$Ni(PPhCl_2)_4 + KSO_2F$	61
	$Ni(CO)_4 + L$	31
$Ni\{P(OC_3H_7\text{-}n)F_2\}_4$	$Ni(CO)_4 + L$	29
$Ni\{P(OPh)F_2\}_4$	$Ni(CO)_4 + L$	29
	$Ni\{P(OPh)Cl_2\}_4 + KSO_2F$	29
$Ni\{P(CF_3)F_2\}_4$	$Ni(CO)_4 + L$	32
	$Ni + L$	16, 20
	$Ni(\pi\text{-}C_5H_5)_2 + L$	125
$Ni\{P(CCl_3)F_2\}_4$	$Ni(\pi\text{-}C_5H_5)_2 + L$	125
	$Ni + L$	20
$Ni\{P(CH_2Cl)F_2\}_4$	$Ni(\pi\text{-}C_5H_5)_2 + L$	20b
$Ni\{P(NMe_2)F_2\}_4$	$Ni(CO)_4 + L$	30
	$Ni + L$	20
	$[Ni(\pi\text{-allyl})Br]_2 + L$	131
	$Ni(PF_3)_4 + NHMe_2$	64

TABLE 2.I *continued*

Compound	Method of synthesis	References
$Ni\{P(NEt_2)F_2\}_4$	$Ni(\pi\text{-}C_5H_5)_2 + L$	125
	$Ni(CO)_4 + L$	39
	$Ni(PF_3)_4 + NHEt_2$	64
$Ni\{P[N(C_3H_7\text{-}n)_2]F_2\}_4$	$Ni(PF_3)_4 + (n\text{-}C_3H_7)_2NH$	64
$Ni\{P(NC_5H_{10})F_2\}_4$	$Ni + L$	20
	$Ni(CO)_4 + L$	30
	$Ni(\pi\text{-}C_5H_5)_2 + L$	125
	$Ni(PF_3)_4 + C_5H_{10}NH$	64
$Ni(PMeCl_2)_4$	$Ni + L$	10
$Ni(PPhCl_2)_4$	$Ni(PMeCl_2)_4 + PPhCl_2$	10
	$Ni(CO)_4 + L$	24
	$Ni(PF_3)_4 + PPhCl_2$	42 *bis*
$Ni\{P(OPh)Cl_2\}_4$	$Ni(CO)_4 + L$	24
$Ni(PMeBr_2)_4$	$Ni + L$	11
$Ni(AsMeBr_2)_4$	$Ni + L$	11
$Ni\{P(CF_3)_2F\}_4$	$Ni(CO)_4 + L$	32
	$Ni(\pi\text{-}C_5H_5)_2 + L$	125
	$Ni + L$	20
$Ni(C_6H_4O_2PF)_4$	$Ni(CO)_4 + L$	29
$Ni(C_6H_4O_2PCl)_4$	$Ni(CO)_4 + L$	39
$Ni(PF_3)_3\{P(CF_3)F_2\}$	$Ni\{P(CF_3)F_2\}_4 + PF_3$	32
$Ni(PF_3)_3\{P(OEt)F_2\}$	$[Ni(PF_3)_3PF_2O]_2Ba + [Et_3O][BF_4]$	63

$Ni(PF_3)_3(PPh_3)$	$Ni(PF_3)_4 + PPh_3$	18, 42 *bis*
$Ni(PF_3)_3(AsPh_3)$	$Ni(PF_3)_4 + AsPh_3$	42 *bis*
$Ni(PF_3)_3(SbPh_3)$	$Ni(PF_3)_4 + SbPh_3$	42 *bis*
$Ni(PF_3)_2\{P(CF_3)F_2\}_2$	$Ni\{P(CF_3)F_2\}_4 + PF_3$	32
$Ni(PF_3)_2(PPh_3)_2$	$Ni(PF_3)_4 + PPh_3$	18, 42 *bis*
$Ni(PF_3)_2(AsPh_3)_2$	$Ni(PF_3)_4 + AsPh_3$	42 *bis*
$Ni(PF_3)_2(SbPh_3)_2$	$Ni(PF_3)_4 + SbPh_3$	42 *bis*
$Ni(PF_3)_2\{C_2H_4(PPh_2)_2\}$	$Ni(PF_3)_4 + C_2H_4(PPh_2)_2$	18
$Ni(PF_3)\{P(OPh)_3\}_3$	$Ni(PF_3)_4 + P(OPh)_3$	18
$Ni(PF_3)(PPh_2Cl)_3$	$Ni(PF_3)_4 + PPh_2Cl$	42 *bis*
$\{Ni(F_2POCH_2CH_2OPF_2)\}_n$	$Ni(CO)_4 + L$	29
$\{Ni(F_2POC_6H_4OPF_2)\}_n$	$Ni(CO)_4 + L$	29
$Ni\{P(NCO)_3\}_4$	$Ni(CO)_4 + L$	25
$Ni\{P(NCS)_3\}_4$	$Ni(CO)_4 + L$	25
$Ni\{P(NH_2)_3\}_4$	$Ni(PF_3)_4 + NH_3$	61
Ni{P(N⌬O)₃}₄	by reduction with metal alkyls	98
Phosphites		
$Ni\{P(OMe)_3\}_4$	$NiX_2 + L + R_3N$	92, 109
	$Ni(PF_3)_4 + NaOMe$	62
	$Ni(PCl_3)_4 + NaOMe$	63 *bis*
	$Ni(CO)_4 + L$	26
$Ni\{P(OEt)_3\}_4$	$Ni(CH_2{=}CHCN)_2 + L$	47
	$NiX_2 + L + R_3N$	47, 92, 93, 109, 111
	$Ni(CO)_4 + L$	26, 35, 38 *bis*
	$NiSO_4 + L + Na_2S_2O_5$	84
	$Ni(PHPh_2)_4 + P(OEt)_3$	113 *bis*
$Ni\{P(OC_3H_7\text{-}n)_3\}_4$	$Ni(CO)_4 + L$	38 *bis*
$Ni\{P(OC_4H_9\text{-}n)_3\}_4$	$Ni(CO)_4 + L$	38 *bis*
$Ni\{P(OCH_2CH_2Cl)_3\}_4$	$Ni(\pi\text{-}C_5H_5)_2 + L$	122
	$NiCl_2 + L + R_3N$	47
	$NiBr_2 + L + AlEt_3$	106

TABLE 2.I *continued*

Compound	Method of synthesis	References
$Ni\{P(OCH_2CCl_3)_3\}_4$	$Ni(CH_2{=}CHCN)_2 + L$	47
$Ni\{tri(2\text{-}ethylhexyl)phosphite\}_4$	$Ni(CO)_4 + L$	36
	$Ni(\pi\text{-}C_5H_5)_2 + L$	122
$Ni\{tri(2\text{-}ethoxyhexyl)phosphite\}_4$	$Ni(\pi\text{-}C_5H_5)_2 + L$	122
$Ni(tridecylphosphite)_4$	$NiBr_2 + L + AlEt_3$	106
$Ni\{P(OCH_2)_3CCH_3\}_4$	$[NiL_6](ClO_4)_2 + NaHCO_3$	112
	$Ni(CO)_4 + L$	37a
	$NiX_2 + L + R_3N$	44
	$Ni(\pi\text{-}C_5H_5)_2 + L$	47
$Ni\{P(OCH_2)_3CC_3H_7\}_4$	$Ni(CO)_4 + L$	92
$Ni\{P(OCH_2)_3CC_2H_5\}_4$	$[NiL_6](ClO_4)_2 + NaHCO_3$	112
$Ni\{P(OCH)_3(CH_2)_3\}_4$	$[NiL_6](ClO_4)_2 + NaHCO_3$	112
	$Ni(CO)_4 + L$	38
	$NiX_2 + L + R_3N$	44
$Ni(diphenyldecylphosphite)_4$	$NiBr_2 + L + AlEt_3$	106
$Ni\{PPh(OMe)_2\}_4$	$NiBr_2 + L$	113
$Ni\{PPh(OEt)_2\}_4$	$NiCl_2 + L$	113, 142
	$Ni(\pi\text{-}C_5H_5)_2 + L$	47
$Ni\{P(OPh)_3\}_3$	$Ni(CH_2{=}CHCN)_2 + L$	48
$Ni\{P(OPh)_3\}_4$	$Ni(PPh_3)_3Cl + P(OPh)_3$	141
	$Ni(CO)_4 + L$	36, 38 *bis*
	$Ni(\pi\text{-}C_5H_5)_2 + L$	93, 122
	$Ni(NO_3)_2 + L + NaBH_4$	43, 83
	$NiBr_2 + L + AlEt_3$	106
	$Ni\{(CH_3)_3CO\}_2 + L$	127
	$Ni(PHPh_2)_4 + P(OPh)_3$	113 *bis*
$Ni\{P(OC_6H_4\text{-}o\text{-}Ph)_3\}_2$	$Ni(acac)_2 + L + AlMe_3$	105
$Ni\{P(OC_6H_4Me\text{-}p)_3\}_4$	$Ni(CO)_4 + L$	36

	$Ni(\pi\text{-}C_5H_5)_2 + L$	122
	$Ni(NO_3)_2 + L + NaBH_4$	43, 83
$Ni\{P(OC_6H_4Me\text{-}o)_3\}_3$	$Ni(NO_3)_2 + L + NaBH_4$	71
$Ni\{P(OC_6H_4Me\text{-}o)_3\}_4$	$NiL_3 + L$	71
$Ni\{P(OC_6H_4Me\text{-}m)_3\}_4$	$Ni(NO_3)_2 + L + NaBH_4$	83
$Ni\{P(OC_6H_4OMe\text{-}p)_3\}_4$	$Ni(CO)_4 + L$	36
	$Ni(\pi\text{-}C_5H_5)_2 + L$	122
	$NiBr_2 + L + AlEt_3$	106
$Ni\{P(OC_6H_4Cl\text{-}p)_3\}_3$	$Ni(NO_3)_2 + L + NaBH_4$	83
$Ni\{P(OPh)_3\}_3(PPh_3)$	$Ni(CO)\{P(OPh)_3\}_3 + PPh_3$	36
$Ni\{P(OPh)_3\}_2\{P(OEt)_3\}_2$	$Ni(CO)\{P(OPh)_3\}_3 + P(OEt)_3$	36
Phosphines and related ligands		
$Ni(PH_3)_4$	$Ni(1,5\text{-cyclooctadiene})_2 + L$	52
$Ni(PHPh_2)_4$	$NiBr_2 + PHPh_2$	113 *bis*
$Ni(PMe_3)_4$	$Ni(1,5\text{-cyclooctadiene})_2 + L$	47
	$Ni(PMe_3)_2Cl_2 + KOH$	94
$Ni(PEt_3)_4$	$Ni(\pi\text{-allyl})_2 + L$	116a
	$Ni(1,5\text{-cyclooctadiene})_2 + PEt_3$	50b
$Ni\{P(C_6H_{11})_3\}_2$ (prepared *in situ*)	$[Ni\{P(C_6H_{11})_3\}_2]_2N_2$ + argon	69
$Ni\{P(C_6H_{11})_3\}_3$	$Ni\{P(C_6H_{11})_3\}_2 + P(C_6H_{11})_3$	69
$Ni(PPh_3)_2$	$NiL_2Cl_2 + NaHg_x$	57
$Ni(PPh_3)_3$. 1/2 toluene	$NiCl_2 + L + Zn$	55
$Ni(PPh_3)_4$	$Ni(C_5H_7O_2)_2 + L + AlEt_3$	47, 99, 107
	by reduction with metal alkyls	98
	$K_4[Ni(CN)_4] + L$	40
$Ni(PMePh_2)_4$	$Ni(1,5\text{-cyclooctadiene})_2 + L$	49
$Ni(AsPh_3)_4$	by reduction with metal alkyls	98
	$K_4[Ni(CN)_4] + L$	40
$Ni(AsMe_2Ph)_4$	$Ni(1,5\text{-cyclooctadiene})_2 + L$	206

TABLE 2.I *continued*

Compound	Method of synthesis	References
$Ni(SbPPh_3)_4$	by reduction with metal alkyls	98
	$K_4[Ni(CN)_4] + L$	40
$Ni\{C_2H_4(PMe_2)_2\}_2$	$Ni(L{-}L)Br_2 + L{-}L$ + sodium naphthalenide	73
$Ni\{C_2H_4(PPh_2)_2\}_2$	$Ni(CO)_4 + L{-}L$	12
	$[Ni(L{-}L)_2](NO_3)_2 + NaBH_4$	73
	$K_4[Ni(CN)_4] + L{-}L$	40
	$Ni(\pi\text{-}C_5H_5)_2 + L{-}L$	102
$Ni\{C_2H_4(AsPh_2)_2\}_2$	$K_4[Ni(CN)_4] + L{-}L$	40
$Ni\{CH_2(PPh_2)_2\}_2$	$K_4[Ni(CN)_4] + L{-}L$	40
	$Ni(\pi\text{-}C_5H_5)_2 + L{-}L$	102
$Ni\{C_3H_6(PPh_2)_2\}_2$	$Ni(acac)_2\,L{-}L + Al(iso\text{-}C_4H_9)_3$	100, 101
	$Ni(\pi\text{-}C_5H_5)_2 + L{-}L$	102
$Ni\{C_4H_8(PPh_2)_2\}_2$	$Ni(acac)_2 + L{-}L + Al(iso\text{-}C_4H_9)_3$	100, 101
$Ni\{CH_3C(CH_2PPh_2)_3\}_2$ (α and β isomers)	$Ni(NO_3)_2 + L{-}L{-}L + NaBH_4$	73
$Ni\{o\text{-}C_6H_4(AsMe_2)_2\}_2$	$[Ni(L{-}L)_2]Cl_2 + NaBH_4$	73
	$Ni(1,5\text{-cyclooctadiene})_2 + L{-}L$	206
$Ni\{o\text{-}C_6H_4(PEt_2)_2\}_2$	$[Ni(L{-}L)_2](NO_3)_2 + NaBH_4$ or NH_2NH_2	73
	$Ni + L{-}L$	12
$Ni\{o\text{-}C_6H_4(PPh_2)_2\}_2$	$Ni + L{-}L$	12

TABLE 2.II

Palladium(0) complexes with tricovalent P, As and Sb derivatives

Compound	Method of synthesis	References
Fluorophosphines		
$Pd(PF_3)_4$	$Pd(CO)_2Cl_2 + L$	114
	$Pd(C_6H_5CN)_2Cl_2 + L$	18, 115
	$Pd + L$	19
	$PdCl_2 + L + Cu$	74
	$PdCl_2 + L$	115
$Pd\{P(NMe_2)F_2\}_4$	$[Pd(\pi\text{-allyl})X]_2 + L$	131
$Pd\{P(NEt_2)F_2\}_4$	$[Pd(\pi\text{-allyl})X]_2 + L$	131
$Pd(PF_3)_3(PPh_3)$	$Pd(PF_3)_4 + PPh_3$	115
$Pd(PF_3)_2(PPh_3)_2$	$Pd(PF_3)_4 + PPh_3$	115
Phosphites		
$Pd\{P(OMe)_3\}_4$	$(\pi\text{-allyl})(\pi\text{-cyclopentadienyl})Pd + L$	120, 121
$Pd\{P(OEt)_3\}_4$	$K_2PdCl_4 + L + R_3N$	92, 93
$Pd\{P(OCH_2)_3CCH_3\}_4$	$Pd(PPh_3)_4 + P(OCH_2)_3CCH_3$	44
	$PdX_2 + L + R_3N$	44
$Pd\{P(OCH)_3(CH_2)_3\}_4$	$Pd(PPh_3)_4 + P(OCH)_3(CH_2)_3$	44
$Pd\{P(OPh)_3\}_4$	$Pd(p\text{-}CH_3C_6H_4NC)_2 + P(OPh)_3$	2
	$(\pi\text{-allyl})(\pi\text{-cyclopentadienyl})Pd + L$	121
	$Pd(PF_3)_4 + P(OPh)_3$	115
$Pd\{P(OPh)_3\}_3$	$(\pi\text{-}C_4H_7PdCl)_2 + P(OPh)_3 + C_6H_5CH_2NH_2$	129
$Pd\{PPh(OMe)_2\}_4$	$PdCl_2 + L$	113
	$(\pi\text{-allyl})(\pi\text{-cyclopentadienyl})Pd + L$	121
$Pd\{PPh(OEt)_2\}_4$	$PdCl_2 + L$	113
$Pd\{P(OPh)_3\}_2(PMePh_2)_2$	$Pd\{P(OPh)_3\}_2\{(CF_3)_2CO\} + PMePh_2$	121

TABLE 2.II *continued*

Compound	Method of synthesis	References
Phosphines and related ligands		
$Pd(PPh_3)_4$	$Pd(NO_3)_2 + L$	2
	$Pd(\pi\text{-allyl})_2 + L$	117
	$Pd\{o\text{-}C_6H_4(PEt_2)_2\}Cl_2 + PPh_3 + NaBH_4$	73
	$(C_5H_5PdC_6H_9) + L$	118
	$K_2PdCl_4 + L + NaBH_4$	78
	$PdL_2X_2 + L + NH_2NH_2$	46
	$(C_8H_{12}acac)Pd(acac) + L$	119
	$Pd(PhCH{=}CHCOCH{=}CHPh)_2 + L$	59
	$PdCl_2 + L + NH_2NH_2$	91
	PdL_2Cl_2 + L + sodium *n*-propoxide	95, 96
	2,2,6,6-tetramethylheptane-3,5-dionato-(phenyl)triphenylphosphinepalladium + $CH_3OH + PPh_3$	97
$Pd(PPh_3)_3$	$(\pi\text{-}C_4H_7PdCl)_2 + PPh_3 + C_6H_5CH_2NH_2$	129
$[Pd(PPh_3)_2]_x$	$Pd(PPh_3)_4 + CH_2{=}CHCH_2Cl$	137
	$Pd(PPh_3)_4 + CH_2{=}C{=}CH_2$	138
$Pd\{P(C_6H_{11})_3\}_2$	$Pd(C_5H_7O_2)_2 + L + AlEt_3$	108
	$Pd\{P(cyclohexyl)_3\}_2(C_2H_4)$ under vacuo	68
$Pd(PHPh_2)_4$	$PdCl_2 + PHPh_2$	113 *bis*
$Pd(PMePh_2)_4$	$(\pi\text{-allyl})(\pi\text{-cyclopentadienyl})Pd + L$	204
$Pd(AsPh_3)_4$	$Pd(p\text{-}CH_3C_6H_4NC)_2 + AsPh_3$	2
	$(C_8H_{12}acac)Pd(acac) + L$	119
$Pd(AsMe_2Ph)_4$	$Pd(acac)_2 + CPh_3^+BF_4^- + L$	139
	$(\pi\text{-allyl})(\pi\text{-cyclopentadienyl})Pd + L$	120
$Pd\{AsMe_2(CH_2Ph)\}_4$	$(\pi\text{-allyl})(\pi\text{-cyclopentadienyl})Pd + L$	121
$Pd(SbPh_3)_4$	$(C_8H_{12}acac)Pd(acac) + L$	119

$Pd\{P(C_6H_4Me\text{-}p)_3\}_3$	$PdCl_2 + L + NH_2NH_2$	55
	$PdL_2I_2 + L + NH_2NH_2$	2
$Pd\{P(C_6H_4Me\text{-}m)_3\}_3$	$PdCl_2 + L + NH_2NH_2$	55
$Pd\{P(C_6H_4Cl\text{-}p)_3\}_4$	$PdL_2Cl_2 + L + NH_2NH_2$	2
$Pd\{P(C_6H_4Cl\text{-}p)_3\}_3$	$Pd(p\text{-}CH_3C_6H_4NC)_2 + P(C_6H_4Cl\text{-}p)_3$	2
$Pd\{PPh_2(C{\equiv}CMe)\}_4$	$K_2PdCl_4 + L + NH_2NH_2$	45
$Pd\{C_2H_4(PMe_2)_2\}_2$	$Pd(L{-}L)Br_2 + L{-}L$ + sodium naphthalenide	73
$Pd\{C_2H_4(PPh_2)_2\}_2$	$[Pd(L{-}L)_2]Br_2 + NaBH_4$	73
	$K_2PdCl_4 + L{-}L + NaBH_4$	78
$Pd\{C_3H_6(PPh_2)_2\}_2$	$[Pd(L{-}L)_2]Cl_2 + NaBH_4$	79
$Pd\{CH_2(PPh_2)_2\}_2$	$Na_2PdCl_4 + L{-}L + NaBH_4$	73
$Pd\{CH_3C(CH_2PPh_2)_3\}_2$ (α and β isomers)	$K_2PdCl_4 + L{-}L{-}L\ NaBH_4$	73
$Pd\{o\text{-}C_6H_4(PEt_2)_2\}_2$	$Na_2PdCl_4 + L{-}L + NaBH_4$	73
	$[Pd(L{-}L)_2]Br_2$ + Na in NH_3 liq.	73
	$Pd + L{-}L$	13
$Pd\{o\text{-}C_6H_4(AsMe_2)_2\}_2$	$[Pd(L{-}L)_2]Cl_2 + NaBH_4$	73
$Pd(PPh_3)_2(SbPh_3)_2$	$Pd(PPh_3)_4 + SbPh_3$	46
	$Pd(PPh_3)_2$(*cis* or *trans*-$MeO_2CCH{=}CHCO_2Me$) + + $SbPh_3$	56
$Pd\{o\text{-}C_6H_4(PEt_2)_2\}\{o\text{-}C_6H_4(AsEt_2)_2\}$	$Pd\{o\text{-}C_6H_4(PEt_2)_2\}Cl_2 + o\text{-}C_6H_4(AsEt_2)_2 + NaBH_4$	73
$Pd\{o\text{-}C_6H_4(PEt_2)_2\}\{o\text{-}C_6H_4(AsMe_2)_2\}$	$Pd\{o\text{-}C_6H_4(PEt_2)_2\}Cl_2 + o\text{-}C_6H_4(AsMe_2)_2 + NaBH_4$	73
$Pd\{o\text{-}C_6H_4(PEt_2)_2\}\{C_2H_4(PPh_2)_2\}$	$Pd\{o\text{-}C_6H_4(PEt_2)_2\}Cl_2 + C_2H_4(PPh_2)_2 + NaBH_4$	73
$Pd\{o\text{-}C_6H_4(PEt_2)_2\}\{CH_3C(CH_2PPh_2)_3\}$	$Pd\{o\text{-}C_6H_4(PEt_2)_2\}Cl_2 + CH_3C(CH_2PPh_2)_3 + NaBH_4$	73
$Pd\{P(C_6H_5)(o\text{-}C_6H_4PEt_2)_2\}(PPh_3)$	$\{Pd(L{-}L{-}L)Cl\}Cl + PPh_3 + NaBH_4$	73
$Pd_2(PPh_3)_4(Ph_2PC{\equiv}CPPh_2)_2$	$Pd(PPh_3)_4 + Ph_2PC{\equiv}CPPh_2$	45

TABLE 2.III

Platinum(0) complexes with tricovalent P, As and Sb derivatives

Compound	Methods of synthesis	References
Fluorophosphines		
$Pt(PF_3)_4$	$Pt + L$	19
	$PtCl_2 + L$	20, 33, 115
	$PtCl_2 + L + Cu$	74
$Pt\{P(CF_3)F_2\}_4$	$PtCl_2 + L$	20, 33
$Pt\{P(CF_3)_2F\}_4$	$PtCl_2 + L$	20, 33
	$K_2PtCl_4 + L$	33
$Pt(PMe_2C_6F_5)_4$	*cis*-$PtL_2Cl_2 + L + NaBH_4$	81
$Pt(PF_3)_3(PPh_3)$	$Pt(PF_3)_4 + PPh_3$	115
$Pt(PF_3)_2(PPh_3)_2$	$Pt(PF_3)_4 + PPh_3$	115
$Pt(PF_3)_2(AsPh_3)_2$	$Pt(AsPh_3)_2(C_2F_4) + PF_3$	58
$Pt\{P(CF_3)F_2\}_2(PMe_2Ph)_2$	$Pt\{P(CF_3)F_2\}_4 + PMe_2Ph$	33
$Pt\{P(CF_3)F_2\}_2(PMePh_2)_2$	$Pt\{P(CF_3)F_2\}_4 + PMePh_2$	33
$Pt\{P(CF_3)F_2\}_2(PPh_3)_2$	$Pt\{P(CF_3)F_2\}_4 + PPh_3$	33
Phosphites		
$Pt\{P(OEt)_3\}_4$	$K_2PtCl_4 + L + KOH$	92, 93
$Pt\{P(OPh)_3\}_3$	$PtL_2I_2 + NH_2NH_2$	3
$Pt\{P(OPh)_3\}_4$	$PtL_2I_2 + L + NH_2NH_2$	3
	$Pt(PPh_3)_{3\ or\ 4} + P(POh)_3$	3, 43
	$K_2PtCl_4 + L + NaBH_4$	80
	$Pt(PF_3)_4 + P(OPh)_3$	115
$Pt\{P(OC_6H_4Cl\text{-}p)_3\}_4$	$PtL_2Cl_2 + L + NH_2NH_2$	3
	$Pt(PPh_3)_4 + P(OC_6H_4Cl\text{-}p)_3$	3
$Pt\{P(OCH)_3(CH_2)_3\}_4$	$Pt(PPh_3)_4 + P(OCH)_3(CH_2)_3$	44
$Pt\{P(OCH_2)_3CCH_3\}_4$	$Pt(PPh_3)_4 + P(OCH_2)_3CCH_3$	44

$Pt\{P(OMe)_2Ph\}_4$	$K_2PtCl_4 + L + NaBH_4$	80
$Pt\{P(OBu\text{-}n)Ph_2\}_4$	$K_2PtCl_4 + L + NH_2NH_2$	90
$Pt\{P(OMe)Ph_2\}_3$	$K_2PtCl_4 + L + NH_2NH_2$	90
$Pt\{P(OEt)Ph_2\}_3$	$K_2PtCl_4 + P(OMe)Ph_2 + NH_2NH_2$ in ethanol	90
$Pt\{P(OPh)_3\}_3$ $\{P(C_6H_4Cl\text{-}p)_3\}$	$Pt\{P(C_6H_4Cl\text{-}p)_3\}_4 + P(OPh)_3$	3
$Pt\{P(OC_6H_4Cl\text{-}p)_3\}_3(PPh_3)$	$Pt(PPh_3)_4 + P(OC_6H_4Cl\text{-}p)_3$	3
Phosphines and related ligands		
$Pt(PMe_3)_4$	$PtB_3H_7L_2 + L$	132
$Pt(PEt_3)_3$	PtL_4 in vacuo	132
$Pt(PEt_3)_4$	$PtB_3H_7L_2 + L$	132
$Pt(PPh_3)_4$	$K_2PtCl_4 + PPh_3 + KOH$	3, 66
	$PtL_2Cl_2 + L + NH_2NH_2$	3, 46
	$K_2PtCl_4 + L + NaBH_4$	78
	$PtL_2(C_2O_4) + L + h\nu$	133
	$PtL_2Cl_2 + L$ + sodium *n*-propoxide	96
$Pt(PPh_3)_3$	$PtCl_2Cl_2 + NH_2NH_2$	3
	$Pt(PPh_3)_4$ in boiling ethanol	66
	$PtL_2(CO_3) + L + h\nu$	135
$Pt(PPh_3)_2$	PtL_2HCl + lithium *n*-butyl	67
$Pt(PMePh_2)_4$	$K_2PtCl_4 + L + NaBH_4$	80
	cis-$PtL_2Cl_2 + L + NaBH_4$	81
	$K_2PtCl_4 + L + KOH$	292
$Pt(PMe_2Ph)_4$	$K_2PtCl_4 + L + NaBH_4$	80
	$K_2PtCl_4 + L + KOH$	81
$Pt\{P(CH_2Ph)Ph_2\}_4$	$K_2PtCl_4 + L + KOH + NH_2NH_2$	292
$Pt(AsPh_3)_4$	$PtL_2Cl_2 + L + NH_2NH_2$	3
	$K_2PtCl_4 + L + NaBH_4$	80
$Pt\{P(C_6H_4F\text{-}p)_3\}_4$	$PtL_2Cl_2 + L + NH_2NH_2$	85
$Pt\{P(C_6H_4Cl\text{-}p)_3\}_3$	$K_2PtCl_4 + L + KOH$	3
	$PtL_2Cl_2 + L + NH_2NH_2$	3
$Pt\{P(C_6H_4Me\text{-}p)_3\}_3$	$PtCl_2 + L + NH_2NH_2$	55

TABLE 2.III *continued*

Compound	Methods of synthesis	References
$Pt\{P(C_6H_4Me\text{-}m)_3\}_3$	$PtCl_2 + L + NH_2NH_2$	55
$Pt\{P(C_6H_4Me\text{-}m)_3\}_4$	$PtCl_2 + L + NH_2NH_2$	55
$Pt\{As(C_6H_4Cl\text{-}p)_3\}_4$	$PtL_2Cl_2 + L + NH_2NH_2$	3
$Pt\{PPh_2(C{\equiv}CMe)\}_4$	$K_2PtCl_4 + L + NH_2NH_2$	45
$Pt\{C_2(PPh_2)_2\}_2$	$K_2PtCl_4 + L + NaBH_4$	82
$Pt\{trans\text{-}C_2H_2(PPh_2)_2\}_3$	$K_2PtCl_4 + L + NaBH_4$	82
$Pt\{P(CH_2CH_2PPh_2)_3\}_3$	$K_2PtCl_4 + L + NaBH_4$	82
$Pt\{C_2H_4(PMe_2)_2\}_2$	$Pt(L{-}L)Cl_2$ + sodium naphthaldenide	77
$Pt\{C_2H_4(PPh_2)_2\}_2$	$Pt(L{-}L)Cl_2 + NaBH_4$	77
	$K_2PtCl_4 + L{-}L + NaBH_4$	78
$Pt\{o\text{-}C_6H_4(AsMe_2)\}_2$	$K_2PtCl_4 + L{-}L + NaBH_4$	80
$Pt\{Ph_2PCH_2CH_2P(C_6H_4CH_3\text{-}m)_2\}_2$	$K_2PtCl_4 + L{-}L + NaBH_4$	82
$Pt(Ph_2PCH_2CH_2AsPh_2)_2$	$K_2PtCl_4 + L{-}L + NaBH_4$	82
$Pt\{cis\text{-}C_2H_2(PPh_2)_2\}_2$	$K_2PtCl_4 + L{-}L + NaBH_4$	82
$Pt(PPh_3)_3(SbPh_3)$	$Pt(PPh_3)_4 + SbPh_3$	46
$Pt_2(PPh_3)_4(Ph_2PC{\equiv}CPPh_2)_2$	$Pt(PPh_3)_4 + Ph_2PC{\equiv}CPPh_2$	45
$Pt_2\{Ph_2PCH_2CH_2P(Ph)CH_2CH_2P(Ph)CH_2CH_2PPh_2\}_2$	$K_2PtCl_4 + L_n + NaBH_4$	82
$Pt_2\{P(CH_2CH_2AsPh_2)_2Ph\}_3$	$K_2PtCl_4 + L_n + NaBH_4$	82
$Pt_3\{P(CH_2CH_2PPh_2)_2Ph\}_4$	$K_2PtCl_4 + L_n + NaBH_4$	82

nearly perfect planar trigonal hybridization [143].* The Pt–P bond lengths lie between 2.25 and 2.28 Å, and are considerably shorter than the sum of the atomic radii (2.42 Å). A distortion of the co-ordination sphere, which allows the packing of two phenyl rings as in graphitic stacking has also been found. In the related carbonyl complexes $Pt(PPh_3)_3(CO)$ [144] and $Pt\{P(C_6H_5)_2(C_2H_5)\}_2(CO)_2$ [145], the Pt–P bond lengths are 2.36 Å. For a second crystalline form of $Pt(PPh_3)_3(CO)$, a slightly shorter mean metal-phosphorus distance was observed [146].

This substantial lengthening of the Pt–P bonds with respect to the non-carbonylated Pt^0 complex, seems to be due to a different degree of metal-to-phosphorus d_π-d_π back bonding as a result of the presence of a strong π-acceptor such as CO [145]. Since the metal-phosphorus distance is the same in the mono and dicarbonyl species (although two different phosphines are present in the two complexes), it was suggested that such distance represents a limiting value, probably corresponding to an essentially pure Pt–P σ-bond. This is in agreement with the non-existence of $Pt(PPh_3)_n(CO)_{4-n}$ ($n = 0, 1$) for which the platinum should not have enough π-electron density to stabilize bonds to the third and fourth CO molecules. It is interesting to point out that recently infrared spectroscopic evidence for the existence of $Pd(CO)_4$ at *ca.* 20°K has been reported [147]. From this point of view, the existence of $Pt(PF_3)_4$ should be attributed to the better σ-donor capacity of PF_3 with respect to CO. The principal structural parameters of $Ni(PF_3)_4$ and $Pt(PF_3)_4$ have been determined by gas-phase electron diffraction [148]. The geometry of PF_3 [149] remains essentially the same both in the free molecule and in the coordinated state in $Ni(PF_3)_4$, whereas there is a marked contraction of the P–F distances in PF_3BH_3 [150], and a lesser but still significant contraction in $Pt(PF_3)_4$ (Table 2.IV).

TABLE 2.IV

Structural parameters of PF_3 and certain of its derivatives.

	r(P–F) Å	< (FPF)°	r(M–P) Å
PF_3	1.569(1)	97.7(2)	–
$Ni(PF_3)_4$	1.561(5)	98.4(8)	2.116(10)
$Pt(PF_3)_4$	1.546(6)	98.9(7)	2.230(10)
PF_3BH_3	1.538(8)	99.8(1)	1.836(12)

Data taken from refs. 148, 149 and 150.

* The compound crystallizes in a triclinic form with two molecules per unit cell on the assumption that the molecular formula is $Pt(PPh_3)_3$. Later it has been shown that it is in fact a mixed compound of formula $2Pt(PPh_3)_3$, $Pt(PPh_3)_3CO$, in a particular crystalline phase different from $Pt(PPh_3)_3$. However the molecular structure of $Pt(PPh_3)_3$ is essentially the same as that considered above (see the note on ref. 292 and V. Albano and P. Bellon, personal communication).

It is expected that pure σ-donation from PF_3 will shorten r(P—F), while π-acceptance into the π-orbital of PF_3 should lengthen r(P—F), because of the reduced possibility of π-bond formation between fluorine and phosphorus.

The structural parameters indicate that σ-donation, which is of course dominant in PF_3BH_3, is in practice compensated by π-back donation in $Ni(PF_3)_4$, while in the platinum complex the σ-donation is probably the more important (see also 1.4).

We have already discussed in Part 1 the most important spectroscopic characteristics of these compounds.

Among the other physico-chemical measurements, the mass spectrum of $Ni(PF_3)_4$ has been reported [151].

The following species have been detected:

$$Ni(PF_3)_4^+ \rightarrow Ni(PF_3)_3^+ \rightarrow Ni(PF_3)_2^+ \rightarrow Ni(PF_3)^+ \rightarrow Ni^+$$

together with other species which originate from processes such as:

$$Ni(PF_3)_3^+ \rightarrow NiP_3F_8^+ + F$$

In another investigation, ions $Ni_2(PF_3)_n$ ($n \geq 6$) were additionally detected in the mass spectrum of $Ni(PF_3)_4$ [152].

A study of $Pt(PPh_3)_3$ by mass spectroscopy has also been attempted, but the mass spectrum showed only peaks due to triphenylphosphine [153]. It is known that, by thermal decomposition of $Pt(PPh_3)_3$, a compound analysing as $[Pt(PPh_3)_2]_x$ (x possibly three) is formed [136] with loss of phosphine. Presumably this cluster compound is not volatile enough to be detected. Moreover the mass spectrum of $Ni\{P(CH_3)_3\}_4$ displays only the fragmentation pattern of the ligand, but no ions from the intact complex [94].

The Raman spectrum of $Ni(PF_3)_4$ has been studied in detail [27, 154]. The Ni—P force constant is 2.37×10^5 [27] or 2.7×10^5 [154] dyn cm^{-1}, a rather low value, close to the Ni–C force constant in $Ni(CO)_4$ ($2.1-2.5 \times 10^5$ dyn cm^{-1}) [155, 156], which does not indicate the presence of a double bond. The i.r., P^{31} and F^{19} n.m.r. spectra for $M(PF_3)_4$ (M = Ni, Pd) have been reported [28, 114]. The PF_3 ligand is less perturbated in the palladium than in the nickel complex, as expected.

The melting and boiling points and decomposition temperatures of $M(PF_3)_4$ (M = Ni, Pd, Pt) are reported below:

	m.p. (°C)	b.p. (°C)	dec. (°C)
$Ni(PF_3)_4$	−55	70	>+70
$Pd(PF_3)_4$	−41	—	−20
$Pt(PF_3)_4$	−15	86	+90

The thermal instability of the palladium derivatives seems again to be related to its high promotion energy, which makes the complex with a strong π-acid such as PF_3 less stable. A more complex situation could be present with more basic ligands such as PPh_3 for which apparently the resistance to oxidation for instance, increases with increasing atomic number, following the increased strength of the σ-ligand–metal bond.

A qualitative M.O. level scheme has been given for a model complex $Ni(PR_3)_4$ of T_d symmetry (Fig. 2.1) [113].

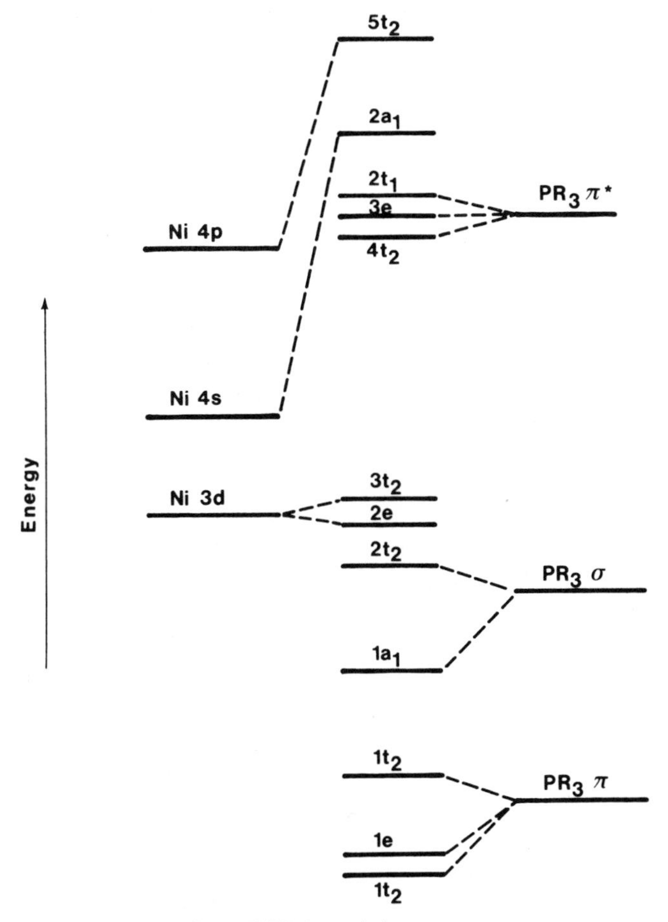

Fig. 2.1. M.O. level scheme for a $Ni(PR_3)_4$ complex.

The predicted ground state electronic configuration for the T_d model is: $\ldots(2e)^4(3t_2)^6 = {}^1A_1$; the $2e$ and $3t_2$ levels are the usual ligand field d-orbital levels. The metal-to-ligand allowed transitions are: $3t_2 \rightarrow 4t_2$ and $2e \rightarrow 4t_2$ (${}^1T_2$ excited states). Since the splitting between t_2 and e levels is less than 5000 cm^{-1}, the two transitions overlap in the electronic spectrum and for

$Ni\{P(OC_2H_5)_3\}_4$, $Ni\{P(OC_2H_5)_2C_6H_5\}_4$ and $Pd\{P(OC_2H_5)_2C_6H_5\}_4$ they were found at 42000, 30000, and 32300 cm^{-1} respectively. Such results have been interpreted as an evidence of the better π-acceptor ability of $P(OC_2H_5)_2C_6H_5$ with respect to $P(OC_2H_5)_3$ (lower energy of the transition) and indicate that the electron transfer from metal to ligand is easier for nickel than for palladium, as expected. This behaviour is in accord with the energies of the electronic transitions for the $M(PPh_3)_3$ complexes (M = Ni, Pd, Pt) which follow the order Ni < Pt < Pd [55].

2.2.3. Coordinative Reactivity

When a neutral molecule reacts with a zerovalent complex, two different types of reaction can take place, namely reactions involving coordinative dissociation and reactions involving coordinative addition [5].

The first step is similar in both cases and corresponds to the formation of a σ-bond with the empty metal orbitals. Some of the electronic density on the metal can be then transferred back to the entering molecule (π-bond). When the coordinating molecule has no π^* antibonding orbitals of low-enough energy (e.g. H_2, RCl, etc.), we can imagine an electron transfer to a σ^* antibonding orbital, with lengthening of the σ bond of the entering molecule, and thus dissociation can take place. On the other hand the coordinative addition is possible when the entering molecule (O_2, CO, olefin, etc.) has low-lying π^* antibonding orbitals. In this latter case the electron transfer from metal to ligand (which can be considered nearly complete with a molecule such as oxygen) causes a reduction of the O–O, C–O, C–C bond order, but does not result in complete dissociation.

The study of the coordinative reactivity of a compound such as $Pt(PPh_3)_4$ should answer first the question of which is the real reactive species in solution. A direct attack at the tetracoordinate complex is unlikely, since a five-coordinate intermediate would require the metal to exceed the 18-electron configuration and would be quite impossible with many of the ligands because of their large size [47]. On the other hand, it has been demonstrated that $Pt(PPh_3)_4$ and some other similar derivatives dissociate in solution into ML_3 and L [2, 3, 47, 55]. The next point to discuss is whether there is further dissociation to ML_2 and L, forming the true reactive two-coordinate species.

In contrast to previous findings [67, 157], it now seems that the three-coordinate complexes are the species involved in the reactions [47, 55, 158] and that in general a coordinative process should be visualized as

$$ML_4 \rightleftharpoons ML_3 + L$$

$$ML_3 + L' \rightarrow ML_3L'$$

which can be followed by

$$ML_3L' \rightarrow ML_2L' + L$$

as a function of the nature of L′.

2.2.3.1. Reaction with Hydrogen

Tris(triethylphosphine)platinum(0), recently synthesized [132], is an extraordinarily strong nucleophile and readily adds hydrogen at room temperature forming the five-coordinate hydride, $H_2Pt\{P(C_2H_5)_3\}_3$, which shows a single platinum–hydrogen stretching frequency at 1766 cm^{-1}. The ^1H n.m.r. spectrum indicates a trigonal bipyramidal structure with the hydrogen nuclei in the axial positions.

The temperature dependence of the hydride spectrum indicates also that above −30°C ligand dissociation is rapid enough via a four-coordinate complex, $H_2Pt\{P(C_2H_5)_3\}_2$, present at very low concentrations. This hydride is thermally unstable and decomposes within several days on standing in hydrocarbon solvents at −40°C. This very remarkable complex represents the first example where hydrogen is activated by a zerovalent Ni, Pd or Pt phosphine complex, with formation of a bis-hydride. A compound previously described as $H_2Pt(PPh_3)_2$ was later shown to be the carbonate complex $Pt(PPh_3)_2(CO_3)$ [159, 160], which is formed from $Pt(PPh_3)_2(O_2)$ and CO_2.

The inertness of $Pt(PPh_3)_3$ towards hydrogen, in contrast with the high reactivity of $Pt\{P(C_2H_5)_3\}_3$, seems then to be due to an electronic effect. The very basic triethylphosphine ligand supplies the platinum with the high electron density required for the activation of hydrogen.

2.2.3.2. Reactions with Acids and Compounds Having Activated Hydrogen

We are considering here the reactions by which an activated hydrogen is abstracted from the entering molecule, at least intermediately. In most cases stable metal(II) hydride derivatives are obtained by this oxidative addition reaction.

The basic character of zerovalent d^{10} phosphine complexes was firstly demonstrated by the reaction of $Pt(PPh_3)_3$ with inorganic acids [161]. The coordinative reaction of the dissociative type takes place in the following way:

$$Pt(PPh_3)_3 \underset{KOH}{\overset{HX}{\rightleftharpoons}} [PtH(PPh_3)_3]X \underset{+PPh_3}{\overset{-PPh_3}{\rightleftharpoons}} trans\text{-}PtHX(PPh_3)_2$$

The ionic hydride complexes can be isolated when the anion X^- is not too nucleophilic ($X^- = ClO_4^-$, HSO_4^-, BF_4^-), while non-ionic hydride complexes are obtained when X = CN^-, NCS^-. With hydrogen chloride it is possible to isolate both ionic and covalent hydride derivatives, depending on the nature of the solvent; the ionic derivatives is favoured by polar solvents. On the contrary, by reaction of hydrogen fluoride with $M(PPh_3)_4$ complexes (M = Pd, Pt), no hydride species were isolated, the products of the reaction being of the type $[PtF(PPh_3)_3][HF_2]$ or $[PdF(PPh_3)_2]_2F_2$, where bridging fluorine atoms are present in the cation, and $Pd(PPh_3)_2F_2$ [162]. Also, by treating $Pt(PPh_3)_4$ with NaN_3 in benzene or ethanol in the presence of sulphuric acid, $Pt(PPh_3)_2(N_3)_2$ is formed [284].

It has been claimed that, in the presence of excess acid, two molecules of HCl will add to $Pt(PPh_3)_4$ to give $Pt(PPh_3)_2H_2Cl_2$ [161]. It was shown later that the suggested platinum(IV) hydride is really the platinum(II) hydride *trans*-$Pt(PPh_3)_2HCl$ in a different crystalline form [163].

The reaction with acids is reversible and $Pt(PPh_3)_3$ can be regenerated by reacting the hydride with KOH under nitrogen [67, 161]. The covalent hydride can also be obtained with $Pt(AsPh_3)_4$. In contrast $M\{C_2H_4(PPh_2)_2\}_2$ complexes (M = Ni, Pd, Pt) with perchloric acid gave only $[M\{C_2H_4(PPH_2)_2\}_2](ClO_4)_2$, and similarly non hydrido species were obtained from $M(PPh_3)_4$ (M = Ni, Pd) by treating the zerovalent species with excess acids [161].

The intermediate hydrides formed react further under these conditions with evolution of hydrogen. In fact the synthesis of *trans*-$PdHCl(PPh_3)_2$ is possible from the stoichiometric reaction between $Pd(PPh_3)_4$ or $Pd(PPh_3)_3(CO)$ and HCl. Excess HCl gives $PdCl_2(PPh_3)_2$ from the hydride [108]. Analogously *trans*-$PdHCl\{P(cyclohexyl)_3\}_2$ has been obtained from $Pd\{P(cyclohexyl)_3\}_2$ and hydrochloric acid [108].

$Ni\{P(C_6H_{11})_3\}_2$, prepared in solution by displacement of nitrogen from $(NiL_2)_2N_2$ [L = $P(C_6H_{11})_3$], has been shown to react with acids such as HCl, CH_3COOH, phenol, pyrrole, forming hydridonickel complexes having the composition $L_2Ni(H)R$ [69].

The related biscoordinated complexes with L = $P(O\text{-}o\text{-}C_6H_4C_6H_5)$, react with acetic acid to give the nickel(I) complex, $NiL_2\{OC(O)CH_3\}$, with evolution of hydrogen [105].

An ionic nickel hydride has been synthesized by the following reaction:

$$Ni\{C_2H_4(PPh_2)_2\}_2 + AlCl_3 + HCl \rightarrow [Ni\{C_2H_4(PPh_2)_2\}_2H]AlCl_4$$

and similar oxidative additions have also been shown to be possible with HBF_4 and HCl [164].

H^1 n.m.r. evidence of the existence of a pentacoordinated hydride such as $[Ni\{P(OC_2H_5)_3\}_4H]^+X^-$ (X = Cl, HSO_4, CF_3COO) in solution obtained from the zerovalent phosphite complex and HX by protonation has also been reported [111]. In this case however the hydride derivatives were not isolated. Such a derivative (X = HSO_4) was later isolated from this reaction as an unstable oily product, and the kinetics and thermodynamics of its formation and decomposition have been reported [165].

Protonation is considered to precede ligand dissociation from the $Ni\{P(OC_2H_5)_3\}_4$ complex; but no comment has been made on the possible free site for the attack of the entering molecule on the coordinatively saturated complex. Equilibrium constants for the reaction:

$$H^+ + NiL_4 \overset{K}{\rightleftharpoons} HNiL_4^+$$

[L = $P(O\text{-}p\text{-}C_6H_4OCH_3)_3$, $P(O\text{-}p\text{-}C_6H_4CH_3)_3$, $P(OMe)_3$, $P(OCH_2CH_2Cl)_3$, $P(OEt)_3$, $PPh(OEt)_2$, PMe_3; 2L = $C_2H_4(PPh_2)_2$]

using H_2SO_4 in CH_3OH have been shown to depend on the electron donor-acceptor character of the ligands L [166].

Addition of H_2SO_4 to solutions of the zerovalent nickel complexes causes the appearance of a symmetrical quintet at high field in the ^1H n.m.r. spectra ($\tau \simeq 23\text{–}27$). The value of J_{P-H} increases as the electron-withdrawing ability of the ligand L increases, while the equilibrium constants decrease as expected.

The protonation of zerovalent platinum(0) complexes having phosphites as ligands has also been studied, with many different acids; although the expected hydrides are formed in solution (n.m.r. evidence), no stable derivatives were isolated, and evaporation of the reaction mixture gives back the zerovalent complex [90].

The protonation of tetrakis(triphenylphosphine)platinum(0) with carboxylic acids has been recently described [167], and a series of ionic hydride complexes $[Pt(PPh_3)_3H]X$ [X = $(CF_3COO)_2H$, $(CF_2ClCOO)_2H$, $(C_2F_5COO)_2H$, $(CF_2COO)_2H$, $(COO)_4H_2$] was obtained. Only with trifluoroacetic acid was a covalent hydride isolated.

Similar complexes of stoichiometry $Pt(PPh_3)_2HX$ (X = $SCOCH_3$, $SCOC_6H_5$, $C_7H_5S_2N$), have been obtained by addition of the acid HX to $Pt(PPh_3)_4$ [168]. This reaction, which does not take place with simple phenols [168], is also possible with picric acid [169]. Both the acidity of the protonic acid and the nephelauxetic effect of the conjugate base (that is its hardness), have been considered responsible for the formation of stable hydrides [168].

The very strong basicity of $Pt\{P(C_2H_5)_3\}_3$ is also well delineated by its interaction with weak acids such as ethanol and water [132]. The tris complex dissolves in both solvents to form the known [170] cationic hydride $[Pt\{P(C_2H_5)_3\}_3H]^+$, which was isolated from the solution as the hexafluorophosphate salt. By simple evaporation of the water solution, the starting tris complex was recovered.

In contrast to the high reactivity towards acids shown by the complexes considered above, the fluorophosphine derivative $Pt\{P(CF_3)F_2\}_4$ does not react with HCl at 60°C [33], and the reactions with PPh_3 and similar ligands gave only partial substitutions. The latter reactions indicate that a free site for the attack of triphenylphosphine is present (or created in solution) on $Pt\{P(CF_3)F_2\}_4$, but the strong electron-withdrawing fluorophosphines do not allow oxidative addition with hydrogen chloride to take place on the presumed unsaturated species.

The protonation of triphenylphosphine derivatives of platinum(0) has also been studied with acids such as H_2M and HMPh (M = S, Se) and the hydride derivatives *trans*-$PtH(MH)(PPh_3)_2$ and *trans*-$PtH(MPh)(PPh_3)_2$ were isolated [171]. In the former case, it has been shown by H^1 n.m.r. spectroscopy that the two species (a) and (b) are present immediately after bubbling hydrogen sulphide or selenide into a solution of $Pt(PPh_3)_3$:

With time the signal of species (a) slowly disappears, while those of the species (b) increases. In the solid state the species (b) is probably the only one present. More recently a series of *para*-substituted benzenethiols have been shown to react with $Pt(PPh_3)_n$ (n, = 3, 4) to form *trans*-$PtH(SC_6H_4Y)(PPh_3)_2$ (Y = NO_2, Br, Cl, F, CH_3, OCH_3, NH_2) and good correlations have been obtained when ν(Pt—H) or J_{Pt-H} were plotted against the Hammett substituent parameter, σ_p [172]. Both increase with increasing σ_p. The reactions with peculiar acids such as imides with $M(PPh_3)_4$ (M = Pd, Pt) and $Pt(AsPh_3)_4$ have also been studied [173].

With the platinum complexes, imido-hydrido complexes of the type PtL_2(H)(imide) have been obtained. $Pd(PPh_3)_4$ reacts with succinimide, but a complex of stoichiometry $Pd(PPh_3)_2(succinimide)_2$ was formed rather than a hydride. Some 5-substituted tetrazoles also react with the zerovalent complexes $M(PPh_3)_4$ (M = Pd, Pt) to form complexes of the type *cis*-$M(PPh_3)_2$-(tetrazolate)$_2$, rather than the hydrides [174]. On the other hand maleimide readily reacts with the palladium and platinum complexes to give compounds which still show an N–H stretch in the i.r. spectrum. They are of stoichiometry ML_2(maleimide), but in this case the imide is coordinated via the olefinic double bond [173].

The cleavage of the C–H bond of the formyl group in aldehydes and esters of formic acid readily occurs on reaction with $Pt(PPh_3)_4$ [175]. Platinum-acyl and -alkoxycarbonyl complexes of the type $Pt(PPh_3)_2(COR)_2$ are obtained, but no hydrides are formed.* β-Dicarbonyl compounds such as acetylacetone and diethyl malonate react with $Pt(PPh_3)_4$ but in this case there is no hydrogen abstraction, and complexes in which the enol form of the organic ligand is bound through the double bond to platinum are obtained [175].

The reactions of $Pt(PPh_3)_4$ with hydroxyacetylenes yields stable platinum complexes, where the organic ligand is η-bonded to platinum; however, 1-ethynylcyclohexanol yields a complex which, from spectroscopic evidence, has been formulated as [176]

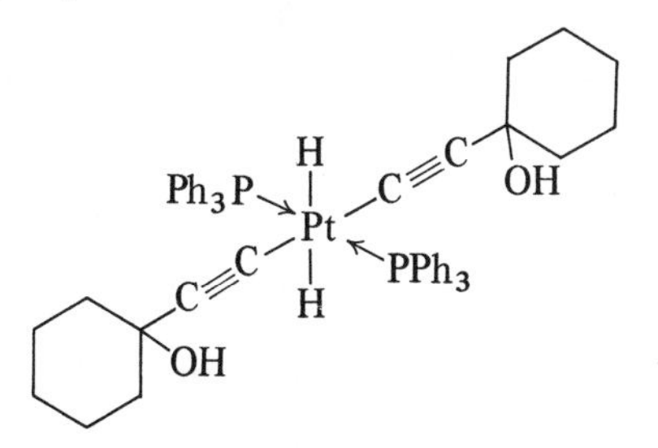

Such compound represents a rare example of a bishydridoplatinum(IV) complex. Hydridic acetylides have been considered the active polymerization catalysts for both linear and cyclic polymerization of acetylenes catalysed by nickel(0) complexes [177]. However, some recent studies on platinum(0) and platinum(II) complexes have shown that a hydridic complex is not necessary to promote the polymerization reaction [178].

Hydrogen abstraction has also been shown to be possible from the carbon atom in the α position with respect to a double bond. By reacting cyclopentadiene with NiL_2 [$L = P(C_6H_{11})_3$], the hydride $NiL(H)C_5H_5$ has been obtained [69a]. Another interesting reaction is given by nitromethane and $Pt(PPh_3)_4$:

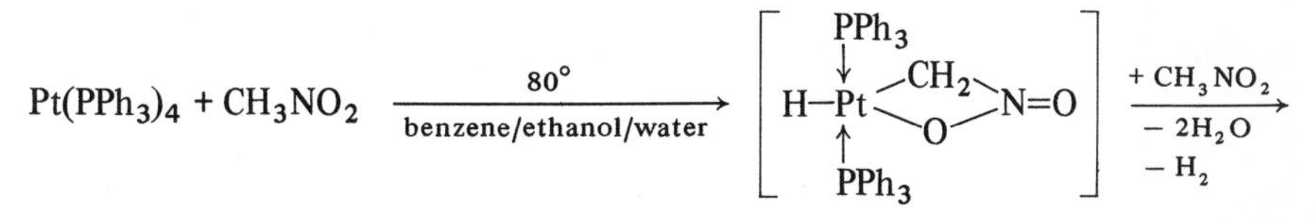

⟶ *trans*-$Pt(PPh_3)_2(CNO)_2$.

It has been assumed that nitromethane reacts as a compound containing an acidic CH group, in a process which corresponds to an oxidative addition reaction [179]. A polar medium is necessary in order that the reaction may

* Tripathy and Roundhill [381] have suggested that these derivatives are in fact the complexes $Pt(PPh_3)_2(COOR)_2$ (R = Et, Ph); their i.r. spectra are identical to those of the compounds obtained from the reactions between $Pt(PPh_3)_2(CO_3)$ and C_2H_5COOH or C_6H_5COOH.

proceed. With nitroethane and 2-nitropropane the reaction does not lead to fulminates, but instead the dinitro complexes $Pt(PPh_3)_2(NO_2)_2$ can be isolated [179].

The complexes $Pt(PPh_3)_2(SiCl_3)_2$, $Pt\{C_2H_4(PPh_2)\}(SiCl_3)_2$, $Pt\{C_2H_4(PPh_2)_2\}(SiHPh_2)_2$ and $Pt(PPh_3)_2H\{Si(C_6H_4X)_3\}$ (X = *meta* and *para*-F or $-CF_3$) have been prepared by oxidative addition of the silicon hydrides Cl_3SiH, Ph_2SiH_2, or $(XC_6H_4)SiH$ to $Pt(PPh_3)_4$ and $Pt\{C_2H_4(PPh_2)_2\}_2$ [180]. All probably have a *cis* configuration. By oxidative addition of methyltrichlorosilane to $Pt(PPh_3)_4$, three types of silicon–platinum complexes are formed, namely $Pt(PPh_3)H(SiCl_2Me)$, $Pt(PPh_3)_2(SiCl_2Me)_2$ and a 1:1 adduct between $Pt(PPh_3)_2(SiCl_2Me)_2$ and $Pt(PPh_3)_2$ [181]; $Pt(PMePh_2)_4$ reacts with Cl_3SiH or $MeCl_2SiH$ to form $Pt(PMePh_2)_2(SiCl_3)_2$ and $Pt(PMePh_2)_2(SiCl_2Me)_2$, respectively [182].

2.2.3.3. *Reactions with Halo and Pseudohalo Derivatives*

A large number of these oxidative addition reactions have been studied over many years [5], and they are very useful for the synthesis of metal(II) complexes, according to the following reaction:

$$ML_n + RX \rightarrow ML_2(R)X + (n-2)L$$

(R = alkyl, aryl, acyl, aroyl, sulphonyl, etc.).

In some cases, and depending on the reaction conditions, the organic part is not present in the reaction product, which is simply the bis-halo derivative ML_2X_2. Also, with acyl and aryl sulphonyl halides, CO or SO_2 can be displaced from the reaction product, for instance with $Pt(PPh_3)_3$, with formation of the corresponding metal–alkyl and -aryl complexes [183]. With benzylsulphonyl halides such as *para*-$NO_2C_6H_4CH_2SO_2Cl$, $Pt(PPh_3)_3$ reacts forming the expected product from the oxidative addition reaction; once formed, such a derivative does not show any tendency to lose HCl by reaction with organic bases [184].

This reaction is typical of benzylsulphonyl halides in organic chemistry and leads to intermediate sulphenes $RCH{=}SO_2$, which cannot be isolated. Organic sulphenes can be trapped in a complex where they are probably bonded as an alkene, when $Pt(PPh_3)_3$ is added to a cooled solution of the sulphonyl halide and triethylamine [185].

The additions of acetyl chloride and ethyl chloroformate to $Pd(PPh_3)_4$, yielding respectively *trans*-$PdCl(COCH_3)(PPh_3)_2$ and *trans*-$PdCl(CO_2C_2H_5)$-$(PPh_3)_2$, have been reported [186].

Similar reactions have been studied in the case of $Ni(PPh_3)_4$ [187]; with benzoyl chloride the aroyl derivative could not be isolated, since it spontaneously loses carbon monoxide giving $NiCl(Ph)(PPh_3)_2$. In contrast, the analogous platinum derivative loses CO only at 210°C [198], as shown by the disappearance of the keto band at 1614 cm^{-1}. With acetyl chloride, the highly unsaturated nickel complex, $Ni(PPh_3)_2$ gives a disproportionation reaction, with formation of $Ni(PPh_3)_2Cl_2$ and $Ni(PPh_3)_2(CO)_2$ [57].

The products $NiCl(CO_2R)(PPh_3)_2$ obtained with ROCOCl (R = Me, Et, CH_2Ph) lose the RCO_2 groups with formation of the nickel(I) species $Ni(PPh_3)_3Cl$ [187].

Some other oxidative addition reactions with RCOCl and RCSCl derivatives to nickel(0) compounds have been reported, according to the scheme [188]:

$$NiL_4 + Cl{-}C\begin{matrix}\nearrow\!\!=A \\ \searrow B\end{matrix} \longrightarrow L_2Ni\begin{matrix}\diagup Cl \\ \diagdown C\begin{matrix}=A \\ \searrow B\end{matrix}\end{matrix}$$

[L = PPh_3, $P(OPh)_3$; 2L = $C_3H_6(PPh_2)_2$, $C_4H_8(PPh_2)_2$] (A = O, S; B = Ph, NMe_2).

Also, carboxamido and thiocarboxamido complexes of formula *trans*-$ML_2(Cl)CONMe_2$ and *trans*-$ML_2(Cl)CSNMe_2$ (M = Pd, Pt; L = PPh_3), have been obtained from ML_4 and $ClCONMe_2$ or $ClCSNMe_2$ [189]. $Pd\{P(OPh)_3\}_4$ and $ClCSNMe_2$ yield the dimeric complex, $[\{P(OPh)_3\}_2Pd_2Cl_2(CSNMe_2)_2]$, in which palladium atoms are bridged by two thiocarboxamido ligands.

Ethyl dithiochloroformate, ClCSSEt, similarly adds to $Pt(PPh_3)_3$ giving $Pt(PPh_3)_2Cl(CSSEt)$ [190], which shows ν(C=S) at 1058 cm^{-1} (1110 cm^{-1} in the free ester) and ν(C–S) at 782 cm^{-1}.

A stable perfluoroacyl complex, $Pt(PMePh_2)_2(COC_3F_7)Cl$ has also been isolated [191].

Chloranil (2,3,5,6-tetrachloro-1,4-benzoquinone) and $Pt(PPh_3)_3$ react with abstraction of a vinylic chlorine, to give:

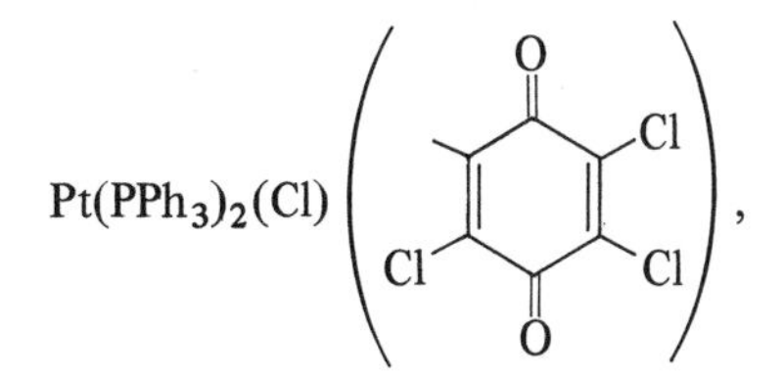

with no interaction between platinum and the double bonds of the quinone moiety in the final product [257].

Oxidative addition of aryl halides to $Ni(PPh_3)_4$ (aryl-X; X = Cl, Br; aryl = C_6H_5, *o*-, *m*-, *p*-$C_6H_4CH_3$, *p*-C_6H_4Cl, *p*-C_6H_4NCO, *p*-$C_6H_4OCH_3$) has been recently studied [187, 192]. In contrast to $Ni(PPh_3)_4$, $Ni(PMePh_2)_4$ and $Ni\{C_2H_4(PPh_2)_2\}_2$ do not react at room temperature with chlorobenzene [192], while $Ni(PEt_3)_4$, and the platinum analogue $Pt(PEt_3)_3$, easily undergo oxidative addition with chlorobenzene [132]. Alkyl halides such as CH_3I or C_2H_5Br readily add to $Ni\{P(C_6H_{11})_3\}_2$, giving the corresponding $Ni\{P(C_6H_{11})_3\}_2(X)R$ complexes [69b]. Analogous reactions occur with $Ni(PEt_3)_4$ [70], and $Pd(PPh_3)_4$ has been shown to undergo oxidative addition with methyl and phenyliodide [186], and with aryl-X (X = I, Br) [193]. In the latter case no reaction was observed with chlorobenzene in benzene solution up to 135°C, although this reaction does occur in neat refluxing chlorobenzene [194].

Aryl chlorides substituted with electron-donating groups are unreactive towards $Pd(PPh_3)_4$, while those with electron-withdrawing substituents give the desired products [193]. It has been suggested that the mechanism of the oxidative addition is similar to that of a bimolecular nucleophilic aromatic displacement reaction, in which breaking of the bond to the leaving group is involved in the rate-determining step [193].

The reactions of $Pt(PPh_3)_3$ or $Pt(PPh_3)_4$ with MeI [195] and MeBr [183] have been studied. With benzyl chloride, polymerization with loss of hydrogen chloride to give *m*-xylene polymers and not the simple oxidative addition reaction was observed [196]. Also, $Pt(PPh_3)_4$ gives *cis*-$Pt(PPh_3)_2Cl_2$ on reaction with chloroform or hexachloroethane; with trichlorobromomethane $Pt(PPh_3)_2Br_2$ is similarly formed, while with carbon tetrachloride what is probably a mixture of *cis* and *trans*-$Pt(PPh_3)_2Cl(CCl_3)$ is obtained [197]. The zerovalent platinum complex gives $Pt(PPh_3)_2Cl(CH_2CN)$ on reaction with chloroacetonitrile [197 *bis*].

Perfluoroalkyl complexes of palladium and platinum have been synthesized by this route [78, 191]. With $Pd\{P(OMe)_3\}_4$ and CF_3I, *cis*-$Pd\{P(OMe)_3\}_2(CF_3)_2$ has been isolated instead of the 1:1 adduct [121], while with C_3F_7I and $Pt(PMePh_2)$, *trans*-$Pt(PMePh_2)_2I(C_3F_7)$ was obtained together with $Pt(PMePh_2)_2I_2(Ph)_2$, which originated from a cleavage of P-aryl bonds [191].

With allyl halides the simple oxidative addition takes place with $Pd(PPh_3)_4$ [186], but with $Pt(PPh_3)_4$ π-allyl compounds of platinum(II) are obtained [130, 198]. Analogously, $Pt(PEt_3)_3$ with C_3H_5Cl gives the π-allyl complex, $[\pi\text{-}C_3H_5Pt(PEt_3)_2]^+$ [132]. Allyl isothiocyanate similarly adds to $Pt(PPh_3)_3$, forming the corresponding allyl derivative [198].

The reactions of halo-vinyl compounds with zerovalent derivatives of nickel, palladium and platinum have been widely explored. This type of oxidative addition reaction proceeds through an intermediate η-olefin complex. This is demonstrated by the reactions studied on the tetrachloroethylene complex, $Pt(PPh_3)_2(CCl_2{=}CCl_2)$ [197]. This derivative in fact isomerizes in boiling ethanol to the halovinyl complex, $Pt(PPh_3)_2(Cl)(CCl{=}CCl_2)$.

The kinetics of this reaction have shown that it proceeds through a carbonium ion intermediate [199], but other mechanisms are possible [200]. $Pt(PPh_3)_4$ and trichloroethylene give only the chlorovinyl complex; with *trans*-dichloroethylene a mixture of $Pt(PPh_3)_2Cl_2$ and $Pt(PPh_3)_2Cl(CH{=}CHCl)$

is obtained, while *cis*-dichloroethylene and hexachloropropene give only $Pt(PPh_3)_2Cl_2$ [197]. The oxidative additions of β-bromo styrene [183], vinyl bromide [201] and α-phenyl-β-bromostyrene [201] to triphenylphosphine derivatives of platinum(0) have been reported. $Pd(PPh_3)_4$ also generally adds chloro-olefins with formation of vinyl complexes [137], but the reaction with vinyl chloride simply results in the formation of $[Pd(PPh_3)_2]_x$. The stereo-chemistry of the products of this type of reaction has been elucidated by studying the reaction between $Pt(PMePh_2)_4$ and *cis*-1,2-dichloroethylene [202]. The product is the *trans* isomer, while there is retention of configuration of the olefin. With *trans*-1,2-dibromoethylene, the product, *trans*-$Pt(PMePh_2)_2Br(CH{=}CHBr)$, in contrast to the thermal stability exhibited by the chloro derivative, loses acetylene at room temperature with formation of $Pt(PMePh_2)_2Br_2$. By carrying out the reaction with *cis*-1,2-dichloroethylene in refluxing ethanol in the presence of ethoxide, the acetylide $Pt(PMePh_2)_2Cl(C{\equiv}CH)$ was obtained; the ethoxide promotes the formation of chloroacetylene, which then undergoes oxidative addition to the zerovalent complex [202].

Perfluorohalo-olefins such as C_2F_3X also react with platinum(0) [191, 203], palladium(0) [120, 204] and nickel(0) [205, 206] derivatives; sometimes the η-olefin intermediate complex cannot be isolated, the oxidative addition products being obtained directly.

The nature of the products from oxidation-addition reactions of halovinyl compounds to the zerovalent derivatives of nickel, palladium, and platinum has generally been elucidated by i.r. and n.m.r. spectroscopy. For example the reaction product between $Pt(PPh_3)_4$ and $CHCl{=}CCl_2$ shows $\nu(C{=}C)$ at 1550 cm^{-1}, $\nu(Pt{-}Cl)$ at 305 cm^{-1}, $\tau_H = 5$ [197]. The n.m.r. signal has platinum satellites with $J_{Pt-H} = 56$ Hz, each line being further split into a triplet owing to coupling with two equivalent phosphorus atoms ($J_{P-H} = 2.2$ Hz) [197].

During these studies [191], an interesting cleavage of the phenyl of the coordinated phosphine was observed. From the reaction between $Pt(PMePh_2)_4$ and C_2F_3Br, $Pt(PMePh_2)_2Br(Ph)$ was obtained, besides the expected product from the oxidative addition reaction. A cleavage of the P-aryl bond has also been observed in the following reaction [194]:

$$Pd(PPh_3)_4 + PdCl_2 \rightarrow Pd(PPh_3)_2Cl_2 + trans\text{-}PdCl(PPh_3)_2(Ph)$$

Haloacetylenes such as C_2I_2 and $C_6H_5C_2I$ add to $Pt(PPh_3)_4$ [183], possibly by way of an intermediate η-complex, as in the case of halo-olefins. Interestingly, the same type of complex has been obtained by oxidative addition of chloroallenes, $ClCH{=}C{=}CR_2$, or haloacetylenes, $CH{\equiv}C{-}CR_2{-}X$, to $Pt(PPh_3)_4$ [201, 207]:

$$Pt(PPh_3)_4 + ClCH{=}C{=}CR_2\ (\text{or}\ CH{\equiv}C{-}CR_2X) \longrightarrow Pt(PPh_3)_2Cl(CH{=}C{=}CR_2)$$

A reaction which can be considered comparable to those reported above is the interaction between tetracyanoethylene and $Pt(PPh_3)_4$, which leads to the cyanide complex, $Pt(PPh_3)_2(CN)_2$ [208] with abstraction of the pseudo-halogen $(CN)_2$. That this type of reaction proceeds through an intermediate η-complex is once more demonstrated by the observation that the dicyano-acetylene complex, $Pt(PPh_3)_2\{C_2(CN)_2\}$, isomerizes to the acetylide $Pt(PPh_3)_2(CN)(C_2CN)$ by irradiation with a sun lamp [209]. Analogously, the reaction of $Pt(PPh_3)_3$ with 1,1,1-tricyanoethane, $MeC(CN)_3$, yields the complex $Pt(PPh_3)_2(CN)\{CMe(CN)_2\}$, which is a rare example of an oxidative addition reaction involving carbon-carbon cleavage in a saturated organic compound [210]. From spectroscopic data the structure

$$\begin{matrix} Ph_3P \searrow & & \nearrow CN \\ & Pt & \\ Ph_3P \nearrow & & \searrow C(CN)_2Me \end{matrix}$$

has been proposed for this compound (ν(CN coordinated) at 2141 cm^{-1}; ν(CN uncoordinated) at 2157 cm^{-1}, unresolved doublet).

By reactions between zerovalent complexes and metal halide derivatives, it is possible to synthesize derivatives with metal–metal bonds. Compounds having covalent Pt–Au, Pt–Hg, Pt–Sn and Pt–Cu bonds have been obtained by this method [211].

Triphenylphosphine derivatives of platinum(0) react according to the following scheme:

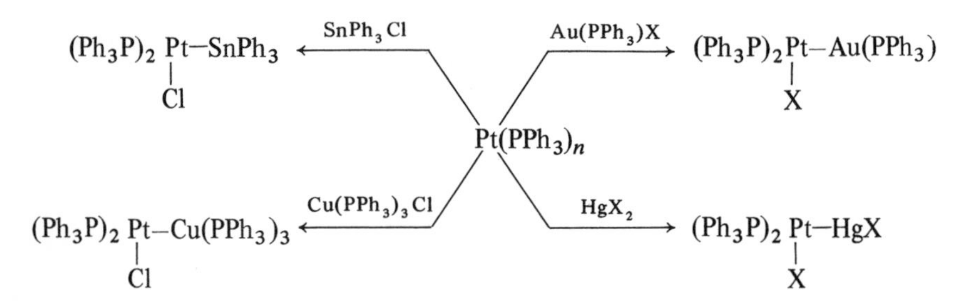

and a kinetic investigation of the reaction of $Pt(PPh_3)_3$ with $SnPh_3Cl$ has been reported [157]. Similarly $Pt(PMePh_2)_4$ reacts with $SnMe_3Cl$ to give *trans*-$Pt(PMePh_2)_2Cl(SnMe_3)$ [81], and $Pt(PPh_3)_4$ with $(CH_3)_3Sn\text{-}Sn(CH_3)_3$ gives *trans*-$Pt(PPh_3)_2(SnMe_3)_2$ [212] [the *trans* configuration has been suggested only on the basis of the unique band found at 420 cm^{-1}, and attributed to ν(Pt–P)]. Trimethylstannane also reacts with $Pt(PPh_3)_4$, the product probably being a mixture of *cis*- and *trans*-$Pt(PPh_3)_2(SnMe_3)_2$ [212]. Trialkyltin acetylides such as $R_3Sn(C{\equiv}CPh)$ (R = Me, Et) oxidatively add to PtL_4 complexes (L = PPh_3, $PMePh_2$), with formation of $PtL_2(SnR_3)(C{\equiv}CPh)$; these compounds have a $\nu(C{\equiv}C)$ at around 2100 cm^{-1}, showing that the triple bond is not coordinated to the metal [213].

However in some cases, as for the reaction of $Pt(PPh_3)_3$ with SiR_3Cl and PbR_3Cl, the reaction product is simply *cis*-$Pt(PPh_3)_2Cl_2$, possibly because an excess of the halo derivative was used [214].

It is possible to prepare compounds with Pt–Pt and Pt–Ni bonds by treating $Pt(PPh_3)_n$ with $Pt(PPh_3)_2I_2$ and $Ni(PPh_3)_2I_2$ respectively [211].

However, no compounds with metal-metal bonds are formed by the addition of $Ni(PPh_3)_2X_2$ to $Ni(PPh_3)_4$, the reaction product being the nickel(I) derivative $Ni(PPh_3)_3X$ [141]. $Ni(PPh_3)_3$ has been recently shown to undergo a double oxidative addition reaction with SnR_3Cl derivatives (R = Me, Ph), with formation of $Ni(PPh_3)_2(SnR_3)_2Cl_2$ [215]. Halogens readily add to triphenylphosphine derivatives of platinum(0), with formation of $Pt(PPh_3)_2X_2$ compounds.

The reaction of $Pt(PPh_3)_3$ with iodine has been used for demonstrating that the former was not a hydride derivative, but a true zerovalent complex [4]. Strangely enough, the reaction of halogens with $Ni(PPh_3)_4$ leads to nickel(I) instead of nickel(II) complexes, $Ni(PPh_3)_3X$ [141]. The addition of cyanogen to ML_4 complexes (M = Ni, Pd, Pt; L = tertiary phosphine) gives the dicyanobisphosphineplatinum(II) derivatives, $ML_2(CN)_2$ [216]. Such reactions are among the few examples where the breaking of a carbon-carbon bond is found. Similar results have been obtained with nickel(0) and palladium(0) complexes having a chelating diphosphine as the ligand [217]. The reaction of cyanogen with $Pt(PPh_3)_4$ has also been reconsidered.

The first product of the reaction is exclusively the *cis* isomer. A slow isomerization to the *trans* isomer in solution then takes place. The two isomers are of comparable stability, and this suggests that the oxidative process primarily involves the formation of a 1:1 adduct. This interaction causes the weakening and breaking of the C–C bond, and synchronous formation of two metal-CN bonds, mutually *cis* when no rearrangements occur [217].

Interestingly, the reaction of cyanogen with $Rh(PPh_3)_3Cl$, which usually readily undergoes oxidative addition reactions, gives at low temperature a cyanogen adduct [217]. In benzene solution at *ca.* 40°C the initial cyanogen adduct [$\nu(CN)_2$, 2240, 2090 cm^{-1}], rearranges to a cyanide complex in about ten minutes [$\nu(CN)$ = 2136, 2133, 2106 cm^{-1}].

The reaction of thiocyanogen, $(SCN)_2$, with $M(PPh_3)_4$ complexes (M = Pd, Pt), with formation of $M(PPh_3)_2(NCS)_2$ has been briefly reported [218].

2.2.3.4. Reactions with Molecules Having Sulphur as the Donor Atom

The poisoning mechanism of a catalytic metal surface by a species such as H_2S could proceed through hydrogen abstraction from the H_2S molecule bonded to the surface. The poisoning could then be due to an unreactive species such as M–SH or M–S formed on the surface [5]. This hypothesis has stimulated some research on the reactivity of low-valent metal complexes with hydrogen sulphide (2.2.3.2) and also with sulphur, and other sulphur-containing molecules. Such research should be considered within the framework of the modern tendency to compare the chemisorption of molecules on active centres of the catalytic surface with the coordination of the same molecules to individual metal atoms in model compounds, usually coordinately unsaturated derivatives with the transition metal in a low oxidation state.

The reaction of triphenylphosphine derivatives of platinum(0) with elemental sulphur in benzene has been studied by two different research groups [171, 219]. The results obtained are substantially different, although the reaction conditions differ only in the temperature employed (room temperature [171] or 0–5°C [219]) and in the molar ratio of the reactants. The use of $Pt(PPh_3)_3$ [171] instead of $Pt(PPh_3)_4$ [219] should not affect the reaction course. At room temperature a pale orange compound analysing as $[Pt(PPh_3)_2S]_n$ has been obtained where n is possibly 2, and the sulphur atoms probably bridge the two platinum atoms:

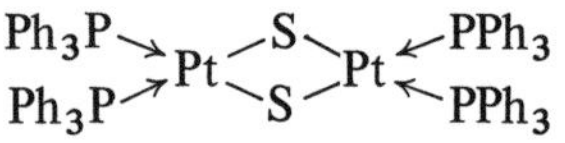

$$\begin{matrix} Ph_3P\searrow & & S & & \swarrow PPh_3 \\ & Pt & & Pt & \\ Ph_3P\nearrow & & S & & \nwarrow PPh_3 \end{matrix}$$

This derivative is too insoluble for molecular weight determinations, and melts at 246°C [171]. An analogous reaction is given by selenium.

The sulphur derivative shows two bands in the far i.r. spectrum at 378 and 317 cm^{-1}, which are absent in the selenium derivative, and which are due to $\nu(Pt{-}S)$.

The platinum-sulphur adduct has also been obtained by reacting $Pt(PPh_3)_2H(SH)$ with oxygen [171]. This is an interesting reaction, since it seems to support the view that the irreversibility of the poisoning of a metal surface with volatile molecules such as H_2S can be attributed to a hydrogen abstraction followed by oxidation to sulphur species which are irreversibly bound.

At lower temperatures, using a larger amount of sulphur, a yellow compound analysing as $Pt(PPh_3)_2S_4$ and with m.p. 138–140°C was obtained [219]. This derivative presumably has the structure

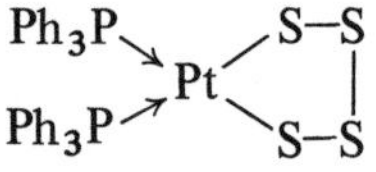

$$\begin{matrix} Ph_3P\searrow & & S{-}S \\ & Pt & \quad | \\ Ph_3P\nearrow & & S{-}S \end{matrix}$$

Similar compounds have been obtained by reacting $Pd(PPh_3)_4$ or $M\{C_2H_4(PPh_3)_2\}_2$ (M = Pd, Pt) with sulphur [219].

A brown compound analysing as $[Pt(PPh_3)S_2]_n$ (n = 4; m.p. = 195–197°C)

was the product of the reaction between $[Pt(PPh_3)_2S]_n$ and excess sulphur in benzene [171].

Both research groups have studied the reaction of sodium sulphide with *cis*-dichlorobisphosphineplatinum(II) derivatives, in order to confirm the nature of the products of the reaction between sulphur and triphenylphosphine–platinum(0) compounds, but again the results are different. In fact while *cis*-$Pt(PMe_2Ph)Cl_2$ reacts with Na_2S giving $[Pt(PMe_2Ph)_2S]_2$ [219], which has a stoichiometry corresponding to the product of the reaction at room temperature [171] (another derivative of formula $[Pt_3(PMe_2Ph)_6S_2]Cl_2$ is also isolated from this reaction), *cis*-$Pt(PPh_3)_2Cl_2$ gives, in ammoniacal ethanol, $[Pt(PPh_3)_2S_{0.5}]_2$ [219] which, strangely enough, has a melting point identical to $[Pt(PPh_3)_2S]_n$ [171], a compound which has been also isolated from the reaction of *cis*-$Pt(PPh_3)_2Cl_2$ and sodium sulphide in benzene solution [171]. The product of the reaction in ethanol may have the structure

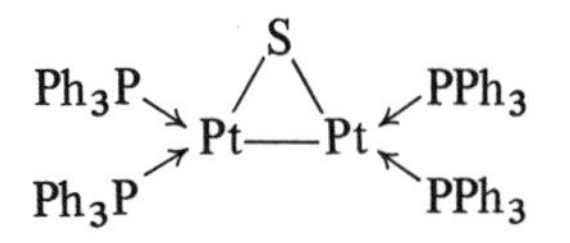

with one bridging sulphur atom while the product obtained in benzene should have two such bridges. Since the reactions of $[Pt(PPh_3)_2S]_n$ [171] and $[Pt(PPh_3)_2S_{0.5}]_2$ [219] with substances such as benzyl halides give different products, these compounds appear to be different.

Support for the structure of the tetrakis (triphenylphosphine) monosulphurdiplatinum derivative comes also from a recent X-ray determination of the structure of $Pt_2S(CO)(PPh_3)_3$, isolated from the thermal decomposition of $Pt(PPh_3)_2(COS)$ (see later) and initially formulated as $Pt_2S(CO)_2(PPh_3)_3$ mainly on the basis of the two CO stretching frequencies present in its i.r. spectrum [198].

As a matter of fact the complex exists in the crystal in two conformationally isomeric forms; the difference between the two isomers is mainly confined to one triphenylphosphine group in which the phenyl groups adopt two quite different configurations with respect to the rest of the molecule [220]. The two platinum and the sulphur atoms form a triangle in which the Pt–Pt distance is 2.65 Å, which corresponds to a simple Pt–Pt bond, in excellent agreement with the sum of Pauling's covalent radii (2.62 Å). It is surprising that among the various products of the reactions between sulphur and the platinum(0) derivatives a compound having a disulphur molecule bound to one metal atom has not been isolated. A compound such as $Pt(PPh_3)_2(S_2)$ may in fact correspond to the well known platinum-dioxygen adduct (see later); it appears that it has been obtained by a different method [221], but its existence has not been confirmed. The X-ray structure of an iridium(I)-disulphur adduct, $[Ir(S_2)\{C_2H_4(PPh_2)_2\}_2]Cl \cdot CH_3CN$ has been reported [222].

Such a cation displays approximately trigonal-bipyramidal coordination and is nearly isostructural with the previously reported $[M(O_2)\{C_2H_4(PPh_2)_2\}_2]^+$ (M = Rh, Ir). The disulphur molecule is η-bonded at an equatorial site, with an average Ir–S distance of 2.41 Å. The S–S distance of 2.066(6) Å is similar to that in octasulphur (2.060(3) Å), but it is significantly longer than the 1.889 Å S–S distance in free disulphur.

The oxidative addition reactions of organic disulphides RSSR (R = C_6H_5, *tert*-C_4H_9, *ortho*- and *meta*-$NO_2C_6H_4$) to $M(PPh_3)_4$ (M = Pd, Pt), have been recently shown to give the mercapto derivatives $M(PPh_3)_2(SR)_2$ [218].

$M(PPh_3)_4$ complexes (M = Pd, Pt) react in oxygen-free benzene solution with sulphur dioxide forming the $M(PPh_3)_3(SO_2)$ adducts [223, 224]. It is reasonable to suppose that in these complexes the SO_2 molecule is coordinated through a sulphur atom, as it was found for a similar iridium(I) compound [225].

The i.r. spectrum of the dark-purple platinum complex shows two bands at 1195 and 1045 cm^{-1}, which can be assigned to the asymmetrical and symmetrical S=O stretching frequencies (see 1.3.1.3). Both platinum and palladium SO_2 adducts react with oxygen giving sulphato-complexes $M(PPh_3)_2(SO_4)$ (M = Pd, Pt) [$\nu(SO_4)$ = 1280, 1155, 885, 665 cm^{-1}]. This reaction indirectly supports the view that catalytic oxidation of sulphur dioxide at platinum surfaces involves chemisorption of the gas on the metal [5]. According to other authors [211] the platinum-SO_2 adduct has a stoichiometry corresponding to $Pt(PPh_3)_2(SO_2)$. A derivative having this stoichiometry has been also obtained by reaction of $Pt(PPh_3)_2(C_2H_4)$ with SO_2; a brown compound is formed, which at 50°C loses SO_2 with formation of the green $Pt(PPh_3)_2(SO_2)$ [ν(S=O) = 1182, 1149, 1035 cm^{-1}] [226]. The nickel adduct, $Ni(PPh_3)_2(SO_2)$, has been briefly mentioned during the study of the reactivity of $Ni(PPh_3)_2$ [57].

By reacting $Pt(PPh_3)_3$ [198] or $Pt(PPh_3)_4$ [227] with carbon disulphide a compound of formula $Pt(PPh_3)_2(CS_2)$ has been obtained. $Pd(PPh_3)_4$ shows the same reactivity towards CS_2, with formation of $Pd(PPh_3)_2(CS_2)$ [198, 227], isomorphous with the platinum derivative [198]. While free CS_2 has ν(C=S) at 1196 and 1117 cm^{-1}, the platinum-CS_2 adduct shows only one ν(C=S) = 1157 cm^{-1} [ν(C=S) = 1193 cm^{-1} in the palladium adduct] and a band at 313 cm^{-1} tentatively assigned to ν(Pt–S) [198]. The i.r. spectra are consistent with the X-ray structure of such compounds.

The molecular structure of $Pt(PPh_3)_2(CS_2)$:

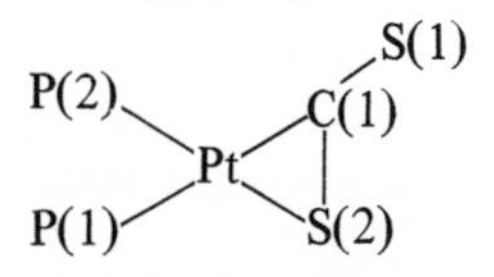

shows that coordination of the carbon disulphide to platinum drastically modifies its geometry. [228]. Free, linear CS_2 has d_{C-S} = 1.554 Å, while the mean C–S bond distance in the coordinated CS_2 is 1.626 Å, close to that observed for the $^3A_2$ excited state of CS_2, where d_{C-S} is 1.64 Å (see also 1.3.1.2). In the complex, CS_2 is no longer linear but bent at an angle of 136.2°; the lowest excited state of CS_2 is in fact bent, the angle being 135°. These values support the view that the electronic density of the metal populates the antibonding levels of the CS_2 molecule through a π-back-donation mechanism [228]. The dihedral angle between the planes P(1)PtP(2) and S(1)C(1)S(2) is only 6°; however, these six atoms are far from coplanar. For example S(1) is displaced 0.23 Å from the mean plane containing the remaining five atoms. The two platinum-phosphorus bond lengths of 2.346 and 2.240 Å are significantly different and are related to the differing *trans*-bond weakening effects of carbon and sulphur. The P(1)PtP(2) bond angle of 107.1° seems unexpectedly large [cf. the angle of 98.5° found in $Pt(PPh_3)_2(CO_3)$ [160]] and may originate from the necessity to reduce repulsions between the phenyl rings [228]. An alternative explanation of such effects has been given, considering that CS_2, but not the carbonato ion, has appreciable π-acidity, so that *d* orbital involvement in the Pt–P bonds in $Pt(PPh_3)_2(CS_2)$ is less than in $Pt(PPh_3)_2(CO_3)$ [228].

The CS_2 geometry is much less perturbed in the palladium complex [229]. First of all there are slight differences between the two C–S bond distances within the CS_2 molecule (1.61 and 1.66 Å respectively) and the C(1)–S(2) bond is notably shorter than in the platinum complex.

The S(1)–C(1)–S(2), S(2)–Pd–C(1) and P(1)–Pd–P(2) bond angles are 140°, 43° and 109° respectively; the molecule is nearly planar, with maximum deviation of 0.09 Å.

The Pd–S(2) and Pd–C(1) bond distances are 2.31 and 2.02 Å. These results are in accordance with the lower tendency of palladium, compared with platinum, to behave as a π-donor (see 1.4).

A nickel compound of formula $[Ni(PPh_3)(CS_2)]_n$ has also been obtained by reacting $Ni(PPh_3)_2(CO)_2$ with CS_2 [198]. Such derivative, which shows a ν(C=S) at 1122 cm^{-1}, is dimeric in chloroform possibly through bridging sulphur:

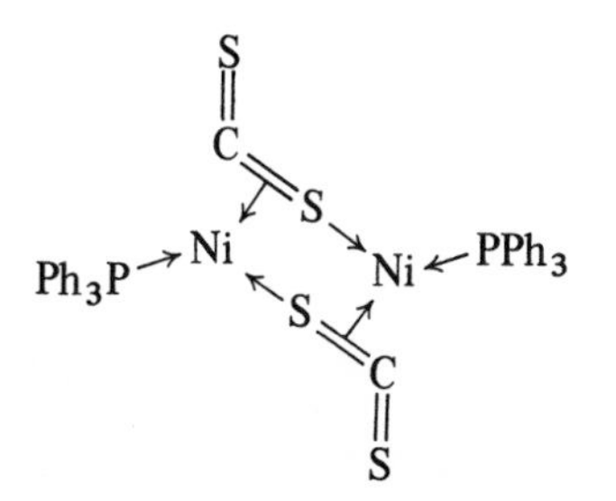

A nickel–CS_2 adduct, probably monomeric, having a chelating diphosphine as ligand, $Ni\{C_3H_6(PPh_2)_2\}(CS_2)$ [ν(C=S), 1170 cm^{-1}] has been very recently obtained from $Ni\{C_3H_6(PPh_2)_2\}_2$ and CS_2 [188]. The reactivity of the linear carbon subsulphide C_3S_2 [(S=C=C=C=S); ν(C=C), 2065 cm^{-1}; ν(C=S), 1019 cm^{-1}], has been studied towards triphenylphosphine platinum(0) derivatives at low temperatures (–15°) [230]. A red-brown compound of formula $Pt(PPh_3)_2(C_3S_2)$ was obtained. This was too unstable for molecular weight determinations, and showed ν(C=C) at 1995 cm^{-1}, ν(C–S) at 935 and 775 cm^{-1}, and a band at 370 cm^{-1} tentatively assigned to ν(Pt–S). On the basis of a probable monomeric structure and the i.r. spectrum structure (A) is preferred to structure (B),

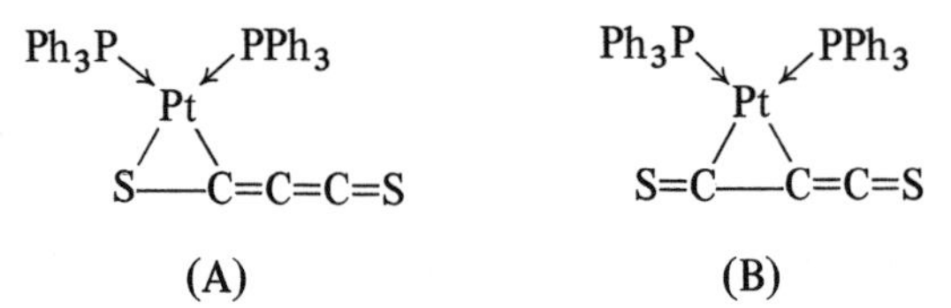

which corresponds to coordination of C_3S_2 as an olefin. In refluxing chloroform, the C_3S_2 adduct transforms into the purple $[Pt(PPh_3)_2(C_3S_2)_2.CHCl_3]_n$, which is believed to be a derivative of a C_3S_2 oligomer [230].

Carbonyl sulphide, COS, has also been reacted with $Pt(PPh_3)_3$ [198]. The reaction product, $Pt(PPh_3)_2(COS)$, showed a ν(C=O) at 1727 cm^{-1}, indicating a coordination through the C=S double bond as in the case of CS_2.

Structurally related η-complexes are obtained from $Pt(PPh_3)_3$ and RNCS (R = Me, Ph). The coordination through the C=S double bond of the isothiocyanates to platinum in $Pt(PPh_3)_2(RNCS)$ is supported by the i.r. spectra which show ν(C=N) at 1643 (R = Ph) and 1653 (R = Me) cm^{-1} and ν(C–S) at 782 cm^{-1} (927 cm^{-1} in free Ph–N=C=S). The methyl derivative shows a single resonance in the ^1H n.m.r. spectrum (τ_{CH_3} = 6.72) with a small coupling with platinum ($J_{Pt-H}) \simeq 1$ Hz) [198].

Hexafluorothioacetone, $(CF_3)_2C$=S, is similarly bonded to platinum in $Pt(PPh_3)_2\{(CF_3)_2C{=}S\}$, obtained from $Pt(PPh_3)_3$ and 2,2,4,4-tetrakis-(trifluoromethyl)-1,3-dithietan [198]. This is of interest because of the high temperature (600°C) normally required to crack the 1,3-dithietan to hexafluorothioacetone.

A series of analogous nickel complexes, $NiL_2\{(CF_3)_2C{=}S\}$ [L = PPh_3, $P(OPh)_3$, $P(OCH_2)_3CEt$, or $L_2 = C_2H_4(PPh_2)_2$], has also been obtained by displacement of 1,5-COD by the ligands L from the derivative Ni(1,5-COD) $\{(CF_3)_2C{=}S\}$ [49]. The latter compound is the product of the reaction between $Ni(1,5\text{-}COD)_2$ and 2,2,4,4-tetrakis(trifluoromethyl)-1,3-dithietan.

A mixture of 3,6-bis-(2,2,2-trifluoro-1-trifluoromethylethylidene)-S-tetrathian(I) and 3,5-bis-(2,2,2-trifluoro-1-trifluoromethylethylidene)-1,2,4-trithiolan(II)

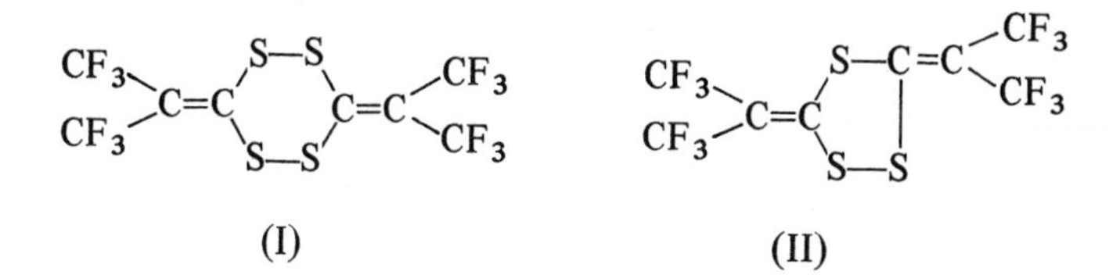

(I) (II)

has been reacted with $Pt(PPh_3)_4$ and $Pt(PMePh_2)_4$ [231]. In the former case two compounds were isolated, to which structures (A) and (B) have been attributed on the basis of i.r. and F^{19} n.m.r. data:

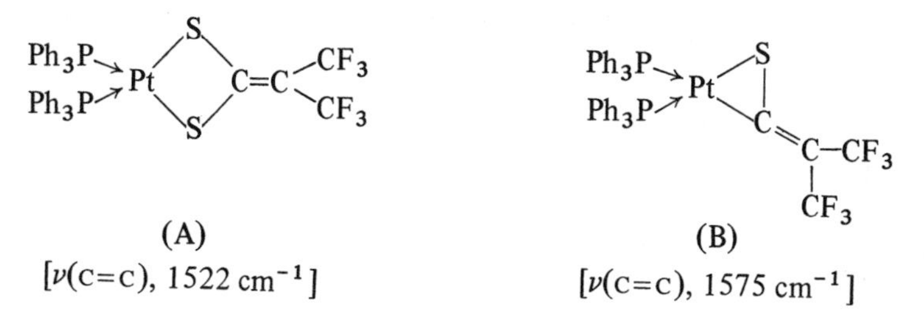

(A) [ν(C=C), 1522 cm^{-1}] (B) [ν(C=C), 1575 cm^{-1}]

In the latter case only the compound corresponding to (A) [ν(C=C), 1517 cm^{-1}] was isolated.

2.2.3.5. *Reactions with Molecules Having Oxygen as Donor Atom*

The activation of molecular oxygen (dioxygen) by transition metal complexes in low oxidation states has received much attention in the last few years. Oxygen activation is a specific reaction only in biological processes, while in heterogeneous and homogeneous catalysis it is never specific and often requires drastic conditions [232]. The difficulty (that is the low reactivity of molecular oxygen) can be directly related to the poor tendency to fill up its degenerate π^* antibonding orbitals. The studies on dioxygen adducts of transition metal complexes have obviously been mostly directed to achieve a better understanding of the nature of the electronic interaction between a transition metal and molecular oxygen and to gain some information on the reactivity of oxygen when bonded to a metal. As far as the compounds we are considering here is concerned, the bonding of dioxygen with transition metal in a low oxidation state can be compared with that of an olefin (Fig. 2.2) (see also 1.3.1.3).

The π-component of the bond thus corresponds to the activation of dioxygen, by transferring electrons into the π^* antibonding orbitals. The amount of this charge transfer depends obviously on the "softness" of the transition

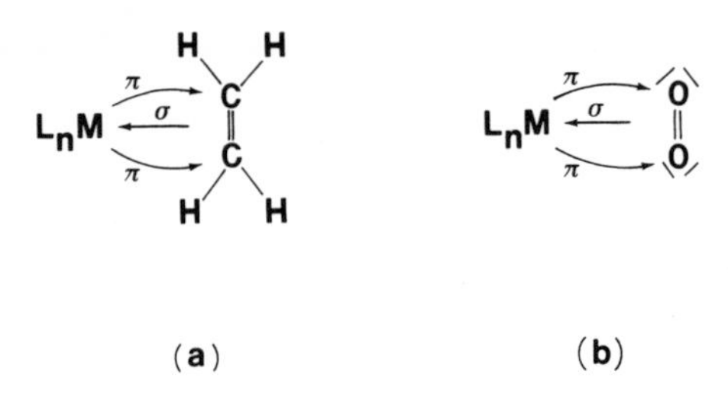

Fig. 2.2. A comparison of the bonding between an olefin (a) or dioxygen (b) and a low valent transition metal complex.

metal atom in the complex, which also depends on the nature of the ligands L. As already discussed, the reactivity of coordinated dioxygen can thus be related to that of its electronically excited state.

Triphenylphosphine complexes of nickel(0), palladium(0) and platinum(0) react with dioxygen to give the following adducts [233, 234, 235, 236, 236 *bis*]:

$$M(PPh_3)_n + \frac{n}{2} O_2 \rightarrow M(PPh_3)_2(O_2) + (n-2)OPPh_3$$

(M = Ni, Pd, Pt; n = 3, 4)

The $M(PPh_3)_2(O_2)$ compounds show in the i.r. spectrum bands at 830 (M = Pt) and 875 (M = Pd) cm^{-1}, due to the three-membered ring M<(O–O) (see 1.3.1.3).

The nickel adduct decomposes above –35°C while the palladium derivative slowly decomposes above 20°C. It has been claimed also that the platinum adduct is unstable at room temperature, and that for all metals the decomposition leads to the metal and triphenylphosphine oxide [233]. However, this seems not to be the case at least for the platinum derivative.

The platinum-dioxygen adduct is in fact much more stable, and decomposes at nearly 120°C in the following way [237]:

$$Pt(PPh_3)_2(O_2) \rightarrow \tfrac{1}{4}[Pt(PPh_3)]_4 + \tfrac{1}{2}O_2 + OPPh_3$$

giving a known cluster compound [136]. In contrast, when the (presumably cluster) derivative of palladium $[Pd(PPh_3)_2]_x$, is reacted with dioxygen, a compound corresponding to $Pd(PPh_3)_2 \cdot \tfrac{3}{2}O_2$ is obtained [137].

The peroxidic nature of the dioxygen molecule, bound to a transition metal in a low oxidation state, is well documented for example by the reaction of $Ni(PPh_3)_2(O_2)$ with water, which gives H_2O_2 [233], and of $Pt(PPh_3)_2(O_2)$ with SO_2, NO_2, NO and CO [226, 238, 239, 240], which react as follows:

$$Pt(PPh_3)_2(O_2) + SO_2 \rightarrow Pt(PPh_3)_2(SO_4)$$
$$Pt(PPh_3)_2(O_2) + 2NO_2 \rightarrow Pt(PPh_3)_2(NO_3)_2$$
$$Pt(PPh_3)_2(O_2) + 2NO \rightarrow Pt(PPh_3)_2(NO_2)_2$$
$$Pt(PPh_3)_2(O_2) + CO \rightarrow Pt(PPh_3)_2(CO_3)$$

The mechanism of the reaction between $Pt(PPh_3)_2(O_2)$ and SO_2 has been studied by i.r. spectroscopy, by using dioxygen labelled with O^{18} [241].

The dioxygen molecule seems to be irreversibly bound to platinum, since $Pt(PPh_3)_2(O_2)$ has no tendency to lose O_2 *under vacuum* [235], unlike the adduct of Vaska's compound $Ir(O_2)Cl(CO)(PPh_3)_2$, but analogously to $Ir(O_2)I(CO)(PPh_3)_2$. This can be considered an electronic effect. In the case of the iridium adducts, the presence of iodine instead of chlorine increases the electronic density on the metal; this allows a greater π-back donation to oxygen and consequently a stronger metal–O_2 bond. This is also reflected in a longer O–O bond distance (1.51 Å) in the dioxygen adduct of the iodo compound [242].

In the platinum-dioxygen adduct, isolated as a benzene solvate the O–O bond distance is 1.45(4) Å, quite close to the value found in the dioxygen adduct of the

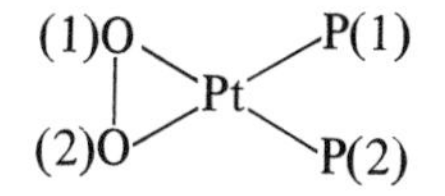

iridium-iodo compound [243]. P(1), P(2), O(1), O(2) and Pt lie almost in the same plane. A much shorter O–O bond distance (1.26 Å) was found in another X-ray structural determination on $Pt(PPh_3)_2(O_2)$, isolated as a toluene solvate [244] but this seems to be a less accurate determination. The kinetics of oxygen absorption by $Pt(PPh_3)_3$ have been studied [53, 232]. The dissociation of $Pt(PPh_3)_3$ into PPh_3 and the active species $Pt(PPh_3)_2$, was considered the first step of the reaction, but direct reaction of molecular oxygen with $Pt(PPh_3)_3$, with formation of $Pt(PPh_3)_2(O_2)$ and PPh_3 cannot be ruled out on the basis of the most recent results on the dissociation of $Pt(PPh_3)_3$ (see 1.3.5).

Nickel(0), palladium(0) and platinum(0) triphenylphosphine complexes are catalysts for the oxidation of triphenylphosphine to triphenylphosphine oxide [46, 53, 232, 233]. Some nickel(0) complexes having a chelating diphosphine as ligand also catalyse the oxidation of the diphosphine to the corresponding oxide [100].

While the platinum complex is generally effective in the oxidation of electron-poor phosphines, the palladium complex is more efficient on the oxidation of electron-rich phosphines such as PBu^t_3 [46]. Both derivatives are catalysts for the oxidation of cyclohexyl isonitrile to the corresponding isocyanate [46].

Two mechanisms have been proposed for the catalytic oxidation of triphenylphosphine to the corresponding oxide, according to the following schemes [232]:

$$Pt(PPh_3)_3 \underset{}{\overset{k_1}{\rightleftharpoons}} Pt(PPh_3)_2 + PPh_3 \qquad (a)$$

$$Pt(PPh_3)_2 + O_2 \xrightarrow{k_2} Pt(PPh_3)_2(O_2) \quad \text{(b)}$$

$$Pt(PPh_3)_2(O_2) + 2PPh_3 \xrightarrow{k_3} Pt(PPh_3)_2 + 2\,OPPh_3 \quad \text{(c)}$$

$$Pt(PPh_3)_2 + OPPh_3 \overset{k_4}{\rightleftharpoons} Pt(PPh_3)_2(OPPh_3) \quad \text{(d)}$$

the last equilibrium corresponds to the poisoning of the catalyst; from catalytic data: $k_2 = 1.92$ and $k_3 = 0.16$ ($mol^{-1}\ l\ s^{-1}$), from one-step investigation: $k_2 = 0.66$ and $k_3 = 0.15$ ($mol^{-1}\ l\ s^{-1}$). The alternative scheme is [53, 245]:

$$Pt(PPh_3)_3 + O_2 \xrightarrow{k_2} Pt(PPh_3)_2(O_2) + PPh_3 \quad \text{(e)}$$

$$Pt(PPh_3)_2(O_2) + PPh_3 \xrightarrow{k_3} Pt(PPh_3)_3(O_2) \quad \text{(f)}$$

$$Pt(PPh_3)_3(O_2) + 2PPh_3 \xrightarrow{\text{fast}} Pt(PPh_3)_3 + 2\,OPPh_3 \quad \text{(g)}$$

$k_2 = 2.6 \pm 0.1$ ($mol^{-1}\ l\ s^{-1}$) and $k_3 = 0.25 \pm 0.001$ ($mol^{-1}\ l\ s^{-1}$). Step (e) seems now to be more probable than steps (a) plus (b). The detailed mechanisms of steps (c) or (f) plus (g) are not yet clear. It has been proposed that the oxidation of phosphine proceeds as follows [53]:

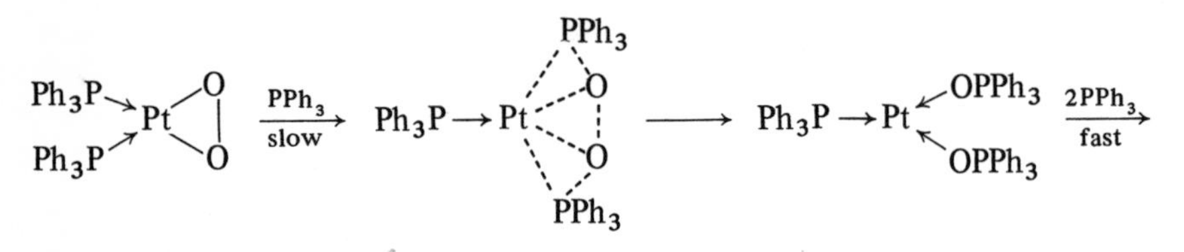

$$\longrightarrow Pt(PPh_3)_3 + 2OPPh_3$$

However the oxygen transfer might also take place with the intermediate formation of a cyclic peroxide [232]:

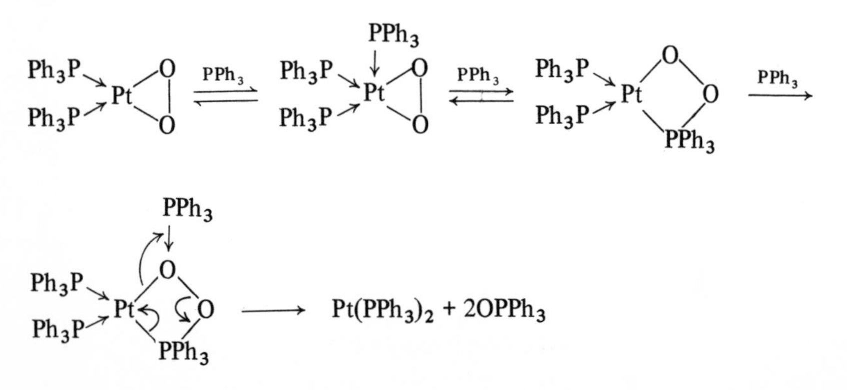

The last hypothesis is in accordance with the great tendency to form cyclic peroxides with many different molecules exhibited by the platinum-dioxygen compound.

In a first paper Wilkinson *et al.* [236] reported that $Pt(PPh_3)_3$ and $Pd(PPh_3)_4$ react with dioxygen in the presence of CO_2 giving the solvated $M(PPh_3)_2(CO_3)$ complexes and phosphine oxide. The platinum complex shows a strong band at *ca.* 1680 cm^{-1} which was assigned to $\nu(C{=}O)$ of the carbonato group. This derivative has been shown to be a good starting material for the synthesis of other platinum(II) compounds. It reacts in fact with acids such as CH_3COOH, CF_3COOH, or PhCOOH with formation of compounds of formula $Pt(PPh_3)_2(RCOO)_2$ [236].

The nature of this carbonato derivative was confirmed by comparison with an authentic sample prepared from *cis*-$Pt(PPh_3)_2Cl_2$ and Ag_2CO_3 [236], and by an X-ray structural determination [160].

Subsequently Wilkinson *et al.* reconsidered the nature of an impure product obtained during the reaction of $Pt(PPh_3)_3$, O_2 and CO_2 [240], which they showed to be the peroxycarbonate:

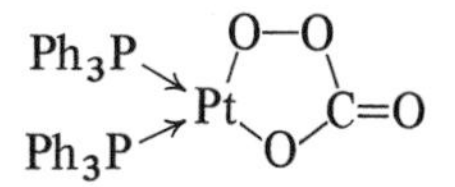

In fact this derivative shows bands at 1678 [$\nu(C{=}O)$], 1250, 815 and 778 [$\nu(O{-}O)$] cm^{-1}, while the corresponding carbonato derivative shows only bands at 1684 [$\nu(C{=}O)$], 1180 and 815 cm^{-1}, but the band due to the peroxo group is absent.

Similar derivatives were obtained also starting from $Pt(AsPh_3)_4$ and $Pt\{C_2H_4(PPh_2)_2\}_2$. The formation of the carbonato derivatives probably proceeds through the peroxycarbonato complexes. This is in accordance with the fact that the peroxocarbonato group is reduced to carbonate by a tertiary phosphine [241]. Such reduction strongly favours the hypothesis of the alternative mechanism, proposed for transferring the dioxygen molecule to the substrate during the catalytic oxidation of PPh_3 to the corresponding oxide, catalysed by triphenylphosphine derivatives of platinum(0) (see above) [232]. As with CO_2, CS_2 adds to $M(PPh_3)_2(O_2)$ (M = Pd, Pt) complexes with formation of the dithiocarbonato derivatives $M(PPh_3)_2(S_2CO)$ [240]. Another interesting reaction of $Pt(PPh_3)_2(O_2)$ was discovered during its recrystallization from acetone [246]. A new compound is formed, analysing as $Pt(PPh_3)_2(Me_2CO_2O)$ which an X-ray structural determination has shown to be the peroxyderivative:

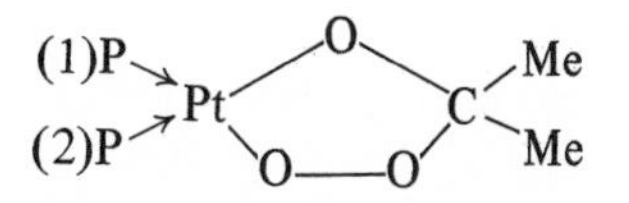

The oxy-peroxy-chelate ring is not completely planar, but the geometry, around the platinum is close to planar.

This reaction has been shown to be possible also with aldehydes such as acetaldehyde and benzaldehyde, but not with amides and esters [246], and later it was extended to acetone oxime and thiourea [240], and also to diphenylketen [246 *bis*]. Hexafluoro- or trifluoro-acetone similarly add to $Pt(PPh_3)_2(O_2)$ [247] with formation of the corresponding peroxy-oxyderivatives, which, in the case of hexafluoroacetone, can be reduced by PPh_3 to:

$$Pt(PPh_3)_2\left\{\begin{matrix} \diagup O \diagdown \\ \diagdown O \diagup \end{matrix} C(CF_3)_2\right\},$$

analogously to the percarbonato-platinum(II) derivative. This reduction is not possible with the acetone adduct of $Pt(PPh_3)_2(O_2)$, and can be attributed to the strong stabilizing effect of the CF_3 groups in the hexafluoroacetone derivative. In the presence of excess hexafluoroacetone, $Pt(PPh_3)_2(O_2)$ reacts with a ring expansion of the peroxo derivative forming [247]:

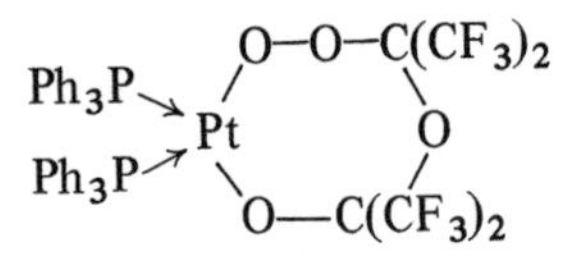

Also α-diketones add to $Pt(PPh_3)_2(O_2)$, with formation of the corresponding peroxo derivative where one C=O group of the original α-diketone is involved in the ring with platinum. With excess $Pt(PPh_3)_2(O_2)$, both C=O groups react forming a compound considered to be [248]:

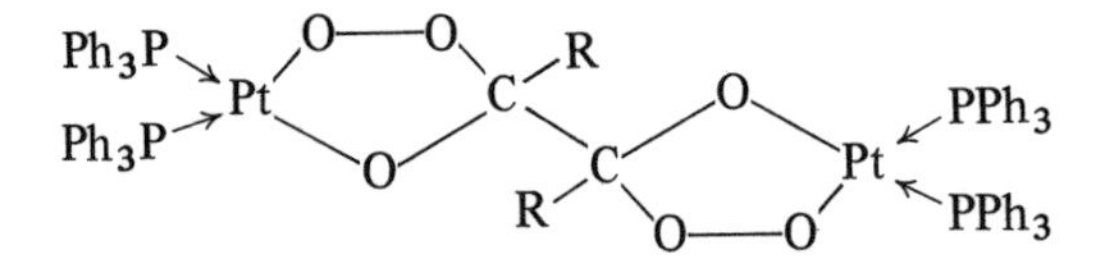

The former compounds decompose in dichloromethane forming the carboxylato derivatives $Pt(PPh_3)_2(OCOR)_2$; moreover, the free keto group reacts with 2,4-dinitrophenylhydrazine with formation of the corresponding hydrazone.

Analogously to α-diketones, *ortho-* and *para*-quinones react with $Pt(PPh_3)_2(O_2)$ giving peroxo compounds, where only one C=O group of the quinone inserts into the peroxy ring [249]. The insertion reaction of double bonds into the peroxy group of $Pt(PPh_3)_2(O_2)$ has also been studied with olefins [232, 249]. Tetracyanoethylene reacts easily to give a stable platinum cyclic peroxide, which is supposed to be:

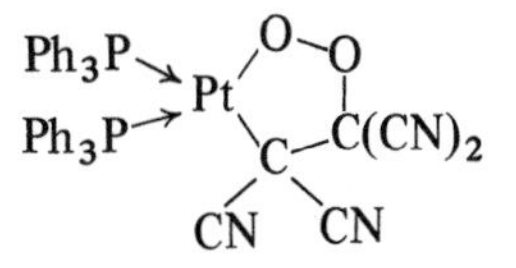

This reaction is more complex with olefins such as acrylonitrile and fumaronitrile [249]. The last reaction is obviously connected with the epoxidation of olefins by oxygen, in both the heterogeneous and the homogeneous phase, which is known to be a very difficult reaction. In the last few years the catalytic oxidation of olefins in the homogeneous phase has been studied. Low-valent complexes, able to form oxygen adducts, have been used as catalysts, since it was expected that, mainly in reversible adducts, the reactivity of coordinated oxygen might be similar to that of singlet oxygen or of the peroxide ion [250]. However, the study of the catalytic oxidation of cyclohexene by $Pt(PPh_3)_3$ or $Pt(PPh_3)_2(O_2)$ and other phosphine metal complexes [250] has shown that the action of the complex does not involve a catalytic oxygen activation which favours the formation of hydroperoxides, but only a radical transformation of hydroperoxides already present in the reaction mixture. In fact α-naphthol inhibits the reaction, while the product distribution is typical of a free radical chain process. The major reaction products are 2-cyclohexen-1-one, 2-cyclohexene-1-ol, cyclohexene oxide and non-volatile polymers. However in some cases (e.g. with catalysts such as $Rh(PPh_3)_3Cl$ [250] or $Ru(PPh_3)_3Cl_2$ [251] it has been found that with peroxy-free cyclohexene only a very small induction period was necessary for the beginning of the reaction, while catalysts such as *trans*-$IrCl(CO)(PPh_3)_2$ require nearly two hours before the oxidation starts [250]. This was interpreted in the latter case as the time required for the formation of small amounts of hydroperoxide by the thermal reaction with oxygen. It follows that catalysts such as $Rh(PPh_3)_3Cl$ or $Ru(PPh_3)_3Cl_2$ have either a large effect on the decomposition of preformed hydroperoxide, or simply do not act as generators of organic radicals. However this problem is not easy to solve.

For example, in the study of the autoxidation of what it was thought to be a peroxide-free cumene, no reaction was detected at 35°C and 1 atm O_2 in the absence of catalysts [252]. On the contrary, it was found that $Pd(PPh_3)_4$, and better $Pd\{P(C_6H_4CH_3\text{-}p)_3\}_3$ are effective catalysts for this oxidation. Also $Pt(PPh_3)_4$ displayed some catalytic activity in cumene autoxidation; however in this case a long induction period was observed, which could be eliminated by addition of preformed cumene hydroperoxide. It appeared then that $Pt(PPh_3)_4$ catalyses hydroperoxide decomposition rather than its formation. For the other cases a mechanism was proposed, which proceeds through the formation of superoxidic oxygen complexes, capable of hydrogen abstraction:

$$Pd(PAr_3)_4 + 2O_2 \rightarrow (Ar_3P)_2Pd(O_2) + 2\ OPAr_3$$
$$(Ar_3P)_2Pd(O_2) + RH \rightarrow (Ar_3P)_2PdO_2H + R.$$
$$R. + O_2 \rightarrow RO_2.$$
$$RO_2. + RH \rightarrow ROOH + R.$$

However the hydride abstraction is also compatible with the reactivity of complexed dioxygen in a singlet state, and this second hypothesis is more consistent with the diamagnetism of the metal-dioxygen adducts [252].

In a subsequent kinetic study of the catalysed autoxidation of cumene, undertaken in order to distinguish whether there was real O_2 activation, it was found that in fact the chain initiation occurred via decomposition of the small quantities of hydroperoxide still present in the cumene [253]. Moreover, $Pd(PPh_3)_4$ was found to catalyse the decomposition of preformed hydroperoxide.

Activated ketones such as $(CF_3)_2CO$ and $(CF_2Cl)_2CO$ have been shown to be able to add directly to complexes such as $M(PPh_3)_4$ (M = Pd, Pt) and $Pt(PMePh_2)_4$ [254], in a manner similar to perfluothioacetone [198]. The ketone is η-bonded to the metal via the $\rangle$C=O bond [the free ketone has ν(C=O) = 1810 cm^{-1}, while the product does not show any band in the 1600–1800 cm^{-1}, region], but only the presence of electronegative groups such as CF_3 can activate the C=O bond for reaction with the zerovalent complexes. These electronegative groups presumably favour the presence of low-lying antibonding orbitals on the C=O group, low enough for accepting π-electron density from the metal.

The possibility that these products might be of the type $Pt(PR_3)_2(CF_3)(COCF_3)$, that is formed by an oxidative addition, was ruled out both by spectroscopic evidence, and by the reaction of the adducts with iodine; such reaction gives in fact $Pt(PR_3)_2I_2$ and the starting hexafluoroacetone [254].

Similar adducts have been obtained with NiL_4 (L = $PMePh_2$, PEt_3) [49] and PdL_4 [L = $P(OPh)_3$, $PMePh_2$] [121] complexes and hexafluoroacetone. Compounds such as PdL_4 [L = $P(OMe)_3$, $P(OMe)_2Ph$, $AsMe_2(CH_2Ph)$], react with hexafluoroacetone with ring expansion, giving derivatives formulated as [121]

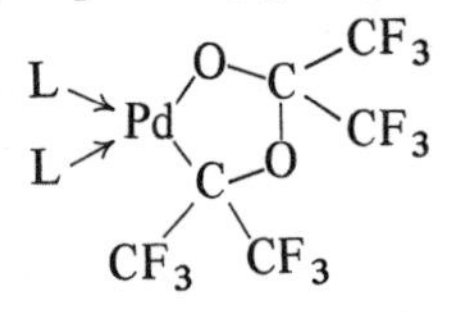

Such ring expansion products can be also obtained from preformed $Pd(PR_3)_2\{(CF_3)_2CO\}$ complexes by reacting them with excess $(CF_3)_2CO$ [121].

An X-ray investigation on $Ni(PPh_3)_2\{(CF_3)_2C{=}O\}$ [255] [obtained from $Ni(PPh_3)_2(C_2H_4)$ [205] and hexafluoroacetone or π-$C_8H_{12}Ni\{(CF_3)_2C{=}O\}$ [49] and PPh_3] has confirmed the presence of the three-membered ring in this type of adduct. The carbonyl group is symmetrically sideways bonded to the nickel and the angle at nickel in the three-membered ring is 41.3(6)°. The $>C{=}O$ bond is lengthened by 0.09(2) Å relative to the bond in the free molecule (1.32 Å). The plane of Ni and $>C{=}O$ is tilted significantly by 6–9° with respect to that defined by the nickel and phosphorus atoms [255].

ortho-Quinones are also able to add to zerovalent phosphine derivatives of platinum and palladium, with formation of diphenolate derivatives such as:

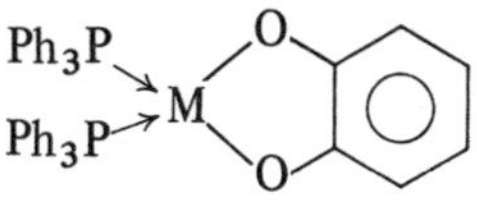

Ph_3P, Ph_3P → M(O,O-C_6H_4) (M = Pd, Pt)

This type of reaction, which requires u.v. activation when an *ortho*-quinone is added to an iridium(I) derivative [256], is in fact spontaneous with the triphenylphosphine derivatives of platinum(0) [256, 257, 258, 259] and palladium(0) [258]. This fact can be related to the stronger basicity of the zerovalent complexes of palladium and platinum compared with the iridium(I) derivative. When $Pt(PPh_3)_3$ is reacted with 3,4,5,6-tetrachloro-1,2-benzoquinone, 1,2-naphthoquinone or 9,10-phenanthrenequinone, the products do not show any band in the 1700–1500 cm^{-1} region attributable to ketonic groups [257]. Some new bands at around 1420–1480 cm^{-1} are present, which can be attributed to the phenolate groups. This type of structure has been confirmed by an X-ray investigation, where it has been found that the molecular structure of the tetrachloroquinone derivative contains simple σ-bonded oxygens [260]. Such a structure is also consistent with the ^1H n.m.r. and electronic spectra of these derivatives [257]. In fact the electronic spectra of the *ortho*-quinone moiety shows a very large dependence on co-ordination. In particular the $n \rightarrow \pi^*$ transition of the $>C{=}O$ groups is shifted to higher energy to a great extent, as expected for this type of co-ordination. This is not the case for analogous derivatives of *para*-benzoquinones, where the co-ordination is via the double bonds (see later) [257].

A nickel derivative of tetrachloro-*o*-benzoquinone, $Ni\{C_2H_4(PPh_2)_2\}(C_6Cl_4O_2)$ has also been reported, being obtained by oxidative addition of the *ortho*-quinone to $Ni\{C_2H_4(PPh_2)_2\}(CO)_2$ [261]. Treatment of $Pt(PPh_3)_4$ with dibenzoylperoxide gives *cis*-dibenzoato*bis*(triphenylphosphine)platinum(II) [262].

An interesting compound has been recently obtained from the reaction between carbon dioxide and the unsaturated complex $Ni\{P(C_6H_{11})_3\}_2$ [69b].

Such compound, which is one of the few examples of carbon dioxide derivatives, has probably the structure:

$$\{P(C_6H_{11})_3\}_2Ni\cdots\overset{\displaystyle O}{\underset{\displaystyle C}{\|}}$$
$$\overset{\displaystyle\|}{\underset{\displaystyle O}{}}\cdots Ni\{P(C_6H_{11})_3\}_2$$

which is substantiated by a band at 1735 cm^{-1} in the i.r. spectrum due to ν(C=O). A linear arrangement such as $L_2Ni(O{=}C{=}O)NiL_2$ seems to be less probable. It seems that only the unsaturation of the starting complex allows the coordination of the hard CO_2 ligand, while it is difficult to discuss the influence of the bulky and basic tricyclohexylphosphine; interestingly, the same starting complex is able to add reversibly to dinitrogen, with formation of a dimer with N_2 bridging the two nickel atoms [69].

2.2.3.6. Reactions with Molecules Having Nitrogen as the Donor Atom

In a first report the products of the reaction between $M(PPh_3)_4$ (M = Pd, Pt) and NO were formulated as binuclear, metal–metal bonded nitrosyl derivatives, $[M(NO)(PPh_3)_2]_2$, on the basis of their diamagnetism and of the results of oxidative titration [263] Subsequent experiments failed to substantiate these findings and led instead to a different formulation of the products [264]. In fact when NO, free from NO_2, was bubbled into a deoxygenated benzene solution of $Pt(PPh_3)_n$ (n = 3, 4), a lemon yellow material separated. It analysed correctly for $Pt(PPh_3)_2(NO)_2$ and it showed i.r. absorptions at 1285, 1240 and 1062 cm^{-1}. From the mother liquor, a white product could occasionally be isolated [264], which was shown to be $Pt(PPh_3)_2(NO_2)_2$, formed when small amounts of oxygen were present in the reaction mixture. The yellow material was shown to be a hyponitrite derivative (see 1.3.1.3) mainly on the basis of i.r. data, the structure

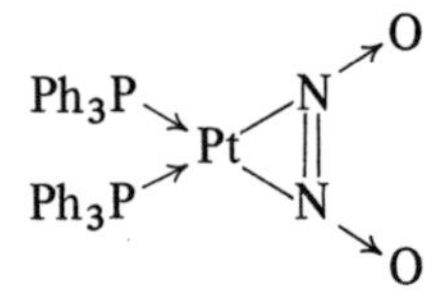

being marginally preferred to other possible structures [264]. The coupling of two NO groups to form a bridging hyponitrite ion has been reported previously to occur for a d^{10} nickel(0) carbonyl-phosphine derivative [256].

The reaction of NO with $Pd(PPh_3)_4$ gave an unstable product; in the presence of oxygen small yields of $Pd(PPh_3)_2(NO_2)_2$ were obtained [264]. Nitrosyl hexafluorophosphate has also been reacted with $Pt(PPh_3)_4$ giving

$PPt(PPh_3)_4]^{++}$ $(PF_6^-)_2$, as the product of this reaction [266]. Such reactions with $Pt(PPh_3)_3$ have been studied with $[NO^+]X^-$ ($X = BF_4, PF_6$) in different solvents, and also in the presence of lithium chloride, with the aim of synthesis of the unknown $Pt(PPh_3)_2(NO)Cl$ [267].

In protic solvents such as ethanol, the formation of the known hydrides $[Pt(PPh_3)_3H]^+(X^-)$ ($X = BF_4, PF_6$) was observed but in no cases were products obtained showing i.r. absorptions in the 1500–2000 cm^{-1} region, indicative of a nitrosyl derivative [267]. This type of oxidative addition of $[NO^+]X^-$ to give nitrosyl derivatives has been observed with $Ni(CO)_2L_2$ ($L = PPh_3$, $PMePh_2$, PMe_2Ph) [268 *bis*]. These compounds react with $NOPF_6$ in methanol/toluene solution, with formation of $[Ni(NO)L_3]^+(PF_6^-)$ complexes [$L = PPh_3$; $\nu(NO) = 1795\ cm^{-1}$].

An interesting reaction is observed between trifluoronitrosomethane, CF_3NO and $Pt(PPh_3)_4$ [203]. An unstable, purple, paramagnetic complex of stoichiometry $Pt(PPh_3)_2(CF_3)(NO)$ is obtained. It shows $\nu(NO)$ at 1660 cm^{-1}, and e.s.r. measurements gave g (on powder) 2.0046. The instability of this compound in solution prevented accurate molecular weight determinations. This compound represents one of the few examples of nitrosyl derivatives of platinum in a low oxidation state.

Its paramagnetism is not easy to explain (although it was considered in accord with the given stoichiometry) [203], since this compound could be considered either a d^{10} platinum(0) derivative (NO as NO^+) or a platinum(II) derivative (NO as NO^-). The related nickel derivatives, $Ni(PPh_3)_2(X)(NO)$ ($X = Br, NO_3$ etc.) are in fact diamagnetic and probably have a quasi-tetrahedral structure, their X-ray powder patterns being very similar to those of the dihalobis(triphenylphosphine)nickel complexes [268], which are known to be tetrahedral. Also, an X-ray structural determination on the analogous $Ni(PPh_3)_2(N_3)(NO)$ [269] has shown that the coordination geometry about the nickel atoms is pseudo-tetrahedral.

Many other molecules containing nitrogen can react with zerovalent phosphine derivatives of nickel, palladium and platinum, with formation of N-bonded complexes.

Hexafluoroisopropylidenimine, $(CF_3)_2C{=}NH$, reacts with zerovalent palladium phosphine complexes in a way similar to the corresponding keto-derivative hexafluoroacetone [120].

In fact $Pd(PPh_3)_4$ gives the three-membered ring complex:

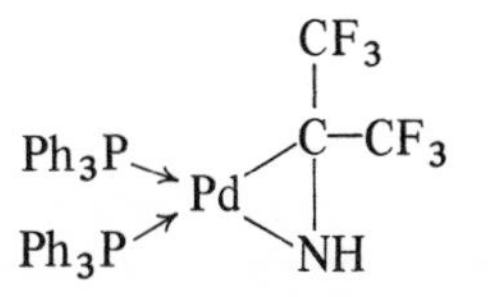

A five-membered ring is formed when $(CF_3)_2C{=}NH$ reacts with a palladium(0) arsine derivative such as $Pd(AsMe_2Ph)_4$. In this case a derivative formulated as

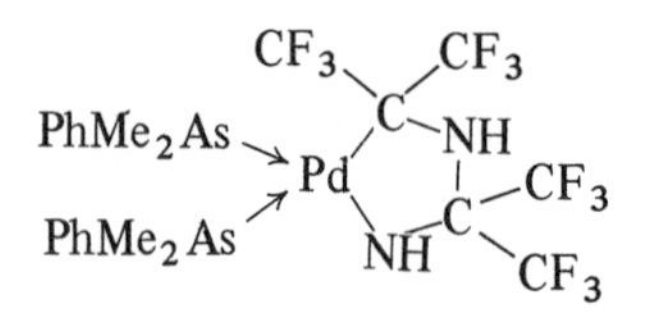

was obtained.

A ring expansion with formation of a keto-imino derivative, was also obtained when $(CF_3)_2C{=}NH$ was reacted with the three-membered $Pd\{C_2H_4(PPh_3)_2\}\{(CF_3)_2C{=}O\}$, giving a derivative for which the following structure has been proposed:

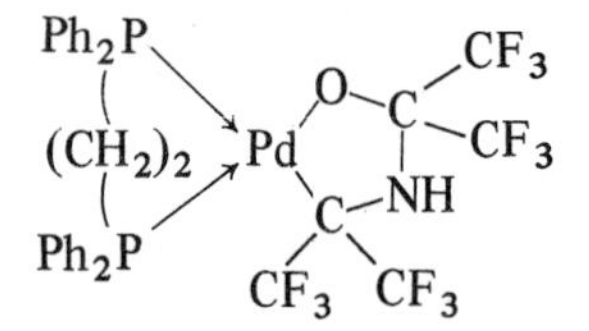

A different structure in which the nitrogen atom is bonded to palladium is also possible. However it was shown by an X-ray investigation that the related nickel derivative $(ButNC)_2\overline{Ni\,.\,C(CF_3)_2NHC(CF_3)_2O}$ has a structure with the metal coordinated to carbon and not to nitrogen [255].

It is interesting to note that among the few nitrile complexes of metals in the zerovalent state, a platinum complex with trifluoroacetonitrile has been reported [270]. This was obtained by displacing *trans*-stilbene with CF_3CN from $Pt(PPh_3)_2$(*t*-stilbene). From spectroscopic evidence a structure containing platinum η-bonded to the C≡N group has been proposed:

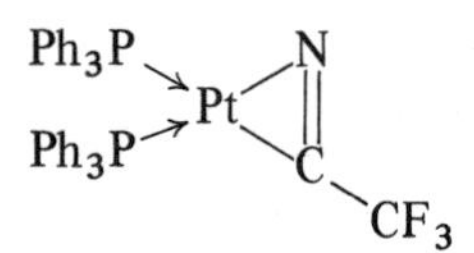

This compound shows in fact an absorption at 1734 cm^{-1}, in the region normally assigned to the C=N stretching mode, and a ^{19}F pattern similar to that of $Pt(PPh_3)_2\{(CF_3)_2CO\}$ [254]. Obviously the strongly electron-withdrawing CF_3 group is responsible for this type of co-ordination. N-bonded nitrile complexes, although not analytically pure, have been obtained with nickel [271]. Addition of a nitrile RCN (R = CH_3, CH_2CH_3, n-C_4H_9 or C_6H_5) to the red-orange toluene solution of NiL_3 [L = $P(O\text{-}o\text{-}C_6H_4CH_3)_3$] causes an immediate loss of colour, attributed to the fast formation of $NiL_3(RCN)$.

The spectroscopic properties of these derivatives in solution have been studied. The benzonitrile derivative shows the larger formation constant, presumably as a consequence of electron delocalization extended to the benzene ring. The i.r. spectrum of the methyl cyanide compound in toluene shows a band at 2266 cm^{-1} due to ν(CN) bonded to the complex in addition to a band at 2256 cm^{-1} due to free CH_3CN. On the other hand the i.r. spectrum of the phenyl cyanide complex shows a shift to lower energies of the ν(CN) absorption [ν(CN) = 2217 and 2229 cm^{-1} for bonded and free benzonitrile respectively]. Owing to the ease with which the nitrile complexes dissociate in solution, impure derivatives were obtained by removal of excess nitrile under vacuum.

From the reaction between $Pt(PPh_3)_4$ and trifluoroacetonitrile in benzene a compound analysing as $Pt(PPh_3)_2(CF_3CN)_2N$ was isolated [270]. Although the X-ray investigation on the complex could not distinguish between carbon and nitrogen, the following structure was considered as the most probable:

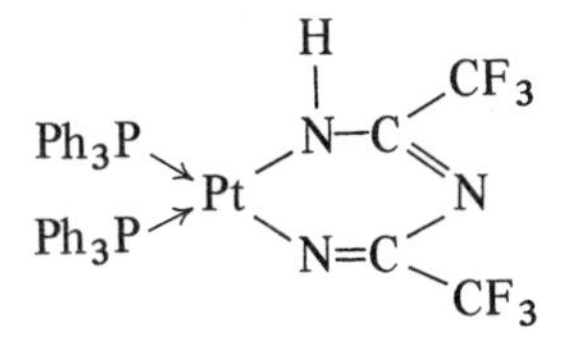

This compound shows a band considered to be ν(NH) at 3358 cm^{-1} in the i.r. spectrum, while the ^{19}F n.m.r. spectrum indicates two non-equivalent CF_3 groups.

A strong peak corresponding to the fragment $(CF_3CN)_2NH$ appeared in the mass spectrum. The source of the NH group was presumably the hydrolytic degradation of CF_3CN, caused by traces of water in the reaction medium [270]. From the reaction above considered, other as yet unidentified complexes were isolated.

The substitution reactions of $Ni\{P(NCO)_3\}_4$ and $Ni(PF_3)_4$ with pyridine have been briefly described [25, 42 *bis*]; also with 2,2′-bipyridyl and 1,10-phenanthroline [42 *bis*]. The aromatic azo group, $\diagdown$N=N$\diagup$ can coordinate to a transition metal through one or both the nitrogen lone pairs. Azobenzene is also known to form 2-(phenylazo)phenyl-metal complexes:

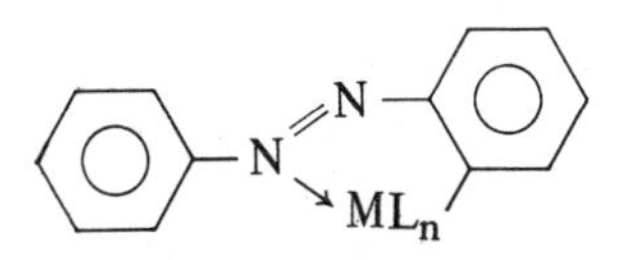

which also contain a σ-bond between the metal and the *ortho* carbon atom of one phenyl ring. A third type of coordination of the aromatic azo group has been recently discovered, by studying the reactivity of azo derivatives towards complexes of metals in a low oxidation state. In these cases the coordination can be considered to be like that of an olefin, in that the η-bonded azo group receives a large amount of π-back-donation from the metal, up to the limiting case where a rigid three-membered ring

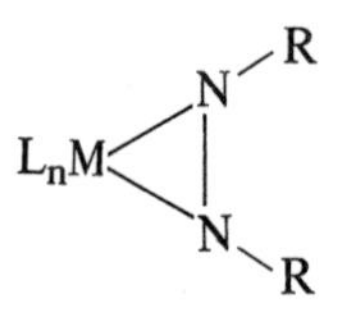

is formed as in the case of fluoro-olefins.

Diethylazodicarboxylate, $EtO_2CN{=}NCO_2Et$, and 4-phenyl-1,2,4-triazoline-3,5-dione,

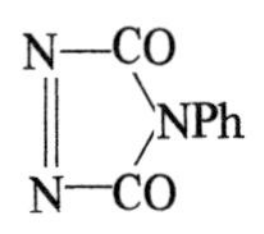

react, with $Pt(PPh_3)_4$ with formation of (A) and (B) respectively [272]:

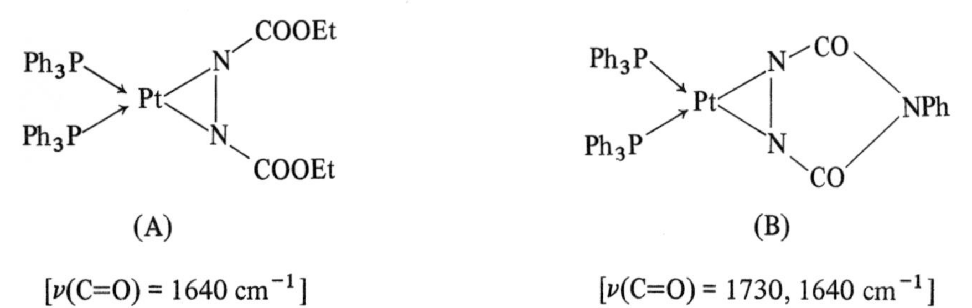

(A) (B)

[ν(C=O) = 1640 cm^{-1}] [ν(C=O) = 1730, 1640 cm^{-1}]

Structures (A) and (B) correspond to systems where the nitrogen atoms remain essentially sp^2-hybridized and the electron-withdrawing substituents lower the energy of the π^* orbitals, with a resultant enhancement in back-bonding from filled d orbitals into the π^* orbitals [272]. On the other hand from the reaction of $Pd(PPh_3)_4$ or $Pt(PMePh_2)_4$ with diethylazodicarboxylate, only the organic 1,2-bis(ethoxycarbonyl)hydrazine was isolated: this corresponds to the formal hydrogenation of the azo group [272]. A dark-red, air-sensitive azobenzene-nickel compound, $Ni(PPh_3)_2(PhN{=}NPh)$ has been obtained by reacting $Ni(PPh_3)_4$ with azobenzene, while $Ni\{P(OMe)_3\}_4$ did not react with azobenzene even at 100°C [273]. $Pt(PPh_3)_3$ did not react with azobenzene at reflux in ethyl alcohol or benzene [274].

Derivatives of formula $Ni(PR_3)_2(PhN{=}NPh)$ (R = Me, Bu, Ph) can be obtained from $Ni(PR_3)_2Cl_2$, azobenzene and metallic lithium in tetrahydrofuran at temperatures below 0°C, or by displacing stilbene from $Ni(PR_3)_2$(stilbene) (R = Ph) with azobenzene [273].

The triphenylphosphine derivative is completely decomposed in solution by oxygen even at –40°C, into metallic nickel, PPh_3, $OPPh_3$ and PhN=NPh. The trimethylphosphine derivative $Ni(PMe_3)_2(PhN{=}NPh)$ shows an interesting reaction in ethanol-water at 60°C; $Ni(PMe_3)_4$ and $Ni(OH)_2$ are formed, while azobenzene is transformed into hydrazobenzene; analogously dimethylglyoxime gives hydrazobenzene, and bis(dimethylglioximato)nickel(II) [273].

Increasing interest has recently arisen in dinitrogen complexes of transition metal ions as possible model systems for nitrogen fixation. A simple analogue of nitrogen reductase has been described by Parshall [275], who suggested that the adduct

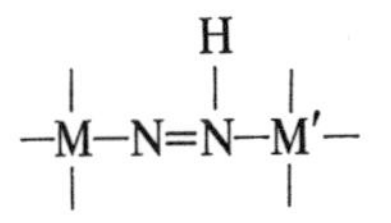

formed from the reaction of a metal hydride with a dinitrogen complex could be the intermediate in the stepwise reduction of molecular nitrogen to ammonia during biological nitrogen fixation. These considerations have stimulated much research on the reactivity of unsaturated organic compounds, such as diazonium salts and azides, towards transition metal complexes in low oxidation states. These nitrogen-containing organic compounds can be considered as models of activated molecular nitrogen [275].

In valence bond terms, a dinitrogen complex (A) and an azo derivative (B) are in fact directly comparable:

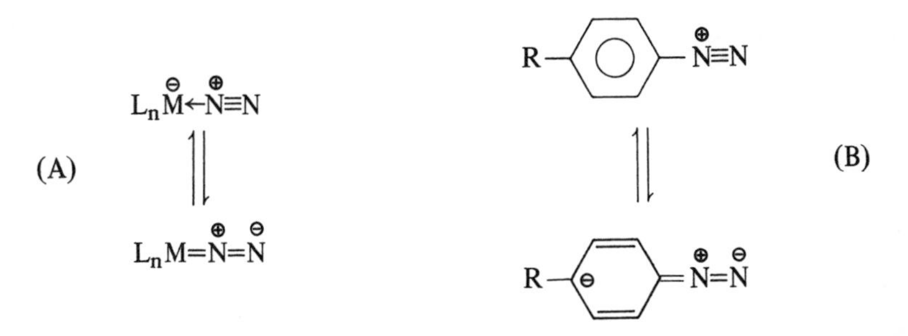

Aliphatic diazo derivatives, which can be considered also as possible sources of electrophilic carbenes, have been reacted with tris or tetrakis(triphenylphosphine)platinum(0).

The simplest diazo derivatives, namely diazomethane CH_2N_2, is rather unreactive towards $Pt(PPh_3)_3$ when the reaction takes place in an ethereal suspension; in benzene solution fast gas evolution was noted but no well defined derivatives could be isolated [274]. A diazo derivative stabilized by electron-withdrawing substituents, such as bis(trifluoromethyl)diazomethane

$(CF_3)_2CN_2$, gave on the other hand a stable compound of stoichiometry $Pt(PPh_3)_2\{(CF_3)_2CN\}_2$ [276, 277].

This derivative is not a simple adduct, but a derivative of the azine, $(CF_3)_2C{=}N{-}N{=}C(CF_3)_2$. In fact with iodine it affords $Pt(PPh_3)_2I_2$ and the azine, while the free azine and $Pt(PPh_3)_2$(stilbene) gives the same derivative. Its structure has been completely elucidated by an X-ray investigation [277]:

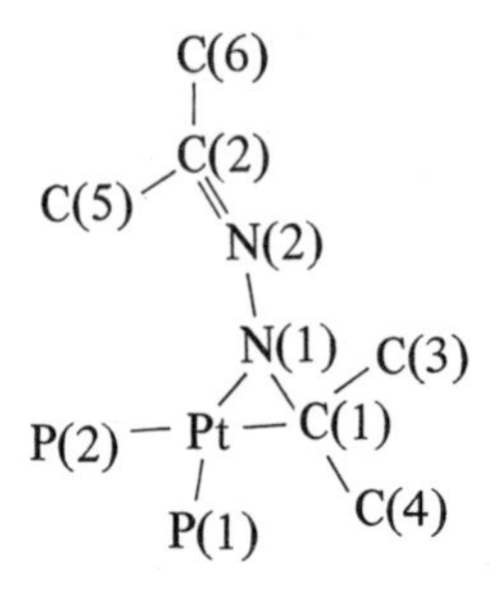

The C(1)N(1) distance is similar to the C–N distance in aziridine (1.488 Å). The interplanar angle is in accordance with an idealized sp^3 hybridization of C(1) and N(1). In the i.r. spectrum a band at 1560 cm^{-1} can be attributed to the N(2)=C(2) stretching. The ^{19}F n.m.r. spectrum shows three resonances at 59.5 (doublet with ^{195}Pt satellites, 6F, J_{P-F} = 10.0 Hz, J_{Pt-F} = 79.0 Hz); 63.5 (multiplet, 3F, J_{FF} = 6.0 Hz, J_{P-F} = 2.5 Hz) and 65.8 (apparent quintet, 3F, J_{P-F} = 6.0 Hz) p.p.m. relative to CCl_3F. Double irradiation of the high-field band collapsed the multiplet at 63.5 p.p.m. to a doublet (J_{FF} = 6.0 Hz), whereas irradiation of the band at 63.5 p.p.m. caused collapse of the high field band to a doublet (J_{P-F} = 2.5 Hz). In order to fit the ^{19}F n.m.r. spectrum with the structure, it has been postulated that in solution rapid inversion occurs at the nitrogen bonded to platinum [277].

Diazonium salts have also been reacted with $Pt(PPh_3)_3$. A series of highly coloured azo-platinum complexes have been prepared [278]:

$$Pt(PPh_3)_3 + [p\text{-}RC_6H_4N_2]^+X^- \rightarrow [(PPh_3)_3Pt{-}N{=}N{-}C_6H_4R{-}p]^+(X^-)$$

$$(R = NO_2, F, H, OMe, Me, X = BF_4; R = NMe_2, NEt_2, X = BPh_4)$$

This reaction follows the general scheme of oxidative addition reactions. The influence of the *para* substituents on the nature, stability and reactivity of the bond between the platinum atom and the azo group has been investigated [278]. The co-ordinated azo group is easily lost as molecular nitrogen in solvents such as chloroform (R = NO_2) or by the action of sunlight in the solid state (R = OMe, F), with formation of complexes which presumably contain an aryl group σ-bonded to platinum. It is interesting to note that the reverse reaction has been postulated during amine formation from molecular nitrogen under reducing conditions [279].

A suspension of the azo-platinum complex in methanol reacts with hydrogen at atmospheric pressure and room temperature to give the known cationic platinum hydride, $[Pt(PPh_3)_3H]^+(BF_4^-)$. It may be interesting to compare this reaction with the reaction of the non-ionic platinum(II) azo-complexes reported by Parshall [275], where PEt_3 was present as ligand. In the latter case the reduction with molecular hydrogen takes place only in the presence of a heterogeneous catalyst, with the formation of amines and hydrazines from the azo-ligand. In the former case no formation of N–H bonds was noted, the organic group being transformed into the corresponding hydrocarbon with evolution of nitrogen:

$$[(PPh_3)_3PtN{=}NC_6H_4R\text{-}p]^+(BF_4^-) + H_2 \xrightarrow[\text{suspension}]{\text{MeOH}} [(PPh_3)_3PtH]^+(BF_4^-) + C_6H_5R + N_2$$

A similar kind of reaction takes place with oxygen at atmospheric pressure and room temperature:

$$[(PPh_3)_3PtN{=}NC_6H_4R\text{-}p]^+(BF_4^-) + O_2 \xrightarrow[\text{suspension}]{\text{MeOH}} [(PPh_3)_2Pt(OH)]_2^{2+}(BF_4^-)_2 + p\text{-}RC_6H_4OH + N_2 + PPh_3$$

Again nitrogen is lost from the azo group. In this case the organic part is converted into the corresponding phenol, while a hydroxo-platinum complex which may contain bridging OH groups is formed. The azo-platinum complexes can be protonated with non-co-ordinating strong acids such as HBF_4 or $HClO_4$:

$$[(PPh_3)_3PtN{=}NC_6H_4R\text{-}p]^+(BF_4^-) + HBF_4 \rightarrow [(PPh_3)_3PtNH{=}NC_6H_4R\text{-}p]^{2+}(BF_4^-)_2$$

The reaction is reversible and with bases the azo-complexes are reformed. The proton is probably on the nitrogen atom directly bound to platinum, as it has been demonstrated for similar derivatives using ^{15}N [275]. Neutral azo-derivatives $(PPh_3)_2Pt(N{=}NC_6H_4R\text{-}p)(X)$ (X = I, N_3) can be obtained by reacting the ionic azo complexes with LiX.

The spectroscopic properties of these derivatives have been investigated. An i.r. absorption of the cationic azo complexes at *ca*. 1600 cm^{-1} has been tentatively assigned to $\nu(N{=}N)$ but this value is much higher than that observed in the related neutral azo-platinum derivatives [275], where triethylphosphine is used as ligand and $\nu(N{=}N)$ occurs at *ca*. 1450 cm^{-1}. If this assignment is correct, a resonance formula such as

$$[R\text{—}\langle\ominus\rangle{=}N\text{—}N{=}\overset{\oplus}{Pt}\text{—}]^+$$

is unlikely to contribute significantly to the electronic structures of these compounds. The position of the $n \rightarrow \pi^*$ transition of the azo group provides further evidence against such a structure [278]. Moreover when R is a very electronegative substituent such as NO_2 which should greatly favour the conjugation of the π-electron system of the metal with the organic azo group the azo cation is so unstable that it instantaneously decomposes in solution. All these facts support the assumption that the –N=N– group has a strong double bond character in these azo-platinum complexes, and in fact no activation towards hydrogenolysis has been found [278].

A series of cyclic azo-platinum compounds has been obtained during attempts to stabilize benzyne by coordination to zerovalent platinum. Thermal decomposition in solution of benzyne precursors, in the presence of $Pt(PPh_3)_4$, does not lead to a benzyne derivative of platinum, since coordination of the precursors to the metal atom is preferred [280]. So 1,2,3-benzothiadiazole-1,1-dioxides

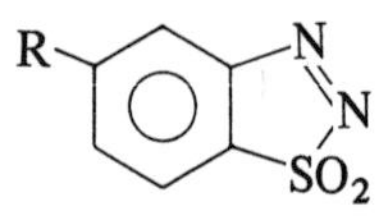

(R = H, Me) reacts with $Pt(PPh_3)_4$ in benzene giving

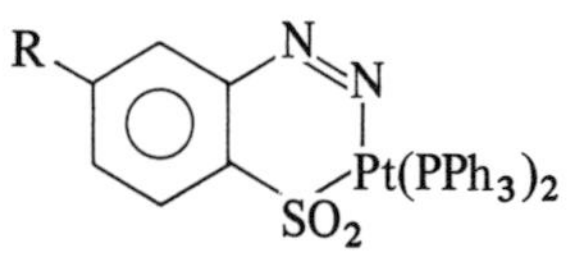

A more detailed investigation on the former complex (R = H) and on similar derivatives prepared by using $Pt(PPh_3)_2(C_2H_4)$ [281] as the zerovalent platinum compound has also appeared. In these compounds the azo group absorbs at 1446 cm^{-1}.

The cyclic platinum-azo derivative (R = H) does not generate benzyne on photolysis but irradiation with a sun lamp, in a mixture of furan and tetrahydrofuran, gives the benzyne adduct, 1,4-dihydronaphthalene endoxide and its rearrangement product, 1-naphthol [280]. Some interesting thermal rearrangements have also been observed [281]:

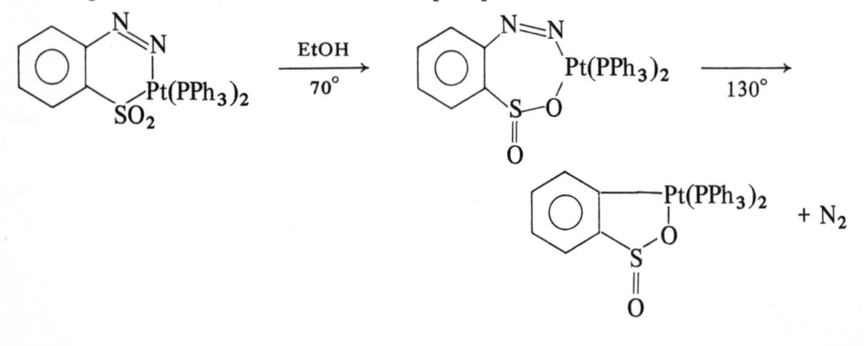

Another benzyne precursor, benzene diazonium-2-carboxylate

R^1 N_2^+ $(R^1 = R^2 = H)$ R^2 CO_2^-

gives at 10° the cyclic azo derivatives of the type

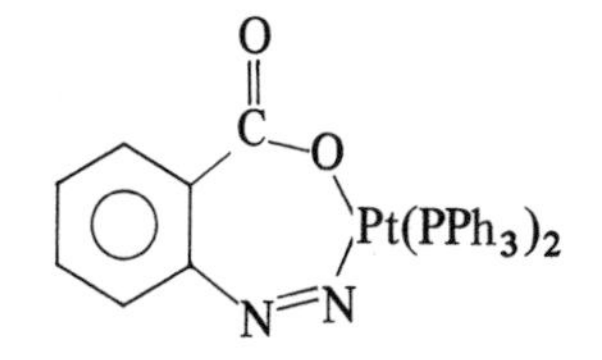

by reacting with $Pt(PPh_3)_2(C_2H_4)$ [281] [ν(N=N) = 1517 cm^{-1}]. This compound loses nitrogen at 130°C giving

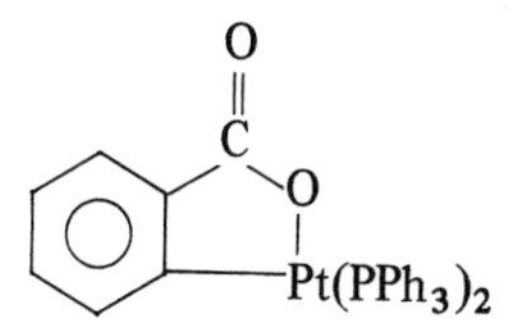

which may be obtained directly from the corresponding benzene diazonium carboxylates (R^1 = H, R^2 = H, Me, Cl, NO_2; R^1 = Cl, R^2 = H) and $Pt(PPh_3)_4$ [280].

The carboxylato compounds are stable to u.v. irradiation, and they are not benzyne precursors as was the sulphone derivative. Their structures have been mainly elucidated by i.r. spectroscopy [281].

The information we have on the reactions of azides indicates them as the best source for the preparation of nitrene complexes. Indeed when electron-withdrawing substituents are attached to the N_3 group, the dissociation $RN_3 \rightarrow RN: + N_2$ is very easy, both thermally and photochemically [282].

The coordination of azide to a metal in a low oxidation state could thus correspond to a photochemical excitation of the azide, through π-back-donation of electron density from the metal into antibonding orbitals of the azide.

The synthesis of fluoroalkylimido (or nitrene) derivatives of palladium and platinum has been successfully accomplished by reacting 2H-hexafluoropropyl azide (R_FN_3) with $M(PMePh_2)_4$ complexes (M = Pd, Pt) in benzene [283]. Deep red compounds of formula $M(PPh_3)_2(NR_F)$ have been obtained. If the imido ligand may be regarded formally as being in the singlet state, then a *p* orbital on the nitrogen would be vacant, and thus able to accept electron density from a filled metal *d* orbital [283]; π-back-bonding would be favoured by the strongly electron-withdrawing fluoroalkyl groups.

Zerovalent triphenylphosphine derivatives of platinum(0) have also been reacted with sulphonyl, aryl, acyl and aroyl azides [284]. The most interesting compounds have been obtained with sulphonyl azides RN_3 ($R = C_6H_5SO_2$, $p\text{-}CH_3C_6H_4SO_2$). In non-polar solvents such as dry benzene pale yellow compounds of formula $Pt(PPh_3)_2(N_4R_2)$ were isolated. A concomitant fast nitrogen evolution was observed while the displaced phosphine can be recovered from the mother liquor as the phosphinimino adduct, $Ph_3P{=}NR$. It was suggested that these compounds could be either tetrazene derivatives or bis-azo compounds [284]. However the azo-platinum complexes above considered show a chemical behaviour completely different from that of $Pt(PPh_3)_2(N_4R_2)$ derivatives. In particular with hydrochloric acid the following reaction has been observed [285].

$$Pt(PPh_3)_2(N_4R_2) + 2HCl \rightarrow Pt(PPh_3)_2Cl_2 + RNH_2 + RN_3$$

This unexpected ring opening not only ruled out a structure corresponding to a bis-azo derivative, but also suggested that the platinum complex is a tetrazene derivative (α), stabilized by coordination to the metal, but in equilibrium with open forms such as (β), in which the chelate ring has been broken:

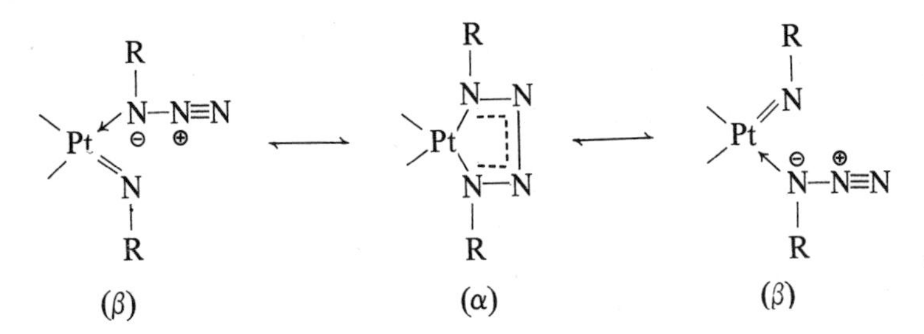

The type of Pt-tetrazene bonding in the closed form (α) has already been discussed (see 1.3.1.2). An analogous type of compound has been found also in the nitrosyl derivatives, $M(NO)(PPh_3)(N_4R_2)$ (M = Rh, Ir), obtained from $M(NO)(PPh_3)_3$ and *p*-toluenesulphonyl azide [285]. In the latter case, the suggestion that open forms such as (β) may be present in solution has been substantiated by variable temperature H^1 n.m.r. studies [285]. In fact the single n.m.r. peak of the methyl group at room temperature collapses at low temperature and, in the case of $Ir(NO)(PPh_3)(N_4R_2)$ ($R = p\text{-}CH_3C_6H_4SO_2$), gives place to two distinct signals at temperatures lower than −50°C.

On raising the temperature, the single peak reappears, as expected for a fast exchange equilibrium between an open and a closed ring. However the n.m.r. spectrum of the platinum tetrazene complex, $Pt(PPh_3)_2(N_4R_2)$, from +40°C to −80°C in methylene chloride showed only a slight broadening of the methyl peak [267], without any splitting of the band. Also, chemical evidence failed to substantiate (for the platinum complex) the presence, in solution, of the open forms (β), where the azide should be weakly bound and present in a reactive form. For example enamines, which are known to react easily with sulphonyl azides at room temperature according to the reaction [282]

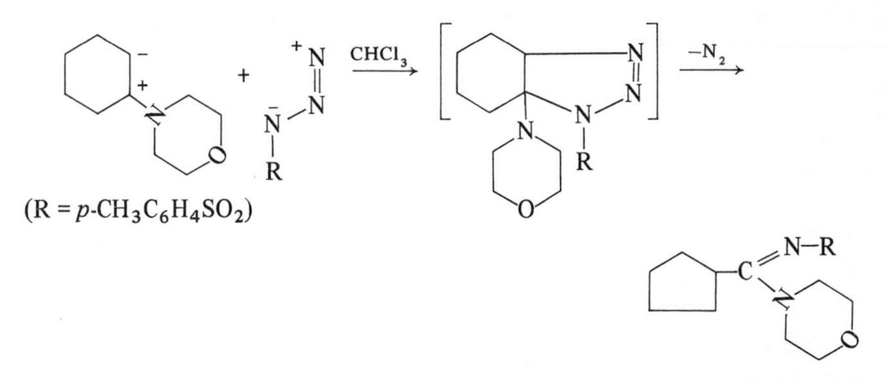

($R = p\text{-}CH_3C_6H_4SO_2$)

did not react with the platinum complex, even in boiling chloroform [267].

With carbon monoxide, the platinum tetrazene complex exchanges one phosphine with CO, but no insertion of carbon monoxide into the tetrazene ring was observed [267]. The last insertion reaction was in principle quite probable, since the reaction of carbonyl–phosphine–platinum(O) and –palladium(O) compounds such as $M(PPh_3)_3(CO)$ (M = Pd, Pt) with RN_3 ($R = p\text{-}CH_3C_6H_4SO_2$) in benzene gives derivatives where the R–N–CO–N–R or R–N–CO–N=N–NR groups are bonded to the metals [286]. At the moment it is not easy to explain the poor reactivity of the tetrazene ligand in these complexes.

Nickel-tetraazadiene derivatives of formula $NiL_2(N_4R_2)$ [L = PPh_3, PPh_2Me, $PPhMe_2$, $P(OMe)_3$; R = C_6F_5] have been obtained from Ni(1,5-COD)(N_4R_2) by displacement of 1,5-cyclooctadiene by the ligands L [287]. The latter compound was isolated from the reaction between Ni(1,5-COD)$_2$ and pentafluorophenyl azide in diethyl ether at −35°C [287].

The reaction of sulphonyl azides with tris- and tetrakis-(triphenylphosphine)-platinum(0) was shown to be strongly dependent on the nature of the solvent employed in the reaction [284]. When benzene containing traces of water was used as solvent, the major product was $Pt(PPh_3)_2(OH)(NHR)$. In the reaction mixture some other by-products were always present; these were $Pt(PPh_3)_2(N_3)_2$, $Pt(PPh_3)_2(NHR)_2$ and $Pt(PPh_3)_2(N_3)(NHR)$. The latter compound was identical to the main product obtained when the reaction was conducted in ethanol [284], in which the derivative $Pt(PPh_3)_2(N_3)(R)$ corresponding to the simple oxidative addition reaction product, was also isolated. Its structure was supported by the absence in the i.r. spectrum of the ν(N–S) absorption, which is always present in the other cases at about 900–940 cm^{-1}. From the reaction between acyl and aroyl azides with phosphine-platinum(0) derivatives, only

$Pt(PPh_3)_2(N_3)_2$ and $Pt(PPh_3)_2(N_3)(NCO)$ have been recovered. The possible routes for the formation of these different compounds have been discussed, and the formation of intermediate nitrenes seems to be the most plausible hypothesis [284]. A preformed nitrene, generated from p-$CH_3C_6H_4SO_2NNaCl . 3H_2O$ (chloramine T), gives rise to $Pt(PPh_3)_2(X)(NHR)$ derivatives (X = OH, Cl), when reacted with $Pt(PPh_3)_4$ in wet benzene.

The reactions of tris(triphenylphosphine) platinum(0) with aryl azides are very slow and difficult. With phenyl azide only intractable oils were obtained [284].

With methyl and ethyl azide in benzene, both $Pd(PPh_3)_4$ and $Pt(PPh_3)_4$ give simply $M(PPh_3)_2(N_3)_2$ complexes [288].

The reaction of $Pt(PPh_3)_3$ with cyclohexane diazide has also been briefly investigated; in benzene no immediate gas evolution was noted and after 16 h a material showing only weak i.r. absorptions at around 2050 cm^{-1}, indicative of azido groups bound to platinum, was recovered [267].

The reactions between 1,3-diaryltriazenes (dtt) with $M(PPh_3)_4$ (M = Pd, Pt) have also been studied. In benzene or alcohols red-orange bis-triazenido derivatives, $M(PPh_3)_2(dtt)_2$, were obtained [289]. The 1,3-diaryltriazenido ligand has a close structural analogy with the carboxylato ligand, R—C(O)(O)—. However at first square planar four-coordinate structures with monodentate triazenido ligands were proposed for these derivatives [289].

Subsequent variable temperature n.m.r. studies revealed the occurrence of dynamic equilibria involving co-ordinated monodentate 1,3-diaryl-triazenido groups through a bidentate form [290]:

$$-M-\overset{*}{N}(Ar)-N{=}N(Ar) \longrightarrow \left[-M\langle N(Ar) \cdots N \cdots N(Ar) \rangle \right] \longrightarrow -M-N(Ar)-N{=}\overset{*}{N}(Ar)$$

The n.m.r. signal of the p-methyl derivative of the platinum complex, τ_{CH_3} = 7.8 at 40°C, broadens on lowering the temperature, and at *ca.* −10°C becomes a sharp doublet. This temperature dependence was also shown by the analogous platinum complex having $C_2H_4(PPh_2)_2$ as the ligand, and by the palladium triphenylphosphine derivative. Some 1,3-diaryltriazenido ligands other than the p-methyl derivative have also been studied [290].

In general triazene and tetrazene derivatives of low valent metals seem to afford new types of fluxional organometallic molecules.

2.2.3.7. *Reactions with Carbon Monoxide and Isocyanides*

The coordinative reactivity of carbon monoxide towards $Pt(PPh_3)_n$ ($n = 3, 4$) derivatives was first studied by Malatesta and Cariello [3]. The platinum(0) derivatives were treated with carbon monoxide under pressure in the solid state. The following reactions took place:

$$Pt(PPh_3)_4 + 2CO \rightarrow Pt(PPh_3)_2(CO)_2 + 2PPh_3$$

$$Pt(PPh_3)_3 + 2CO \rightarrow Pt(PPh_3)_2(CO)_2 + PPh_3$$

The phosphine displaced cannot be removed by crystallization of the reaction mixture, since in solution rapid gas evolution occurs according to the reaction:

$$Pt(PPh_3)_2(CO)_2 + PPh_3 \rightarrow Pt(PPh_3)_3(CO) + CO$$

The same reactions have been studied with $Pt\{P(C_6H_4Cl\text{-}p)_3\}_3$ [3]. In the latter case the bis-carbonyl product of the carbonylation reaction seems to lose carbon monoxide both in the solid state or in solution, and the interesting coordinatively unsaturated complex, $Pt\{P(C_6H_4Cl\text{-}p)_3\}_2(CO)$ was formed.

In view of the interest on the chemical behaviour of phosphine derivatives of platinum(0), these reactions were reconsidered later in more detail [291, 292]. When $Pt(PPh_3)_3$ or $Pt(PPh_3)_4$ are allowed to react with carbon monoxide at atmospheric pressure and room temperature in suspension of saturated hydrocarbons such as *n*-hexane or cyclohexane, or in methanol, a fast formation of the mono carbonyl derivative $Pt(PPh_3)_3(CO)$ [$\nu(CO)$, 1908 cm^{-1}] was observed [291]. It was affirmed that with time this derivative gave $Pt(PPh_3)_2(CO)$ [$\nu(CO)$, 1942 cm^{-1}]. Later, however, it was shown that these are simply two different crystalline forms of the trisphosphinemonocarbonyl derivative [292], a pale yellow form (A) [$\nu(CO)$, 1940 cm^{-1}] which can be obtained directly from $Pt(PPh_3)_4$ and carbon monoxide in acetone, and a colourless form (B) [$\nu(CO)$, 1908 cm^{-1}] which can be obtained from $Pt(PPh_3)_3$ and carbon monoxide in ethanol–tetrahydrofuran mixtures.

Both forms in tetrahydrofuran absorb at 1930 cm^{-1}. This point emphasizes the importance of determining $\nu(CO)$ in solution and not in the solid state since two different crystalline forms of the same compound may give, as in this case, a $\Delta\nu(CO)$ of as much as 32 cm^{-1}, a shift which is usually associated with strong electronic effects.

Other $Pt(PR_3)_3(CO)$ derivatives have been obtained similarly from $Pt(PMePh_2)_4$ and $Pt\{PPh_2(CH_2Ph)\}_3$ and carbon monoxide [292]. With longer reaction times, $Pt(PPh_3)_n$ ($n = 3, 4$) and carbon monoxide give the bis-carbonyl derivative, $Pt(PPh_3)_2(CO)_2$ [$\nu(CO)$, 1982 and 1950 cm^{-1}], if the solvent is periodically removed in order to withdraw the displaced phosphines from the reaction [291]. Mono and bis carbonyl derivatives of platinum having phosphines other than PPh_3 have also been obtained, but through syntheses in which the reduction of a platinum(II) complex has been applied [292]. The X-ray structures of some of these carbonyl derivatives have already been discussed (2.2.2). The far i.r. spectra of some of these compounds have also been reported, and Pt–C stretching frequencies tentatively assigned [291]. However, we have already pointed out how these data could be incorrect (1.3.1.1). Molecular weights have shown considerable dissociation of $Pt(PPh_3)_3(CO)$, $Pt(PMePh_2)_3(CO)$ and $Pt(PPh_3)_2(CO)_2$ [292] and the following set of equilibria in solution has been proposed:

$$PtL_3(CO) \rightleftharpoons PtL_3 + CO \quad (1)$$

$$PtL_3 \rightleftharpoons PtL_2 + L \quad (2)$$

$$PtL_2 + 2CO \rightleftharpoons PtL_2(CO)_2 \quad (3)$$

$$5\ PtL_2(CO)_2 \rightleftharpoons Pt_3L_4(CO)_3 + 2PtL_3 + 7CO \quad (4)$$

Equilibrium (1) is consistent with the formation of $Pt(PPh_3)_3$ by boiling $Pt(PPh_3)_3(CO)$ for a long time in ethanol [291, 292].

With a shorter reaction time a different intermediate which analysed as $2Pt(PPh_3)_3 \cdot Pt(PPh_3)_3(CO)$ was also obtained [292]; this shows in the i.r. $\nu(CO)$ at 1958 cm^{-1} in the solid state, but in solution it absorbs at the same frequency as $Pt(PPh_3)_3(CO)$. By treatment with carbon monoxide it reverts to $Pt(PPh_3)_3(CO)$. We have already discussed the controversial existence of equilibrium (2) which is in any case considered to lie strongly to the left [292].

Equilibrium (4) is responsible for the reddish colour of benzene solutions of $Pt(PPh_3)_2(CO)_2$; the trimeric phosphine carbonyl derivative $Pt_3(PPh_3)_4(CO)_3$ has in fact a red colour [291, 292].

This reaction has been shown to be possible also in the solid state [291]. It has been observed that increasing steric requirements of the group R of the ligand PPh_2R increases the stability of the $Pt(PPh_2R)_2(CO)_2$ compounds [292].

The compounds $Pt(PR_3)_3(CO)$ and $Pt(PR_3)_2(CO)_2$ react quantitatively *in vacuo* with 1,2-bis(diphenylphosphino)ethane giving $Pt\{C_2H_4(PPh_2)_2\}_2$ [292]. Similarly, when R = Ph, the reaction with triphenylphosphine gives $Pt(PPh_3)_4$, while the interaction with acids causes loss of carbon monoxide [291] with formation of the same compounds obtained directly from $Pt(PPh_3)_4$ [161].

The $\nu(CO)$ frequencies in $Pt(PPh_3)_n(CO)_{4-n}$ ($n = 2, 3$) are not very different from those of the related nickel carbonyl complexes which are stable towards carbon monoxide dissociation and it is not clear why platinum(0) carbonyl complexes should so easily lose carbon monoxide [5, 291]. This confirms that the position of the CO bands is definitely not a decisive parameter for comparing the strength of metal–carbon interactions (1.3.1). The stability of the species in the case of phosphine carbonyl platinum(0) derivatives seems not to be a matter of steric requirements. Loss of carbon monoxide from compounds such as

$M(PPh_3)_3(CO)$ (M = Ni, Pt) should in any case relieve the steric hindrance in the complex. Moreover from this point of view the loss of one phosphine instead of CO should be more favourable for the complex and it would be interesting to know the kinetics of a reaction such as $Pt(PPh_3)_3(CO) + CO = Pt(PPh_3)_2(CO)_2 + PPh_3$.

The reaction of carbon monoxide with palladium(0)–phosphine complexes has been briefly studied. It was observed that $Pd(PPh_3)_4$ suspended in *n*-hexane gives the pale yellow $Pd(PPh_3)_3(CO)$ [ν(CO), 1955 cm^{-1} in nujol mull] and in good yield [293].

Owing to the ease with which the starting material $Pd(PPh_3)_4$ can now be prepared, this seems to be a more convenient preparation of $Pd(PPh_3)_3(CO)$ than those already reported [294]. The previous method of synthesis required the use of aluminum alkyls; also, the reduction of $Pd(PPh_3)_2Cl_2$ with $NaBH_4$ in the presence of phosphine and carbon monoxide at low temperature [294], gives $Pd(PPh_3)_3(CO)$, but the product is sometimes contaminated by some of the unreacted starting material [293].

The reaction between $Pd(PPh_3)_4$ and CO does not take place in benzene, while addition of PPh_3 to a solution of $Pd(PPh_3)_3(CO)$ immediately gives $Pd(PPh_3)_4$ with CO evolution [294].

In contrast to the corresponding platinum complex, $Pd(PPh_3)_3(CO)$ loses phosphine when heated in ether suspension giving the trimeric, red cluster $Pd_3(PPh_3)_4(CO)_3$ [294]. From molecular weight determinations $Pd(PPh_3)_3(CO)$ appears to dissociate in solution to $Pd(PPh_3)_2(CO)$ and phosphine. The reactions of $Pd(PPh_3)_3(CO)$ with methyl iodide, allyl chloride and vinyl chloride, lead to the acyl complexes *trans*-$Pd(PPh_3)_2Cl(COR)$ (R = Me, $CH_2CH{=}CH_2$, $CH{=}CH_2$), which show $\nu(C{=}O)$ at around 1665–1690 cm^{-1}.

With methyl iodide $Pd(PPh_3)_2I(Me)$ has also been isolated [294]. The reactions of carbon monoxide with $Ni(PPh_3)_4$ and with analogous derivatives having a chelating diphosphine as ligand, have been studied recently, mainly by means of i.r. spectroscopy [295].

At room temperature and atmospheric pressure, in benzene or dichloromethane solution, $Ni(PPh_3)_2(CO)_2$ is rapidly formed [ν(CO) 1995 and 1940 cm^{-1} in benzene]; then $Ni(PPh_3)(CO)_3$ [ν(CO), 2065 cm^{-1}] begins to appear. Under the same conditions $Ni\{C_2H_4(PPh_2)_2\}_2$ and carbon monoxide give place in *ca.* 60 min to $Ni\{C_2H_4(PPh_2)_2\}_2(CO)$ [ν(CO), 1920 cm^{-1}], in which one of the diphosphines is believed to be bonded as a monodentate ligand. With longer times small amounts of the known $Ni\{C_2H_4(PPh_2)_2\}(CO)_2$ [ν(CO), 2000 and 1940 cm^{-1}] appear in solution. Similarly $Ni\{C_3H_6(PPh_2)_2\}_2$ gives a mixture of $Ni\{C_3H_6(PPh_2)_2\}_2(CO)$ [ν(CO), 1910 cm^{-1}] and $Ni\{C_3H_6(PPh_2)\}(CO)_2$ [ν(CO), 2000 and 1935 cm^{-1}]. With $Ni\{C_4H_8(PPh_2)_2\}_2$ the stepwise substitution is more differentiated and each reaction product could be isolated and characterized. The first step is fast and complete in about 2 min, with formation of $Ni\{C_4H_8(PPh_2)_2\}_2(CO)$ [ν(CO), 1910 cm^{-1}], where only one phosphine is acting as a chelating ligand. The uptake of the second molecule of CO is slow and is complete in about 3 h, with formation of $Ni\{C_4H_8(PPh_2)\}(CO)_2$ [ν(CO), 1995 and 1940 cm^{-1}]; from very concentrated solutions this violet compound precipitated, but even in these conditions it could not be isolated in a pure state. Strangely enough, the rate of CO uptake and the number of moles of carbon monoxide adsorbed per mole of nickel(0) complex did not depend appreciably on the presence of free diphosphine [295].

The absorption of carbon monoxide has been also studied with $Ni\{P(OPh)_3\}_4$ [296]. By using stoichiometric amounts of carbon monoxide $Ni\{P(OPh)_3\}_3(CO)$ [ν(CO), 2020 cm^{-1} in CCl_4] and $Ni\{P(OPh)_3\}_2(CO)_2$ [ν(CO), 2053 and 2016 cm^{-1} in CCl_4] have been isolated. On the contrary phosphite complexes of platinum(0) did not show any reactivity towards carbon monoxide [3].

The formation of $NiL_2(CO)_2$ [L = $P(O\text{-}o\text{-}C_6H_4C_6H_5)_3$ [105] or = PPh_3 [57]] from NiL_2 and carbon monoxide has been briefly mentioned.

Similarly $Pt(PPh_3)_2(CO)_2$ has been obtained both from preformed $Pt(PPh_3)_2$ and carbon monoxide [67] and by the reaction of CO on $Pt(PPh_3)_2$ prepared *in situ* via the photochemical decomposition of $Pt(PPh_3)_2(CO_3)$ [135]. The latter reaction has also been used for the synthesis of the corresponding triphenylarsine derivative.

The trifluorophosphine derivatives, $Pd(PF_3)(CO)(PPh_3)_2$ and $Pt(PF_3)_2(CO)_2$, have been obtained by reaction of $M(PF_3)_2(PPh_3)_2$ complexes (M = Pd, Pt) with carbon monoxide [115]. We briefly mention here that among the very few isocyanide complexes of platinum(0), $Pt(PF_3)_3(C_6H_{11}NC)$ has been isolated during the kinetic studies on the reaction of $Pt(PF_3)_4$ with cyclohexyl isocyanide (see also 1.3.5) [297].

The analogous nickel complex has also been isolated.

These two derivatives show ν(CN) at 2169 (Ni) and 2152 (Pt) cm^{-1}, that is at a frequency higher than the free ligand [ν(CN) = 2138 cm^{-1}]. Usually the coordination of an isocyanide to a transition metal in a low oxidation state causes a lowering of the frequency of the CN stretching vibration, as a consequence of electron charge displacement from the metal *d* orbitals into the π^* orbitals of the ligand [298]. The observed increase in ν(CN) was interpreted as an indication that the isocyanide ligand was simply acting as a σ donor [297].

The increased electron density on the metal upon substitution of a trifluorophosphine by an isocyanide determines an increase of the bond order between the metal and the phosphorus of the unsubstituted PF_3 ligands, and a shift to lower energy of the phosphorus-fluorine vibration [M = Ni, $\Delta\nu$(P–F) = 44 cm^{-1}; M = Pt, $\Delta\nu$(P–F) = 28 cm^{-1}]. This explains why the substitution of a second PF_3 group by $C_6H_{11}NC$ occurs much more slowly than the replacement of the first group [297]. On the other hand, the reaction of $C_6H_{11}NC$ with

$Ni\{P(OEt)_3\}_4$ did not allow the isolation of the monosubstituted compound, and only $Ni\{P(OEt)_3\}_2(C_6H_{11}NC)_2$ was obtained [299]. In this compound ν(CN) are at 2070 and 2050 cm^{-1}. This is due to the fact that phosphites are much poorer π-acceptors than PF_3. Such an interpretation also explains the faster substitution of the second PF_3 in comparison to the first [299].

The reactions of zerovalent nickel, palladium and platinum(0) derivatives with isocyanides have not been studied to any great extent. This could be due to easy polymerization of isocyanides by the catalytic action of low valent complexes, but probably much more research would be successful in this field.

2.2.3.8. Reactions with Alkenes and Related Ligands

As we will see, an extraordinary amount of work has been done on the chemistry of alkene (and alkyne, 2.2.3.9) phosphine derivatives of nickel, palladium and platinum. The purpose of this research was twofold and interdependent. The nature of the interaction between the metal and the unsaturated hydrocarbon has received much attention, and we have already discussed how this type of bond can be interpreted (1.3.1.2). At the same time, some of these complexes have been used in homogeneous catalysis, for instance in the oligomerization of alkenes. Owing to the great development of the chemistry in this field, we will subdivide this paragraph and the following into separate sections. Also, we will consider not only the derivatives obtained directly from the interaction of an alkene or alkyne with a zerovalent phosphine derivatives of nickel, palladium and platinum, but also similar compounds obtained by different routes. The popularity of this research is reflected in the notable number of reviews appeared on the field [1a, 5, 116b, 300]. However in the last few years many new and interesting results appeared, and we will focus our attention on them.

(*a*) *Methods of synthesis.* The ligand exchange reaction:

$$M(PR_3)_n\ (n = 3,4) + \text{olefin} \rightarrow M(PR_3)_2(\text{olefin}) + (n - 2 \text{ or } 1)PR_3$$

is a very useful method of synthesis, particularly for platinum complexes. However it can be in general applied only when the alkene possesses strongly electron-withdrawing substituents, and the metal–olefin bond is strong. According to the most recent results (1.3.5) the reaction scheme is:

$$M(PR_3)_4 \rightleftharpoons M(PR_3)_3 + PR_3 \quad (1)$$

$$M(PR_3)_3 + \text{olefin} \rightarrow M(PR_3)_3(\text{olefin}) \quad (2)$$

$$M(PR_3)_3(\text{olefin}) \rightarrow M(PR_3)_2(\text{olefin}) + PR_3 \quad (3)$$

When the metal–alkene bond is not strong enough, the $M(PR_3)_3$ or $M(PR_3)_4$ complex is isolated from the reaction mixture even when the alkene derivative is clearly detected in solution [301]. A complex of stoichiometry $M(PR_3)_3$(olefin), which is now presumed to be the intermediate in the reaction, has been isolated in only one case, from the displacement-disproportionation reaction of $Ni(PPh_3)_2(C_2H_4)$ and tetrafluoroethylene [302]. This complex is very sensitive to air, and it is possibly tetrahedral. Tetrafluoroethylene is only slightly perturbed on coordination, as shown for instance by the small downfield shift of the ^{19}F resonance of the olefin in the complex with respect to free tetrafluoroethylene.

The ^{31}P spectrum shows a single peak at −20.3 p.p.m. relative to 85% H_3PO_4 as external standard, and this could be due to dissociation of phosphine in solution. Presumably this compound can be isolated only because it is insoluble in the solvent used in the reaction (toluene).

In order to avoid the reverse reaction between $M(PR_3)_2$(olefin) and the displaced PR_3, dicoordinated complexes such as $Pt(PPh_3)_2$, both preformed [67] or prepared *in situ* by reduction of $Pt(PPh_3)_2(O_2)$, by sodium borohydride [235] can be used. In these cases even a weak ligand such as ethylene can be firmly coordinated to the complex. The equilibrium constants at 25°C in benzene for the reactions:

$$M(PPh_3)_3 + C_2H_4 \overset{K}{\rightleftharpoons} M(PPh_3)_2(C_2H_4) + PPh_3$$

(M = Ni, Pd, Pt) have been measured [55]. They are $K = 3 \times 10^2$ (M = Ni), K = 0.12 (M = Pd) and $K = 0.013$ (M = Pt) (see also 1.3.5), and indicate that at least for nickel it might be possible to isolate the ethylene adduct from the solution.

Another route to prepare $Pt(PPh_3)_2$ *in situ* is the photochemical decomposition of $Pt(PPh_3)_2(C_2O_4)$ [133] or $Pt(PPh_3)_2(CO_3)$ [135]; in the presence of alkenes and alkynes the corresponding $Pt(PPh_3)_2$ (unsaturated hydrocarbon) complexes have been obtained. Displacement reactions can be conveniently used by replacing an olefin in a preformed complex with a different olefin forming a stronger bond with the metal, as for example in [301]:

$$Pt(PPh_3)_2(C_2H_4) + CH_3CH{=}CHCHO \rightarrow Pt(PPh_3)_2(CH_3CH{=}CHCHO) + C_2H_4$$

This reaction is particularly useful when the entering alkene is not able to give a stable complex with the $M(PPh_3)_n$ ($n = 3, 4$) complexes because of the presence of excess phosphine, as in the case considered above.

This type of displacement reaction has been also used to prepare the tetracyanoethylene complex $Pt(PPh_3)_2\{C_2(CN)_4\}$ both from $Pt(PPh_3)_2(C_2H_4)$ [303] and from $Pt(PPh_3)_2(PhC_2H)$ [208], and this last reaction represents a rather rare example of an alkene able to replace an alkyne in a complex. Alkynes usually form a stronger bond with the metal than alkenes, and only powerful complexing reagents such as tetracyanoethylene, which is extremely activated by the four CN groups, are preferentially coordinated rather than an alkyne.

On the other hand when tetracyanoethylene reacts with $Pt(PPh_3)_4$, an oxidative addition, giving $Pt(PPh_3)_2(CN)_2$, takes place [208].

Complexes having a strong π-acid, such as tetrafluoroethylene, coordinated to the metal, may exchange the other ligands. For example the series of derivatives $PtL_2(C_2F_4)$ [L = $PMePh_2$, PMe_2Ph, PEt_2Ph, PBu^n_3 or L_2= $C_2H_4(PPh_2)_2$] has been obtained from $Pt(AsPh_3)_2(C_2F_4)$ and the ligands L [58]. The related nickel complex, $Ni(PPh_3)_2(C_2F_4)$, has been similarly obtained by displacing CDT by PPh_3 from Ni(CDT) (C_2F_4) (CDT = 1,5,9-cyclododecatriene) [206 *bis*].

A potentially very useful method of synthesis is the reaction of $M(dba)_2$ complexes (M = Pd, Pt; dba = dibenzylideneacetone) with alkenes in the presence of ligands such as a phosphines or phosphites. This double displacement reaction has been recently applied to the synthesis of $Pd\{P(OR)_3\}_2$ (maleic anhydride) (R = Me, Ph) [304].

Ligand addition reactions have also been used for the synthesis of olefin phosphine complexes of nickel(0). By adding for example, triphenylphosphine, to bisacrylonitrilenickel(0), $Ni(PPh_3)_n(CH_2CH{=}CN)_2$ (n = 1,2) have been isolated [305].

Nickel-η-1,2-5,6-9,10-(1,5,9-cyclododecatriene), Ni(CDT), similarly adds phosphines giving nickel-η-1,2-5,6-9,10-(1,5,9-cyclododecatriene) phosphine, $Ni(PR_3)(CDT)$; this type of reaction has been studied in order to elucidate the mechanism of cyclooligomerization of butadiene [116b].

Other methods of synthesis are those related to reduction reactions. The reduction of *cis*-$Pt(PPh_3)_2Cl_2$ with hydrazine in the presence of the unsaturated ligand was applied by Chatt *et al.* [306] for the first time; $Pt(PPh_3)_2$(olefin) complexes (olefin = *trans*-stilbene, *trans*-4,4-dinitrostilbene, acenaphthylene) were obtained in this way. The reduction with hydrazine was supposed to lead to the unsaturated $Pt(PPh_3)_2$ intermediate, which then reacts with the alkene. However, more likely is the suggestion that the dehydrodiimideplatinum(II) complex, isolated from the reduction of *cis*-$Pt(PPh_3)_2Cl_2$ and hydrazine, is the reactive intermediate in these reactions (2.2.1.e) [176]. Obviously the most serious limitation to this method of synthesis is that its use is precluded with alkenes bearing substituents sensitive to hydrazine.

Other reducing agents usefully applied to the synthesis of phosphine–alkene derivatives of palladium, and particularly of nickel, are the aluminium alkyls. For example by reduction of $Ni(acac)_2$, with $Al(OEt)Et_2$ in the presence of triphenylphosphine, $Ni(PPh_3)_2(C_2H_4)$ has been isolated [103]. The source of ethylene is the aluminium alkyl: at first this had been overlooked and this compound was thought to be $Ni(PPh_3)_2$. By conducting the reaction with a saturated solution of ethylene, the yields of the reaction are increased from 75% to 90–95% [103]. Similarly $Pd(PR_3)_2(C_2H_4)$ complexes have been synthesized [68]. The use of aluminium alkyls, which are rather difficult to handle, puts a serious limitation on this method of synthesis, although in some cases no other method is available.

Very recently the use of lithium metal as reducing agent has been reported with $Ni(PR_3)_2Cl_2$ (R = Ph, Bu) in the presence of *trans*-stilbene, for the synthesis of $Ni(Pr_3)_2$(*trans*-stilbene) [273]. We briefly mention here that in few cases a hydrido-platinum(II) complex such as $Pt(PR_3)_2HX$ (X = Cl, Br, etc.) can be reduced to the platinum(0) derivative $Pt(PR_3)_2$(alkene), with displacement of HX simply by reaction with the alkene. However this reaction seems to be limited to very strong π-acids such as tetracyanoethylene [208, 307].

In the Tables 2.V, 2.VI and 2.VII the known $M(PR_3)_2$(alkene) complexes (M = Ni, Pd, Pt) are reported together with their method of synthesis. Also, related complexes such as the derivatives of *para*-quinones, which can be considered as 1,4-substituted dienes, allenes, etc., which are obtained by routes analogous to those above considered, are reported.

Among the most recently isolated compounds we note the platinum derivative of hexafluorobicyclo[2,2,0]hexa-2,5-diene ("Dewar hexafluorobenzene") [308, 309] obtained from $Pt(PPh_3)_4$ and the fluorodiene, in which only one double bond is coordinated to the metal.

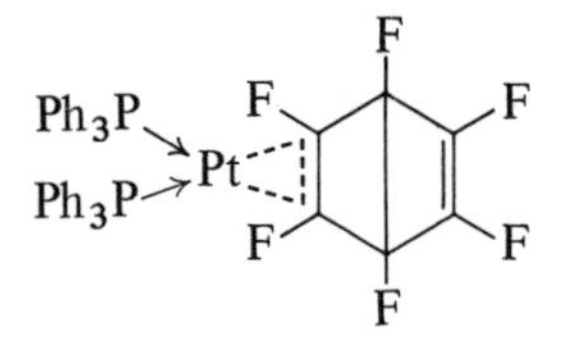

The nickel complex $Ni(PCy_3)(C_2H_4)_2$, obtained from $Ni(PCy_3)$(CDT) and ethylene at −20°C, represents one of the few examples where two alkene molecules are coordinated to a M–L moiety (M = Ni, Pd, Pt; L = phosphine) [310] and it can be considered as a model for the mechanism of oligomerization of olefins catalysed by metal complexes. From this point of view, the alkene complex (A) appears to be very important. This has been shown by ^1H n.m.r.

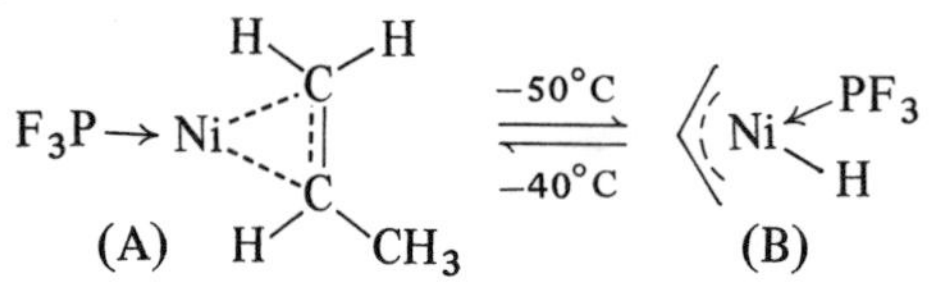

spectroscopy to be in equilibrium with the π-allylhydrido nickel complex (B) [311]. This is the first direct evidence for the so-called "π-allyl mechanism" of olefin isomerization at a transition metal.

A derivative of acraldehydeimine, $CH_2{=}CH{-}CH{=}NH$, a ligand comparable to butadiene, has been obtained through the following route [312]:

$$CH_2{=}CHCN + AlH(2\text{-butyl})_2 \rightarrow CH_2{=}CH{-}CH{=}NAl(2\text{-butyl})_2 \xrightarrow{Ni(PPh_3)_2(C_2H_4)}$$

CH–CH
CH_2 NAl(2-butyl)$_2$ $\xrightarrow{\text{acac}}$ CH_2 NH
Ni Ni
Ph_3P PPh_3 Ph_3P PPh_3

The reaction of the final nickel complex with triphenylphosphite allows the isolation of free acraldehydeimine, a compound which is difficult to purify when the uncomplexed N-aluminium substituted acraldehydeimine is directly reacted with acetylacetone.

Very recently some unstable organic molecules have been prepared *in situ*, and trapped by coordination on platinum. By this way thermally unstable cyclic allenes such as

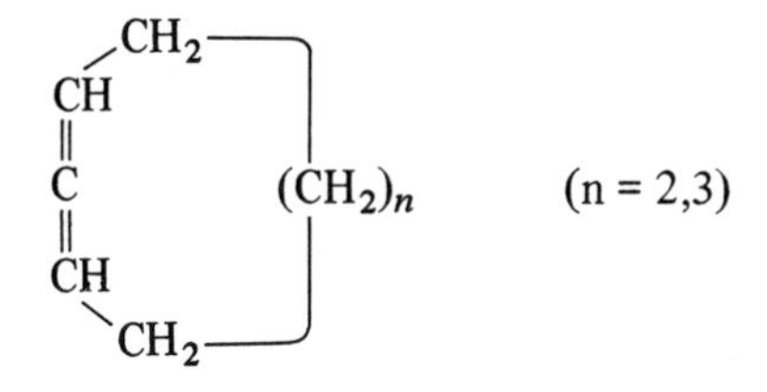

prepared *in situ* in the presence of $Pt(PPh_3)_2(C_2H_4)$ give the $Pt(PPh_3)_2$(allene) complexes [313].

(*b*) *Collateral reactions.* A metallacyclobutenone complex has been shown to be the product of the reaction between $Pt(PPh_3)_4$ and

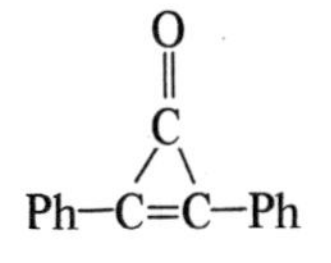

in dichloromethane [342].

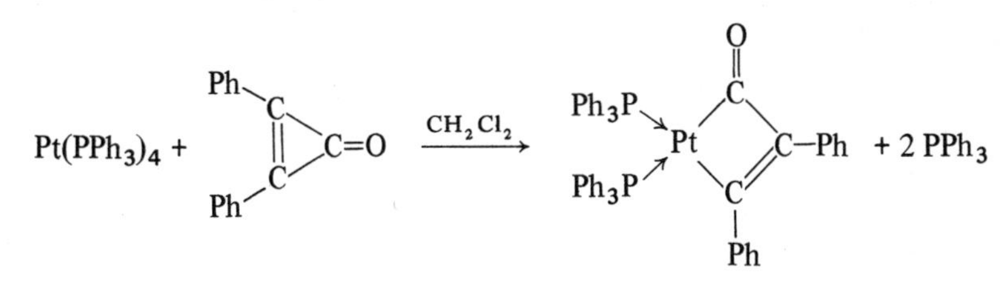

This derivative shows ν(C=C) at 1652 cm^{-1}, and its X-ray structure corresponds to that of a distorted square planar platinum(II) derivative, with normal values for the bond distances in the ring. Several mechanisms have been suggested for the hydrocarboxylation of alkynes to form $\alpha\beta$-unsaturated acids, and in one case the intermediacy of a metallacyclobutenone derivative has been invoked [342].

Another interesting feature of this compound is that it arises from an oxidative addition reaction where a single carbon–carbon bond has been broken. Reaction of methylcyclopropenone with $Pd(PPh_3)_2(C_2H_4)$ at −65°C proceeds by exchange with ethylene to yield the expected alkene complex. However in solution at −30°C the platinum atom inserts into a carbon–carbon single bond to yield the metallacyclobutenone complex [343]. Dimethylcyclopropenone gives the insertion product directly. The coordination and insertion reactions described above were not observed with $Pd(PPh_3)_4$ [343].

We shall also consider here some side reactions which can occur when a fluoro-olefin reacts with a zerovalent nickel complex, according to one of the methods of synthesis previously considered. The reaction between $Ni(PPh_3)_2(C_2H_4)$ and $CF_2{=}CFH$ gives a nickel(II) derivative having a five membered C_4Ni ring [205], probably as a mixture of isomers.

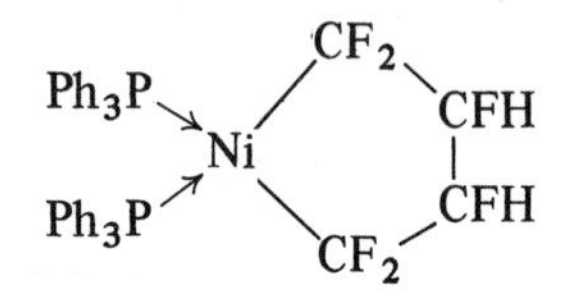

Related octafluoronickelacyclopentane complexes, $NiL_2(CF_2)_4$ [L = PEt_3, PBu^n_3, $PMePh_2$, $P(OMe)_3$, $\frac{1}{2}C_2H_4(PPh_2)_2$] have been obtained from the reaction of C_2F_4 with NiL_2(1,5-COD) (L = PEt_3, PBu^n_3), NiL_4 [L = PEt_3, $PMePh_2$, $P(OMe)_3$], and $Ni(PBu^n_3)_2\{C_2H_4(PPh_3)_2\}$ from which tributylphosphine is selectively displaced [50]. The corresponding triphenylphosphine derivative has been prepared by the following route:

$$Ni(PBu^n_3)_2(CF_2)_4 + 1{,}5\text{-COD} \xrightarrow{ZnBr_2} Ni(1{,}5\text{-COD})(CF_2)_4 \xrightarrow{PPh_3} Ni(PPh_3)_2(CF_2)_4$$

where $ZnBr_2$ acts as a scavenger for PBu^n_3 [50].

Similarly $Ni(AsMe_2Ph)_4$ and $Ni\{o\text{-}C_6H_4(AsMe_2)_2\}_2$ react with C_2F_4 and C_2HF_3 with formation of the corresponding five membered ring derivatives [206].

Compounds such as $NiL_2(CF_2{=}CF_2)$ are probably the precursors of the nickelacyclopentane derivatives, and in fact $Ni(PPh_3)_2(C_2F_4)$ reacts with C_2F_4 or C_2F_3H giving $Ni(PPh_3)_2(CF_2)_4$ and $Ni(PPh_3)_2(C_4F_7H)$ respectively [206 *bis*].

Hexafluorobuta-1,3-diene, which reacts with $Pt(PPh_3)_4$ as a mono-alkene forming $Pt(PPh_3)_2$ (CF_2=CF–CF=CF_2, η²-bonded through CF_2=CF) [203], gives a hexafluoronickelacyclopent-3-ene derivative with $Ni(AsMe_2Ph)_4$ [206]:

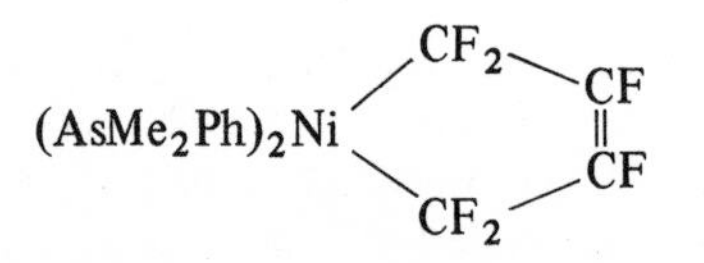

The reaction probably proceeds through an anionic intermediate:

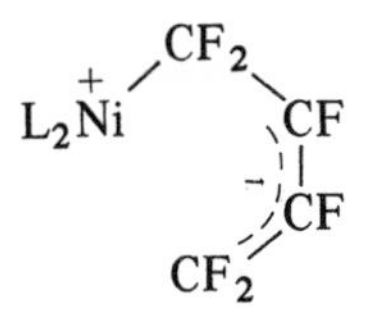

and this reaction should thus correspond to a new type of oxidative addition reaction.

(*c*) *X-ray structures.* As a result of X-ray structure determinations on $Ni(PPh_3)_2(C_2H_4)$ [344] and $Pt(PPh_3)_2(C_2H_4)$ [344b] the plane containing the metal and the two phosphorus atoms and the plane containing the metal and the two carbon atoms form a small, but definite dihedral angle (θ = 5.0°, M = Ni; θ = 1.3°, M = Pt) [344].

Free ethylene has a carbon–carbon bond distance of 1.344 Å. This distance is lengthened in the complexes to 1.46 Å (M = Ni) and 1.43 Å (M = Pt) [344]. The lengthening led to the suggestion that a significant π-back-donation had taken place in spite of the high energy of the π^* orbitals of alkenes such as ethylene, which does not have electron withdrawing substituents. However, it is rather improper to discuss these variations in terms of electronic effects, owing to the difficulty of determining accurately these molecular parameters. Moreover a high degree of σ-character in the metal–olefin bonding should also contribute to the lengthening of the carbon–carbon double bond.

In a concurrent X-ray structure of $Ni(PPh_3)_2(C_2H_4)$ [345] a carbon–carbon bond distance of 1.41 Å was determined and it was found that the ethylene carbon atoms are approximately coplanar with nickel and the two phosphorus atoms as expected for the metal in the dsp^2 hybridization state. Mean values of metal–carbon and metal–phosphorus bond distances were found to be: d_{Ni-C} = 1.99, d_{Pt-C} = 2.11, d_{Ni-P} = 2.15, d_{Pt-P} = 2.27 Å [344].

A distorted, square planar co-ordination geometry has also been found in $Pt(PPh_3)_2\{(CN)_2C{=}C(CN)_2\}$ [346]. (See Fig. 2.3 on page 150)

The planes determined by P(1)PtP(2) and C(1)PtC(2) form a dihedral angle θ of 8.3°. The $C_1{-}C_2$ bond distance of 1.49(5) Å is 0.18 Å longer than the corresponding distance in the free ligand and very close to that of single bond. On considering that the average of the C–C distances in a number of metal–olefin complexes examined by X-ray methods is about 0.1 Å longer than in the uncomplexed olefin [346a] one comes to the conclusion that a very strong π-back-donation is present in the tetracyanoethylene complex, in accord with the effect foreseen for the presence of the four cyano groups. The C(1)PtC(2) angle is very small [41.5° (1.3)] so that a "platinacyclopropane" structure would have a considerable strain [346a]. Other significant molecular parameters are: $d_{Pt-C(1)}$ = 2.12(3), $d_{Pt-C(2)}$ = 2.10(3), $d_{Pt-P(1)}$ = 2.291(9), $d_{Pt-P(2)}$ = 2.288(8) Å; P(1)PtP(2) = 101.4° (0.3).

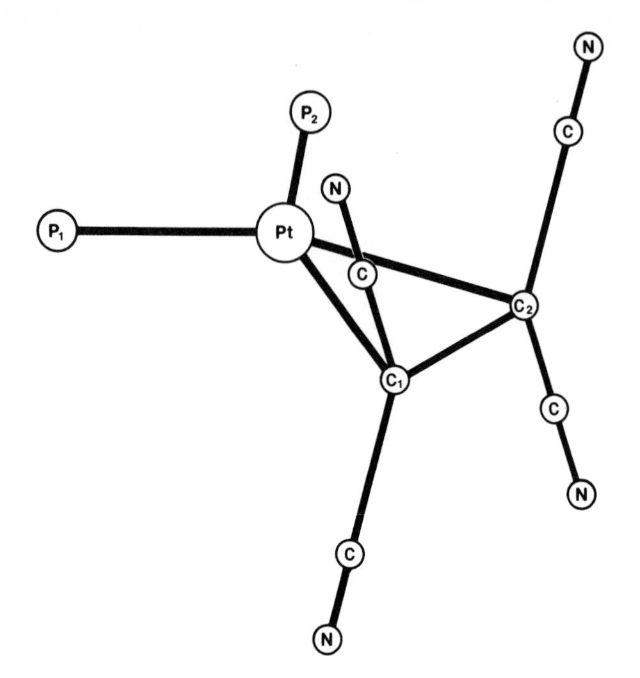

Fig. 2.3.

The essential geometry of the coordinated carbon atoms is pyramidal.

Now, there is some evidence that an excited state of the ethylene molecule has a pyramidal configuration, in accordance with the geometry found in $Pt(PPh_3)_2\{(CN)_2C{=}C(CN)_2\}$ and interpreted in terms of participation of the excited state of the alkene in the bonding (1.3.1.2).

However in the X-ray structure of the triphenylphosphineplatinum derivative of 1,2-dimethylcyclopropene:

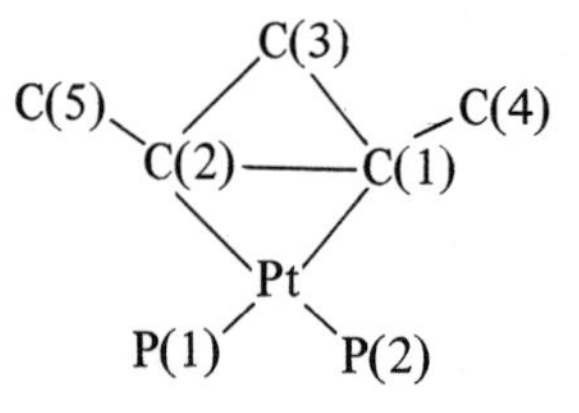

that is an alkene which should exhibit only a very small tendency to behave as a π-acid, a strong perturbation of the alkene upon coordination has been found [340], although the cyclopropene ring remains intact on complexing, and the metal has not inserted into the ring.

The double bond is lengthened from 1.30 to 1.50 Å, a distance very close to that found in the tetracyanoethylene complex. The methyl groups are bent out of the plane of the ring, away from the metal. P(1), P(2), C(1) and C(2) lie in a distorted square planar conformation about platinum, the angle between the cyclopropene ring and the plane defined by Pt, P(1) and P(2) being 116°(1). C(3) is 2.83 Å away from platinum. Other molecular parameters are: $d_{Pt-P(1)}$ = 2.28, $d_{Pt-P(2)}$ = 2.29, $d_{Pt-C(1)}$ = 2.12, $d_{Pt-C(2)}$ = 2.11, $d_{C(1)-C(4)}$ = $d_{C(2)-C(5)}$ = 1.54, $d_{C(1)-C(3)}$ = 1.54, $d_{C(2)-C(3)}$ = 1.55 Å; C(2)PtC(1) = 41.6, C(3)C(1)C(4) = 126, C(3)C(2)C(5) = 128°.

A long olefinic bond (1.53 ± 0.036 Å) in a distorted square planar configuration about platinum, has also been observed in $Pt(PPh_3)_2$(CNCH=CHCN) [346 *bis*]. Two non-equivalent Pt–C distances [2.05 and 2.16 Å] and the usual dihedral angle θ [5.2°] were determined. The angle between the best lines containing the two nitrile groups is 147°, while the C(1)PtC(2) angle is 42.6°.

Also for the series of platinum complexes, $Pt(PPh_3)_2$(alkene) (alkene = $Cl_2C{=}CCl_2$, $Cl_2C{=}CF_2$, $ClFC{=}CF_2$), the X-ray structure determinations have shown that the olefin has lost its planar geometry, the substituents being bent back from the central metal atom [347].

The tetrachloroethylene complex has a mean platinum-carbon distance of 2.03(3) Å, one of the shortest distances ever observed in this type of complex. The chlorine atoms are pushed back from the olefinic carbon atoms more than the cyano groups in the corresponding tetracyanoethylene derivative. The olefinic carbon-carbon distance in the complex is 1.62(3) Å, with very marked lengthening of this bond on complexation. However, some difficulties have been found on refining the anisotropic thermal parameters for the olefinic carbons. Other molecular parameters are: PPtP = 100.6(2) and CPtC = 47.1° (1.0); the planes defined by PPtP and CPtC are twisted by 12.3° (1.5). The platinum-phosphorus bond distances are comparable with the distances determined in similar compounds. In $Pt(PPh_3)_2(Cl_2C{=}CF_2)$ a rotational disorder about the carbon-carbon bond has been found [347]. Assuming that a dipolar intermediate is involved in the formation reaction:

$$Pt(PPh_3)_n + X_2C{=}CY_2 \longrightarrow (Ph_3P)_2\overset{+}{Pt}{-}CX_2{-}\overset{-}{C}Y_2 \longrightarrow (Ph_3P)_2Pt\langle CX_2{-}CY_2 \rangle$$

in the case of $Cl_2C{=}CF_2$ the negative charge should be localized on the CF_2 moiety (Y = F). Such an intermediate would then exist for a sufficient time to allow the observed rotation about the C–C bond before ring closure [347]. The platinum-CCl_2 bond distance [2.05(3) Å] is again very short, even shorter than the platinum-CF_2 bond. The PPtP angle is 103.4°(2), while the platinum-phosphorus bond distances are 2.303(6) and 2.314(5) Å. For $Pt(PPh_3)_2\{Cl_2C{=}C(CN)_2\}$:

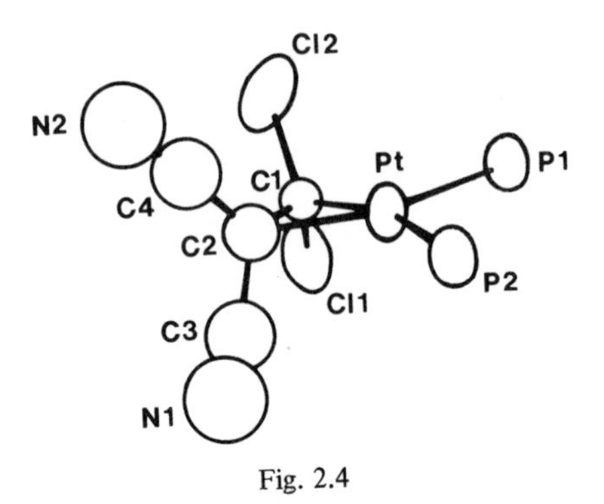

Fig. 2.4

d_{Pt-CCl_2} [2.00(2) Å] is significantly shorter than $d_{Pt-C(CN)_2}$ [2.10(2) Å] [338]. The C(1)–C(2) distance is 1.42(3) Å, shorter than the same distance in the tetrachloroethylene and tetracyanoethylene complexes; however, high standard deviations were observed. The chlorine atoms are bent back from the metal twice as much as the cyano groups. This could be due to the π interaction between the p_π orbitals of the carbon atom and the cyano group [338]. Surprisingly, C(1), C(2), Pt, P(1) and P(2) lie in the same plane. Other molecular parameters are: $d_{Pt-P(1)}$ = 2.260(6), $d_{Pt-P(2)}$ = 2.339(6) Å; P(1)PtP(2) = 102.0(2) and C(1)PtC(2) = 40.6°(9). The longer Pt–P(2) distance in respect to Pt–P(1) could be due to a *σ-trans* influence being P(2) *trans* to C(1), that is to the carbon with the shorter distance to platinum [338].

We can now compare the most significant molecular parameters observed in the structures considered above. The data are summarized in Table 2.VIII. The lengthening of the C(1)–C(2) bond does not seem to follow any regular variation in function of the nature of the substituents of the alkene, neither does it show any correlation with the metal–carbon bond distances. Moreover in $Pt(PPh_3)_2\{Cl_2C{=}C(CN)_2\}$ the shorter bond is that between Pt and CCl_2, while one would expect a shorter distance with the carbon having the more electron-withdrawing cyano groups as substituents.

The same anomalous result is observed when these distances in $Pt(PPh_3)_2(C_2Cl_4)$ and $Pt(PPh_3)_2\{C_2(CN)_4\}$ are compared.

Ibers *et al*. [338] considered that the π and π^* orbitals of tetracyanoethylene have such low energies that σ-donation to the metal is of very little importance in practice. On the other hand the amount of π-back donation from the metal has a limiting value, which is not reached with a very strong π-acid such as tetracyanoethylene, particularly because of the lack of the synergic reinforcement due to σ-donation. In this way the very poor σ-donation is not fully compensated by the π-back-bonding. On the other hand with tetrachloroethylene, which is a better σ-donor but poorer π-acceptor than tetracyanoethylene, the π-back-donation from the metal more than compensates the higher, though still scarce, σ-donation of the alkene bearing four electronegative substituent groups such as chlorine. The long metal–carbon bond distances and the relatively small lengthening of the C(1)–C(2) bond upon complexation for the ethylene derivatives, are explained by the poor π-acidity of ethylene which in this case is not compensated by a strong enough σ-donation to the metal.

Finally the lengthening of the C(1)–C(2) bond in the platinum–cyclopropene derivative is quite unexpected from the above considerations and could be solely ascribed to the necessity of relieving, at the olefinic carbon atoms, the strain in the cyclopropene ring.

The X-ray structure of the related allene complex has been recently reported [348]:

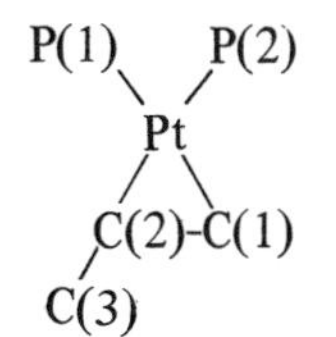

The coordination around platinum is approximately square planar with the allene coordinated through only one double bond. For the related tetramethylallene derivative of platinum(II), $[Cl_2Pt(C_7H_{12}) . CCl_4]_2$, the ligand acts as a η-bonded monolefin, but the double bond is approximately orthogonal to the plane.

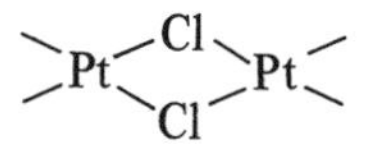

of the complex [349].

The situation recalls the behaviour of the simple alkene derivatives of platinum(0) and platinum(II). Naturally, the C(1)–C(2) bond distance in the complex [1.48(5) Å] was found longer [1.31(5) Å] than that in the uncoordinated C(2)–C(3) double bond. The bonded allene is no longer linear, the C(1)C(2)C(3) angle being 142.3°. The planes P(1)PtP(2) and C(1)PtC(2) from the usual dihedral angle (9°). Other molecular parameters are: $d_{Pt-C(1)}$ = 2.13(3), $d_{Pt-C(2)}$ = 2.03(3), $d_{Pt-P(1)}$ = 2.278(9), $d_{Pt-P(2)}$ = 2.286(9) Å; P(1)PtP(2) = 107.1 (0.3), C(1)PtC(2) = 41°(2).

The X-ray structure of the very interesting diethylene nickel derivative $Ni(PCy_3)(C_2H_4)_2$, shows an overall planarity of the bis(diethylene)(heavy atom) framework [350]:

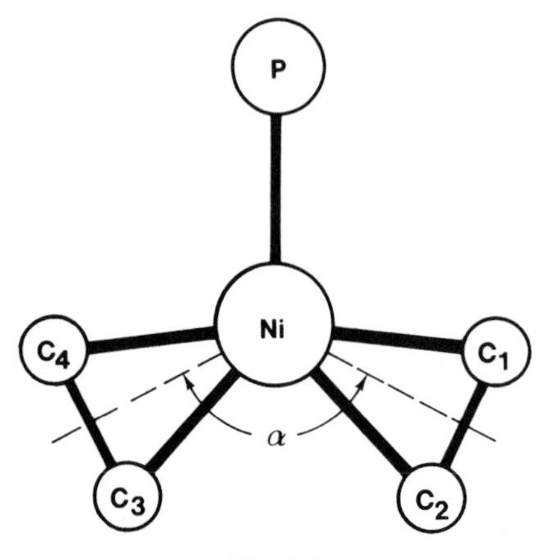

Fig. 2.5.

The nickel has a slightly distorted trigonal environment. Steric repulsion between the olefinic moieties may be the cause of this distortion; also, the P–C(1), 3.146 Å and P–C(4), 3.123 Å distances are shorter than the sum of the van der Waals radii (1.9 Å + 1.7 Å). The carbon–carbon bond distances are: $d_{C(1)-C(2)} = 1.387$ and $d_{C(3)-C(4)} = 1.415$ Å. Other molecular parameters are: $d_{Ni-P} = 2.196$, $d_{Ni-C(1)} = 2.009$, $d_{Ni-C(2)} = 2.042$, $d_{Ni-C(3)} = 1.987$, $d_{Ni-C(4)} = 2.017$ Å; C(1)NiC(2) = 40.0, C(3)NiC(4) = 41.4, $\alpha = 126.9°$.

In our opinion this compound represents further evidence that the metal–alkene bond should be interpreted, for complexes in low oxidation states, along the lines of the Dewar–Chatt–Duncanson model, while an interpretation in terms of a rigid cyclopropane ring is not satisfactory.

(*d*) *Spectral properties.* Very few carbon–carbon and metal–carbon stretching frequencies have been assigned for the alkene derivatives of nickel, palladium and platinum, owing to the difficulty of analysing the region of the spectrum where ligands such as triphenylphosphine absorb.

In practice only for some ethylene derivative an absorption at around 1485 cm^{-1} (Tables 2.V and 2.VI) has been attributed to ν(C=C). Free ethylene absorbs at 1623 cm^{-1} (Raman) with a lowering of about 140 cm^{-1} of ν(C=C) in the complexes, in accord with what has been already discussed (1.3.1.2).

For $Pd(PPh_3)_2(C_2H_4)$ and $Pd(PCy_3)_2(C_2H_4)$ two absorptions at 388 and 350 cm^{-1} respectively have been assigned to the palladium–carbon stretching frequencies [68].

TABLE 2.V

Nickel(0) complexes with alkenes and related ligands

Compound	Method of synthesis	Infrared absorptions (cm^{-1})	References
$Ni(PPh_3)_2(C_2H_4)$	$Ni(acac)_2 + Al(OEt)Et_2 + PPh_3 + C_2H_4$		103
	$Ni(acac)_2 + AlEt_3 + PPh_3 + C_2H_4$		314
$Ni(PEt_3)_2(C_2H_4)$	$Ni(acac)_2 + Al(OEt)Et_2 + PEt_3 + C_2H_4$		103
$Ni(PCy_3)_2(C_2H_4)$	$Ni(acac)_2 + Al(OEt)Et_2 + PCy_3 + C_2H_4$		103
	$\{Ni(PCy_3)_2\}_2N_2 + C_2H_4$		315
$Ni\{P(NC_5H_{10})_3\}_2(C_2H_4)$	$Ni(acac)_2 + Al(OEt)Et_2 + P(NC_5H_{10})_3 + C_2H_4$		103
$Ni\{P(NC_4H_8O)_3\}_2(C_2H_4)$	$Ni(acac)_2 + Al(OEt)Et_2 + P(NC_4H_8O)_3 + C_2H_4$		103
$Ni\{P(O\text{-}o\text{-}C_6H_4Me)_3\}_2(C_2H_4)$	$Ni(acac)_2 + AlEt_3 + P(O\text{-}o\text{-}C_6H_4Me)_3 + C_2H_4$	$\nu(C{=}C)$, 1487	54
	$Ni\{P(O\text{-}o\text{-}C_6H_4Me)_3\}_3 + C_2H_4$		71, 316
$Ni\{P(O\text{-}o\text{-}C_6H_4OMe)_3\}_2(C_2H_4)$	$Ni(acac)_2 + Al(OEt)Et_2 + P(O\text{-}o\text{-}C_6H_4OMe)_3 + C_2H_4$		103
$Ni\{P(O\text{-}o\text{-}C_6H_4 . C_6H_5)_3\}_2(C_2H_4)$	$Ni(acac)_2 + Al(OEt)Et_2 + P(O\text{-}o\text{-}C_6H_4 . C_6H_5)_3 + C_2H_4$		103
$Ni(PPh_3)_2(C_2H_3F)$	$Ni(PPh_3)_2(C_2H_4) + C_2H_3F$		205
$Ni\{P(O\text{-}o\text{-}C_6H_4Me)_3\}_2(C_2H_3CN)$	$Ni(CH_2{=}CHCN)_2 + P(O\text{-}o\text{-}C_6H_4Me)_3$		317, 317 *bis*
$Ni(PCy_3)_2(C_2H_5CH{=}CH_2)$	$Ni(acac)_2 + AlBu^i_3 + PCy_3 +$ *cis* or *trans*-$CH_3CH{=}CHCH_3$		318
$Ni(PPh_3)_2(PhCH{=}CH_2)$	$Ni(PPh_3)_2(C_2H_4) + PhCH{=}CH_2$		103
$Ni(PPh_3)_2(CH_2{=}CF_2)$	$Ni(PPh_3)_2(C_2H_4) + CH_2{=}CF_2$		205
$Ni\{P(O\text{-}o\text{-}C_6H_4 . C_6H_5)_3\}_2(CH_2{=}C(CH_2)_2)$	$Ni\{P(O\text{-}o\text{-}C_6H_4 . C_6H_5)_3\}_2 + CH_2{=}C(CH_2)_2$		105
$Ni(PPh_3)_2(PhMeC{=}CH_2)$	$Ni(PPh_3)_2(C_2H_4) + PhMeC{=}CH_2$		103
$Ni(PCy_3)_2(cis\text{-}CH_3CH{=}CHCH_3)$	$Ni(PCy_3)_2 + cis\text{-}CH_3CH{=}CHCH_3$		69b
$Ni(PCy_3)_2(trans\text{-}CH_3CH{=}CHCH_3)$	$Ni(PCy_3)_2 + trans\text{-}CH_3CH{=}CHCH_3$		69b
$Ni(PCy_3)_2(C_6H_{10})$	$Ni(PCy_3)_2 + C_6H_{10}$		69b

TABLE 2.V *continued*

Compound	Method of synthesis	Infrared absorptions (cm^{-1})	References
$Ni(PPh_3)_2$(*trans*-PhCH=CHPh)	$Ni(PPh_3)_2Cl_2$ + Li + *trans*-PhCH=CHPh		273
	$Ni(PPh_3)_2(C_2H_4)$ + *trans*-PhCH=CHPh		103
$Ni(PBut_3)_2$(*trans*-PhCH=CHPh)	$Ni(PBut_3)_2Cl_2$ + Li + *trans*-PhCH=CHPh		273
$Ni(PPh_3)_2$(*cis*-MeOOCCH=CHCOOMe)	$Ni(PPh_3)$(*cis*-MeOOCCH=CHCOOMe) (Br) $\xrightarrow[20^\circ]{CH_3OH}$ disproportionation reaction	ν(C=C), 1690	319
$Ni(PPh_3)_2$(*trans*-MeOOCCH=CHCOOMe)	{Ni(*trans*-MeOOCCH=CHCOOMe)Br}$_n$ + PPh_3 $\longrightarrow$ disproportionation reaction	ν(C=C), 1666	319
	$Ni(PPh_3)_2$(*cis*-MeOOCCH=CHCOOMe) + *trans*-MeOOCCH=CHCOOMe		319
	$Ni(PPh_3)_3Br$ + *cis*-MeOOCCH=CHCOOMe → isomerization and disproportionation reaction		319
$Ni(PBut^n_3)_2\{(CN)_2C{=}C(CN)Ph\}$	$[Ni\{(CN)_2C{=}C(CN)(Ph)\}]_n + PBut^n_3$		320
$Ni\{C_3H_6(PPh_2)_2\}\{(CN)_2C{=}C(CN)_2\}$	$Ni\{C_3H_6(PPh_2)_2\}_2 + (CN)_2C{=}C(CN)_2$		101
$Ni\{C_4H_8(PPh_3)_2\}\{(CN)_2C{=}C(CN)_2\}$	$Ni\{C_4H_8(PPh_2)_2\}_2 + (CN)_2C{=}C(CN)_2$	ν(C≡N), 2220	101
$Ni(PPh_3)_2(C_2F_4)$	$Ni(CDT)(C_2F_4) + PPh_3$		206 *bis*
$Ni(PPh_3)_3(C_2F_4)$	$Ni(PPh_3)_2(C_2H_4) + C_2F_4$		302
$Ni(PPh_3)_2(CF_3CF{=}CF_2)$	$Ni(PPh_3)_2(C_2H_4) + CF_3CF{=}CF_2$		205
	$Ni(1,5\text{-}COD)(CF_3CF{=}CF_2) + PPh_3$		50
	$Ni(PPh_3)_2 + CF_3CF{=}CF_2$		57
$Ni(PEt_3)_2(CF_3CF{=}CF_2)$	$Ni(PEt_3)_2(1,5\text{-}COD) + CF_3CF{=}CF_2$		50
	$Ni(1,5\text{-}COD)(CF_3CF{=}CF_2) + PEt_3$		50
$Ni(PPh_3)_2$ (CF—CF_2 ‖ | CF—CF_2)	$Ni(PPh_3)_2(C_2H_4)$ + CF—CF_2 ‖ | CF—CF_2		205
$Ni(PPh_3)_2(C_2H_3CN)_2$	$Ni(CH_2{=}CHCN)_2 + PPh_3$	ν(C≡N), 2167	321
$Ni(PPh_3)_2(C_2H_3CHO)_2$	$Ni(CH_2{=}CHCHO)_2 + PPh_3$		321
$Ni(PPh_3)(C_2H_4)_2$	$Ni(PPh_3)(CDT) + C_2H_4$		310
$Ni(PCy_3)(C_2H_4)_2$	$Ni(PCy_3)(CDT) + C_2H_4$	ν(C=C), 1490	310
$Ni(PPh_3)(C_2H_3CN)_2$	$Ni(CH_2{=}CHCN)_2 + PPh_3$	ν(C≡N), 2180	305, 322
	$Ni(PPh_3)(CO)_3 + CH_2{=}CHCN$		322

$Ni(PPh_3)(C_2H_3CHO)_2$	$Ni(C_2H_3CHO)_2 + PPh_3$		305
$Ni(PCy_3)(CH_3CH{=}CH_2)_2$	$Ni(PCy_3)(CDT) + CH_3CH{=}CH_2$	$\nu(C{=}C)$, 1485	310
$Ni(PCy_3)(\text{CH=CH}_2 \ldots)_2$	$Ni(PCy_3)(CDT) + $ (CH=CH$_2$ …)		310
$Ni(PPh_3)(\ldots)$	$Ni(PPh_3)(CDT) + \ldots$		310
$Ni(PPh_3)(CH_2{=}CH{-}CH{=}CH_2)_2$	$Ni(PPh_3)(CDT) + CH_2{=}CH{-}CH{=}CH_2$		310
$Ni\{P(O\text{-}o\text{-}C_6H_4 . C_6H_5)_3\}_2(CH_2{=}CH{-}CH{=}CH_2)$	$Ni\{P(O\text{-}o\text{-}C_6H_4 . C_6H_5)_3\}_2 + CH_2{=}CH{-}CH{=}CH_2$		105
$Ni\{P(O\text{-}o\text{-}C_6H_4 . C_6H_5)_3\}_2(\ldots)$	$Ni\{P(O\text{-}o\text{-}C_6H_4 . C_6H_5)_3\}_2 + \ldots$		105
$Ni(PPh_3)_2(\ldots)$	$Ni(acac)_2 + AlR_3 + PPh_3 + 1,5\text{-COD}$		98
$Ni(PBut^n_3)_2(\ldots)$ (O, H_3C, CH_3, H_3C, CH_3, O)	$Ni(CO)_4 + Dq + PBut^n_3$		300d
	$Ni(1,5\text{-COD})(Dq) + PBut^n_3$	$\nu(C{=}O)$, 1540	51
	$Ni(COT)(Dq) + PBut^n_3$		51
$Ni(PCy_3)_2(CDT)$	$\{Ni(PCy_3)_2\}_2N_2 + CDT$		69b
$Ni\{P(O\text{-}o\text{-}C_6H_4 . C_6H_5)_3\}$(*cis, cis,cis*-CDT)	Ni(*cis,cis,cis*-CDT) + $P(O\text{-}o\text{-}C_6H_4 . C_6H_5)_3$		324
$Ni\{P(O\text{-}o\text{-}C_6H_5 . C_6H_5)_3\}$(*trans,trans,trans*-CDT)	Ni(*trans,trans,trans*-CDT) + $P(O\text{-}o\text{-}C_6H_4 . C_6H_5)_3$		116b
$Ni(PPh_3)_2(CH_2{=}C{=}CH_2)$[a]	$Ni(PPh_3)_2 + CH_2{=}C{=}CH_2$		57
	$Ni(PPh_3)_2Br_2 + NaHg_x + CH_2{=}C{=}CH_2$		57
$Ni\{P(O\text{-}o\text{-}C_6H_4 . C_6H_5)_3\}_2$ (1,2–cyclononadiene)	$Ni\{P(O\text{-}o\text{-}C_6H_4 . C_6H_5)_3\}_2 + \ldots$		326

[a] This compound could be a derivative of the allene trimer C_9H_{12}, of formula $Ni(PPh_3)(C_9H_{12})$ (see ref. 325).

TABLE 2.VI

Palladium(0) complexes with alkenes and related ligands

Compound	Method of synthesis	Infrared absorptions (cm^{-1})	References
$Pd(PPh_3)_2(C_2H_4)$	$Pd(acac)_2 + Al(OEt)Et_2 + PPh_3 + C_2H_4$	$\nu(C{=}C)$, 1488 $\nu(Pd{-}C)$, 388	68
$Pd(PCy_3)_2(C_2H_4)$	$Pd(acac)_2 + Al(OEt)Et_2 + PCy_3 + C_2H_4$	$\nu(C{=}C)$, 1483 $\nu(Pd{-}C)$, 350	68
$Pd\{P(O\text{-}o\text{-}C_6H_4Me)_3\}_2(C_2H_4)$	$Pd(acac)_2 + Al(OEt)Et_2 + P(O\text{-}o\text{-}C_6H_4Me)_3 + C_2H_4$		68
$Pd(PPh_3)_2$(*cis*-MeOOCCH=CHCOOMe)	$Pd(PPh_3)_4$ + *cis*-MeOOCCH=CHCOOMe		56
$Pd(PPh_3)_2$(*trans*-MeOOCCH=CHCOOMe)	$Pd(PPh_3)_2$ + *trans*-MeOOCCH=CHCOOMe		56
$Pd(PPh_3)_2$(*trans*-EtOOCCH=CHCOOEt)	$Pd(PPh_3)_4$ + *trans*-EtOOCCH=CHCOOEt		137
$Pd(PPh_3)_2\left(\begin{matrix}CH{-}CO\\ \Vert\\ CH{-}CO\end{matrix}\!\!>\!O\right)$	$Pd(PPh_3)_4 + \begin{matrix}CH{-}CO\\ \Vert\\ CH{-}CO\end{matrix}\!\!>\!O$		56, 137
$Pd\{P(OMe)_3\}_2\left(\begin{matrix}CH{-}CO\\ \Vert\\ CH{-}CO\end{matrix}\!\!>\!O\right)$	$Pd(dba)_2 + P(OMe)_3 + \begin{matrix}CH{-}CO\\ \Vert\\ CH{-}CO\end{matrix}\!\!>\!O$		304
$Pd\{P(OPh)_3\}_2\left(\begin{matrix}CH{-}CO\\ \Vert\\ CH{-}CO\end{matrix}\!\!>\!O\right)$	$Pd(dba)_2 + P(OPh)_3 + \begin{matrix}CH{-}CO\\ \Vert\\ CH{-}CO\end{matrix}\!\!>\!O$		304
$Pd\{C_2H_4(PPh_2)_2\}\left(\begin{matrix}CH{-}CO\\ \Vert\\ CH{-}CO\end{matrix}\!\!>\!O\right)$	$Pd\{C_2H_4(PPh_2)_2\}_2 + \begin{matrix}CH{-}CO\\ \Vert\\ CH{-}CO\end{matrix}\!\!>\!O$		56
	$Pd(PPh_3)_2\left(\begin{matrix}CH{-}CO\\ \Vert\\ CH{-}CO\end{matrix}\!\!>\!O\right) + C_2H_4(PPh_2)_2$		56
$Pd(PPh_3)_2\left(\begin{matrix}CH{-}CO\\ \Vert\\ CH{-}CO\end{matrix}\!\!>\!NH\right)$	$Pd(PPh_3)_4 + \left(\begin{matrix}CH{-}CO\\ \Vert\\ CH{-}CO\end{matrix}\!\!>\!NH\right)$		173
$Pd(PPh_3)_2$(CNCH=CHCN)	$Pd(PPh_3)_4$ + CNCH=CHCN	$\nu(C{\equiv}C)$, 2206	327

$Pd(PPh_3)_2(CH_3-C=CH-CO-O-CO)$ [ring] 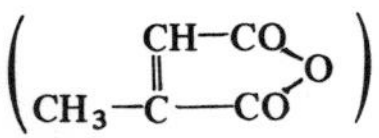	$Pd(PPh_3)_4 + CH_3-C=CH-CO-O-CO$ [ring] 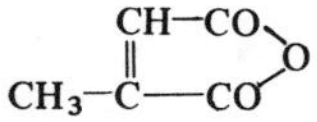		56
$Pd(PPh_3)_2(CF_3CF{=}CFCF_3)$	$Pd(PPh_3)_4 + CF_3CF{=}CFCF_3$		328
$Pd(PPh_3)_2\{(CN)_2C{=}C(CN)_2\}$	$Pd(PPh_3)_4 + (CN)_2C{=}C(CN)_2$	$\nu(C{\equiv}N)$, 2230	137
$Pd\{C_2H_4(PPh_2)_2\}\{(CN)_2C{=}C(CN)_2\}$	$Pd(PhNC)_2\{(CN)_2C{=}C(CN)_2\} + C_2H_4(PPh_2)_2$	$\nu(C{\equiv}N)$, 2221(sh), 2216(vs), 2208(vs)	329
$Pd\{P(OC_3H_7^i)_3\}_2\{(CN)_2C{=}C(CN)_2\}$	$Pd(PhNC)_2\{(CN)_2C{=}C(CN)_2\} + P(OC_3H_7^i)_3$	$\nu(C{\equiv}N)$, 2222(s), 2213(sh)	329
$Pd(PMePh)_2\{(CF_3)_2C{=}C(CN)_2\}$	$Pd\{P(OMe)_3\}_2\{(CF_3)_2C{=}C(CN)_2\} + PMePh_2$		120
$Pd(AsMe_2Ph)_2\{(CF_3)_2C{=}C(CN)_2\}$	$Pd(AsMe_2Ph)_4 + (CF_3)_2C{=}C(CN)_2$		120
$Pd\{P(OMe)_3\}_2\{(CF_3)_2C{=}C(CN)_2\}$	$Pd\{P(OMe)_3\}_4 + (CF_3)_2C{=}C(CN)_2$		120
$Pd(PEt_3)_2\{(CF_3)_2C{=}C(CN)_2\}$	$Pd\{P(OMe)_3\}_2\{(CF_3)_2C{=}C(CN)_2\} + PEt_3$		120
$Pd\{C_2H_4(PPh_2)_2\}\{(CF_3)_2C{=}C(CN)_2\}$	$Pd\{P(OMe)_3\}_2\{(CF_3)_2C{=}C(CN)_2\} + C_2H_4(PPh_2)_2$		120
$Pd(PPh_3)_2(R^1C{=}CR^2SO_2)$ [ring] (R^1 = H, R^2 = Me; R^1 = Me, R^2 = Me; R^1 = Ph, R^2 = Me; R^1 = Ph, R^2 = Ph) 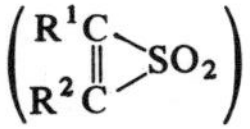	$Pd(PPh_3)_4 + R^1C{=}CR^2SO_2$ [ring] 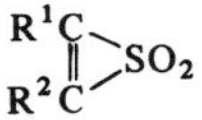		330
$Pd(PPh_3)_2(O{=}C_6H_4{=}O)$ [p-benzoquinone] 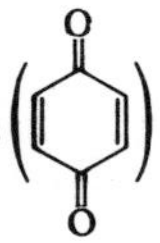	$Pd(PPh_3)_4 + O{=}C_6H_4{=}O$ [p-benzoquinone] 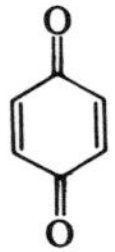		56
$Pd(PPh_3)_2(C_{10}H_6O_2)$ [1,4-naphthoquinone] 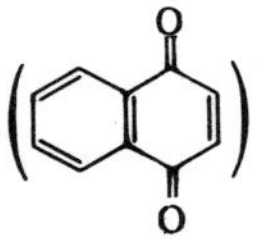	$Pd(PPh_3)_4 + C_{10}H_6O_2$ [1,4-naphthoquinone] 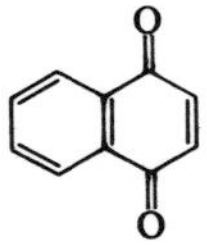		56

TABLE 2.VII

Platinum(0) complexes with alkenes and related ligands

Compound	Method of synthesis	Infrared absorptions (cm^{-1})	References
$Pt(PPh_3)_2(C_2H_4)$	$Pt(PPh_3)_2(O_2) + NaBH_4 + C_2H_4$		235
	$Pt(PPh_3)_2(CO_3) + C_2H_4 + h\nu$		135
	$Pt(PPh_3)_2 + C_2H_4$		67
$Pt(PPh_3)_2(CH_2{=}CHCN)$	$Pt(PPh_3)_3 + CH_2{=}CHCN$	$\nu(C{\equiv}C)$, 2195	301
$Pt(PPh_3)_2(CH_2{=}CClCN)$	$Pt(PPh_3)_4 + CH_2{=}CClCN$		331
$Pt(AsPh_3)_2(CH_2{=}CClCN)$	$Pt(AsPh_3)_4 + CH_2{=}CClCN$		331
$Pt(PPh_3)_2(CH_2{=}CHCOMe)$	$Pt(PPh_3)_3 + CH_2{=}CHCOMe$	$\nu(C{=}C)$, 1630	301
$Pt(PPh_3)_2$(*cis*-$CHCl{=}CHCl$)	$Pt(PPh_3)_2$(*trans*-$PhCH{=}CHPh$) + *cis*-$CHCl{=}CHCl$		332
	cis-$Pt(PPh_3)_2Cl_2$ + *cis*-$CHCl{=}CHCl + NH_2{-}NH_2$		332
$Pt(PPh_3)_2$(*trans*-$CHCl{=}CHCl$)	$Pt(PPh_3)_2$(*trans*-$PhCH{=}CHPh$) + *trans*-$CHCl{=}CHCl$		332
	cis-$Pt(PPh_3)_2Cl_2$ + *trans*-$CHCl{=}CHCl + NH_2{-}NH_2$		332
$Pt(PPh_3)_2(CHCN{=}CHCN)$	$Pt(PPh_3)_2(C_2H_4) + CNCH{=}CHCN$	$\nu(C{\equiv}N)$, 2213	303
$Pt(PPh_3)_2(CH_3CH{=}CHCHO)$	$Pt(PPh_3)_2(C_2H_4) + CH_3CH{=}CHCHO$	$\nu(C{=}O)$, 1630	301
$Pt(PPh_3)_2\left(\begin{array}{l}CH{-}CO\\ \|\\ CH{-}CO\end{array}\!\!>O\right)$	$Pt(PPh_3)_3 + \begin{array}{l}CH{-}CO\\ \|\\ CH{-}CO\end{array}\!\!>O$	$\nu(C{=}O)$, 1800–1725	301
	$Pt(PPh_3)_4 + \begin{array}{l}CH{-}CO\\ \|\\ CH{-}CO\end{array}\!\!>O$		56
$Pt(PPh_3)_2\left(\begin{array}{l}CH{-}CO\\ \|\\ CH{-}CO\end{array}\!\!>NH\right)$	$Pt(PPh_3)_4 + \begin{array}{l}CH{-}CO\\ \|\\ CH{-}CO\end{array}\!\!>NH$		173
$Pt(AsPh_3)_2\left(\begin{array}{l}CH{-}CO\\ \|\\ CH{-}CO\end{array}\!\!>NH\right)$	$Pt(AsPh_3)_4 + \begin{array}{l}CH{-}CO\\ \|\\ CH{-}CO\end{array}\!\!>NH$		173
$Pt(PPh_3)_2$(*trans*-$MeOOCCH{=}CHCOOMe$)	$Pt(PPh_3)_3$ + *trans*-$MeOOCCH{=}CHCOOMe$	$\nu(C{=}O)$, 1680	301
	$Pt(PPh_3)_4$ + *trans*-$MeOOCCH{=}CHCOOMe$		56
	$Pt(PPh_3)_2(PhC_2Ph)$ + *trans*-$MeOOCCH{=}CHCOOMe$		56
$Pt(PPh_3)_2(PhCH{=}CHCHO)$	$Pt(PPh_3)_3 + PhCH{=}CHCHO$	$\nu(C{=}O)$, 1630	301
$Pt(PPh_3)_2$(*trans*-$PhCH{=}CHCOOEt$)	$Pt(PPh_3)_2(C_2H_4)$ + *trans*-$PhCH{=}CHCOOEt$	$\nu(C{=}O)$, 1670	301
$Pt(PPh_3)_2$(*trans*-$PhCH{=}CHPh$)	*cis*-$Pt(PPh_3)_2Cl_2 + NH_2{-}NH_2$ + *trans*-$PhCH{=}CHPh$		306

$Pt(PPh_3)_2(\textit{trans-p}\text{-}NO_2C_6H_4CH{=}CHC_6H_4NO_2\text{-}p)$	*cis*-$Pt(PPh_3)_2Cl_2 + NH_2{-}NH_2 + \textit{trans-p}\text{-}NO_2C_6H_4CH{=}CHC_6H_4NO_2\text{-}p$	306
$Pt(PPh_3)_2\{C_{10}H_6(CH)_2\}$	*cis*–$Pt(PPh_3)_2Cl_2 + NH_2{-}NH_2 + C_{10}H_6(CH)_2$ (acenaphthylene)	306
$Pt(PPh_3)_2(MeC(OH){=}CH{-}C(Me){=}O)$	$Pt(PPh_3)_4 + CH_3COCH_2COCH_3$	175
$Pt(PPh_3)_2(EtOC(OH){=}CH{-}C(OEt){=}O)$	$Pt(PPh_3)_4 + EtO_2CCH_2CO_2Et$	175
$Pt(PPh_3)_2(CHCl{=}CCl_2)$	$Pt(PPh_3)_2$(*trans*-$PhCH{=}CHPh$) + $CHCl{=}CCl_2$	332
$Pt(PPh_3)_2(C_2F_4)$	$Pt(PPh_3)_4 + C_2F_4$	197, 203
	$Pt(PPh_3)_2(OCOCF_3)_2 + C_2F_4$	333
	$[Pt(PPh_3)_3H]^+(BF_4^-) + C_2F_4$	335
$Pt(AsPh_3)_2(C_2F_4)$	$Pt(AsPh_3)_4 + C_2F_4$	58
$Pt(PMePh_2)(C_2F_4)$	$Pt(AsPh_3)_2(C_2F_4) + PMePh_2$	58
$Pt(PMe_2Ph)_2(C_2F_4)$	$Pt(AsPh_3)_2(C_2F_4) + PMe_2Ph$	58
$Pt(PEt_2Ph)_2(C_2F_4)$	$Pt(AsPh_3)_2(C_2F_4) + PEt_2Ph$	58
$Pt(PBu^n_3)_2(C_2F_4)$	$Pt(AsPh_3)_2(C_2F_4) + PBu^n_3$	58
$Pt\{C_2H_4(PPh_2)_2\}(C_2F_4)$	$Pt(AsPh_3)_2(C_2F_4) + C_2H_4(PPh_2)_2$	58
$Pt(PPh_3)_2(C_2F_3Cl)$	$Pt(PPh_3)_4 + C_2F_3Cl$	203
$Pt(PMePh_2)_2(C_2F_3Cl)$	$Pt(PMePh_2)_4 + C_2F_3Cl$	203
$Pt(PEt_2Ph)_2(C_2F_3Cl)$	$Pt(AsPh_3)_2(C_2F_3Cl) + PEt_2Ph$	334
$Pt(PPh_3)_2(C_2F_3Br)$	$Pt(PPh_3)_4 + C_2F_3Br$	203
$Pt(PMePh_2)_2(C_2F_3Br)$	$Pt(PMePh_2)_4 + C_2F_3Br$	191
$Pt(PPh_3)_2(C_2F_3CF_3)$	$Pt(PPh_3)_4 + C_2F_3CF_3$	203
$Pt(PEt_2Ph)(C_2F_3CF_3)$	$Pt(AsPh_3)(C_2F_3CF_3) + PEt_2Ph$	334
$Pt(PPh_3)_2(CF_3CF{=}CFCF_3)$	$Pt(PPh_3)_n(n{=}3,4) + CF_3CF{=}CFCF_3$	328
$Pt(AsPh_3)_2(CF_3CF{=}CFCF_3)$	$Pt(AsPh_3)_4 + CF_3CF{=}CFCF_3$	328
$Pt(PPh_3)_2(CFCl{=}CFCl)$ (mixture of *cis* and *trans* olefin isomers)	$Pt(PPh_3)_2$(*trans*-$PhCH{=}CHPh$) + $CFCl{=}CFCl$ (mixture of *cis* and *trans* isomers)	336
$Pt(PPh_3)_2(C_2Cl_4)$	$Pt(PPh_3)_4 + CCl_2{=}CCl_2$	197
	cis-$Pt(PPh_3)_2Cl_2 + CCl_2{=}CCl_2 + NH_2{-}NH_2$	332
	$Pt(PPh_3)_2L$ (L = *trans*-$PhCH{=}CHPh$, PhC_2Ph) + C_2Cl_4	197
$Pt\{P(OPh)_3\}_2(C_2Cl_4)$	*cis*-$Pt\{P(OPh)_3\}_2Cl_2 + NH_2{-}NH_2 + C_2Cl_4$	332
$Pt(PPh_3)_2\{CCl_2{=}C(CN)_2\}$	$Pt(PPh_3)_2$(*trans*-$PhCH{=}CHPh$) + $CCl_2{=}C(CN)_2$	338

TABLE 2.VII *continued*

Compound	Method of synthesis	Infrared absorptions (cm^{-1})	References
$Pt(PPh_3)_2(CNBrC{=}CCNBr)$	$Pt(PPh_3)_2(CNC{\equiv}CCN)$ + $C_5H_5NHBr_3$ (pyridinium perbromide)		337
$Pt(PPh_3)_2\{(CN)_2C{=}C(CN)_2\}$	$Pt(PPh_3)_2(PhC{\equiv}CH) + (CN)_2C{=}C(CN)_2$	$\nu(C{\equiv}N)$, 2220	208
	$Pt(PPh_3)_2(C_2H_4) + (CN)_2C{=}C(CN)_2$		303
	trans-$Pt(PPh_3)_2H(X)$ (X = Cl, Br, CN) + $(CN)_2C{=}C(CN)_2$		208, 307
	$Pt(PPh_3)_2(CNC{\equiv}CCN) + (CN)_2C{=}C(CN)_2$		337
	$Pt(PPh_3)_2$(*cyclo*-C_6H_8) + $(CN)_2C{=}C(CN)_2$		339
$Pt(PEt_3)_2\{(CN)_2C{=}C(CN)_2\}$	$Pt(PEt_3)_2H(X)$ (X = Cl, Br, I, NO_2, NCO) + $(CN)_2C{=}C(CN)_2$	$\nu(C{\equiv}N)$, 2245(sh), 2220(vs) 2165(sh), 2140(sh)	208, 307
$Pt(PPh_3)_2$(CF=CF–CF₂–CF₂ ring: $CF{-}CF_2$ / $CF{-}CF_2$)	$Pt(PPh_3)_4$ + $CF{=}CF{-}CF_2{-}CF_2$ (cyclic)		203
$Pt(PPh_3)_2(CF_2{=}CF{-}CF{=}CF_2)$	$Pt(PPh_3)_4 + CF_2{=}CF{-}CF{=}CF_2$	$\nu(C{=}C)$, 1760 (only one double bond is involved in the co-ordination)	203
$Pt(PPh_3)_2$(hexafluorocyclohexa-1,3-diene, C_6F_6)	$Pt(PPh_3)_4$ + C_6F_6 (hexafluorocyclohexadiene)		203
$Pt(PPh_3)_2$(hexafluorobicyclo[2,2,0]hexadiene)	$Pt(PPh_3)_4$ + C_6F_6 (hexafluorobicyclo[2,2,0]hexa-2,5 diene)	$\nu(C{=}C)$, 1752 (for the uncoordinated double bond)	308, 309
$Pt(PPh_3)_2(CH_3C{=}CCH_3)$ (cyclopropene, CH_2 bridge)	$Pt(PPh_3)_2(C_2H_4) + CH_3C{=}CCH_3$ (cyclopropene, CH_2 bridge)		340
$Pt(PPh_3)_2(CH_3C{=}CH)$ (cyclopropenone, C=O bridge)	$Pt(PPh_3)_2(C_2H_4) + CH_3C{=}CH$ (cyclopropenone, C=O bridge)		342

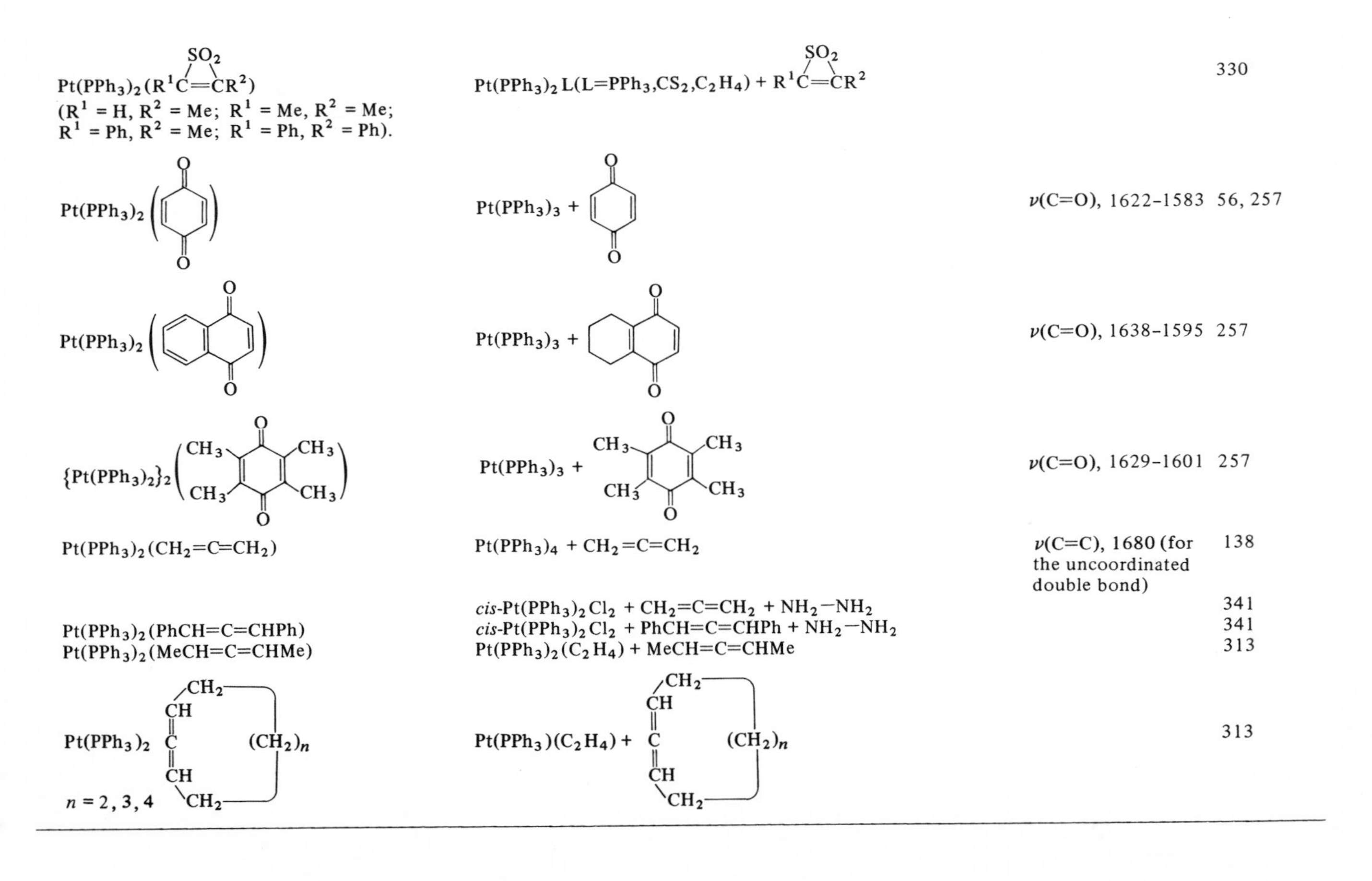

$Pt(PPh_3)_2(R^1C{=}CR^2)$ (SO_2 bridged) ($R^1 = H, R^2 = Me$; $R^1 = Me, R^2 = Me$; $R^1 = Ph, R^2 = Me$; $R^1 = Ph, R^2 = Ph$).	$Pt(PPh_3)_2L(L{=}PPh_3,CS_2,C_2H_4) + R^1C{=}CR^2$ (SO_2 bridged)		330
$Pt(PPh_3)_2$	$Pt(PPh_3)_3$ +	$\nu(C{=}O)$, 1622–1583	56, 257
$Pt(PPh_3)_2$	$Pt(PPh_3)_3$ +	$\nu(C{=}O)$, 1638–1595	257
$\{Pt(PPh_3)_2\}_2$	$Pt(PPh_3)_3$ +	$\nu(C{=}O)$, 1629–1601	257
$Pt(PPh_3)_2(CH_2{=}C{=}CH_2)$	$Pt(PPh_3)_4 + CH_2{=}C{=}CH_2$	$\nu(C{=}C)$, 1680 (for the uncoordinated double bond)	138
	cis-$Pt(PPh_3)_2Cl_2 + CH_2{=}C{=}CH_2 + NH_2{-}NH_2$		341
$Pt(PPh_3)_2(PhCH{=}C{=}CHPh)$	*cis*-$Pt(PPh_3)_2Cl_2 + PhCH{=}C{=}CHPh + NH_2{-}NH_2$		341
$Pt(PPh_3)_2(MeCH{=}C{=}CHMe)$	$Pt(PPh_3)_2(C_2H_4) + MeCH{=}C{=}CHMe$		313
$Pt(PPh_3)_2$ $n = 2, 3, 4$	$Pt(PPh_3)(C_2H_4)$ +		313

In Tables 2.V, 2.VI and 2.VII some of the stretching frequencies of the substituents of the alkenes bound to the metal are also reported. The limited lowering of these absorptions upon complexation are a good indication of the coordination of the alkene through the double bond. An accurate study by i.r. and Raman spectroscopy of $Ni(PPh_3)_n(alkene)_2$ complexes (n = 1, 2; alkene = $CH_2{=}CHCN$, $CH_2{=}CHCHO$) has been reported [351].

For most of the derivatives reported in Tables 2.V -VII the ^1H or ^{19}F n.m.r. spectra have been studied. We will discuss here only some of these spectra and also when reported, the ^{31}P n.m.r. spectra. $Ni(PPh_3)_2(C_2H_4)$ in C_6D_6 shows a sharp singlet at $\tau = 7.43$; this signal moves downfield, still remaining sharp, when ethylene is added to the solution, which indicates a very fast exchange between free and complexed ethylene [55]. Free ethylene absorbs at $\tau = 4.75$. On cooling to −80°C, no P–H coupling was observed, only a slight broadening of the band [344b].

On this basis, and from the observation that by passing argon into the $Ni(PPh_3)_2(C_2H_4)$ solution the signal due to ethylene disappears and a dark brown precipitate is formed, a dissociation of the complex has been postulated [344b]. However, a reaction with traces of oxygen or some unknown decomposition reactions cannot be excluded as the origin of these effects [55]. In fact the experimental molecular weight of the nickel complex as well as of the homologous platinum derivative, does not indicate the slightest dissociation. On the other hand, single sharp resonances, without coupling with phosphorus, have also been observed in nickel complexes such as $Ni(PPh_3)_2(alkene)$ (alkene = *cis* and *trans*-MeOOCCH=CHCOOMe) [319] and in $Pt(PPh_3)_2$(*trans*-MeOOCCH=CHCOOMe) [301].

The ^{31}P spectrum of the nickel ethylene complex shows also a singlet at −31 p.p.m., with no indication of coupling with other nuclei present in the complex [55].

The platinum analogue shows, on the contrary, a 1 : 4 : 1 triplet at $\tau = 7.43$, with $J_{Pt-C_2H_4} = 60$ Hz and unresolved fine structure due to P–H coupling [55]. The coupling with phosphorus is very small, $J_{cis\text{-}P-H} + J_{trans\text{-}P-H} \simeq 3$ Hz [344b], which in any case indicates that there is no free rotation of the ligand about the metal–alkene bond. By adding ethylene to the solution, a broad new resonance in the position of free ethylene is observed. In this case the exchange is too slow on the n.m.r. time scale to give a unique, averaged signal, but enough to broaden the resonance of free ethylene.

By increasing the amount of added ethylene, the resonance at $\tau = 7.43$ also tends to broaden. This behaviour requires an associative mechanism with a small K_{eq} for the diethylene intermediate [55]:

$$^*C_2H_4 + Pt(PPh_3)_2(C_2H_4) \rightleftharpoons Pt(PPh_3)_2(C_2H_4)(^*C_2H_4) \rightleftharpoons C_2H_4 + Pt(PPh_3)_2(^*C_2H_4)$$

By studying this reaction at various temperatures, the following activation parameters were determined: $\Delta H^{\ddagger} = 12$ kcal/mole and $\Delta S^{\ddagger} = -14$ e.u., in accord with an associative process [344b].

Addition of PPh_3 to the solution gives a single averaged resonance, at τ intermediate between 4.75 and 7.43 [55, 67], as required by the reaction [55]:

$$Pt(PPh_3)_2(C_2H_4) + PPh_3 \rightleftharpoons Pt(PPh_3)_3(C_2H_4) \rightleftharpoons Pt(PPh_3)_3 + C_2H_4$$

The ^{31}P spectrum of the platinum-ethylene complex shows a 1 : 4 : 1 triplet at -32 p.p.m., with $J_{Pt-P} = 3360$ Hz.

The nickel-ethylene derivative, $Ni\{P(O\text{-}o\text{-}C_6H_4CH_3)_3\}_2(C_2H_4)$ shows a single ^1H resonance at $\tau = 8.06$ and ^{31}P resonance at -139.7 p.p.m. [54]. By adding different amounts of phosphite to the solution, these species were observed: NiL_4, NiL_3, $NiL_2(C_2H_4)$, $L[L = P(O\text{-}o\text{-}C_6H_4CH_3)_3]$ and free C_2H_4.

In the presence of ethylene, the signals due to phosphorus are averaged via the equilibria:

$$L + NiL_2(C_2H_4) \xrightleftharpoons{\text{fast}} NiL_3 + C_2H_4$$

$$NiL_3 + L \xrightleftharpoons{\text{slow}} NiL_4$$

The solutions do not contain detectable amounts of NiL_2, $NiL_2(C_2H_4)_2$ or $NiL_3(C_2H_4)$ species [54]. The $NiL_2(C_2H_4)$ complex follows Beer's law, which is a further indication that no dissociation occurs in solution.

The ^1H n.m.r. spectra of $Ni(PCy_3)_2(C_2H_4)$ and $Ni(PCy_3)(C_2H_4)_2$ have simple resonances at τ 7.94 and 7.22, respectively [310]. The ^{19}F n.m.r. spectrum of $Pt(PPh_3)_2(C_2F_4)$ has been fully analysed; the spectrum is typical of the X resonance of an $X_2AA'X'_2$ system:

(A)P—Pt(—P(A'))—CF$_2$(X$_2$)—CF$_2$(X$'_2$) [(A)P and (A')P bonded to Pt; Pt bonded to both carbons of CF$_2$(X$_2$)—CF$_2$(X$'_2$)]

centred at 131.1 p.p.m. with satellite multiplets due to coupling with ^{195}Pt [203]. The spectrum is field and temperature invariant, and these observations are consistent with a rigid coordination of tetrafluoroethylene to platinum.

In $L_2Ni(CF_{3(1)}{-}CF_{(2)}{=}C\langle^{F_{(3)}}_{F_{(4)}})$ (L = PEt_3, PPh_3) complexes, $J_{F(3)F(4)}$ is *ca.* 170–190 Hz, a value close to that present in substituted cyclopropanes and cyclobutanes [50].

Moreover $J_{F(3)F(4)} = 168$ Hz ($L = PPh_3$) is lower than $J_{F(3)F(4)} = 193.6$ Hz ($L = PEt_3$) and this can be associated to the stronger π-back-donation from the metal when the ligand is the more basic triethylphosphine.

$J_{F(2)F(3)}$ decreases in the complexes compared to the free ligand, with a value which is similar to that predicted for sp^3 hybridized carbons.

For $M(PPh_3)_2(CF_3CF{=}CFCF_3)$ complexes (M = Pt, Pd), the ^{19}F n.m.r. spectrum shows the absorptions reported under [328]:

	δ p.p.m. (relative to benzotrifluoride)		J_{Pt-F} (Hz)	J_{P-CF_3} (Hz)
$CF_3CF{=}CFCF_3$	+ 6.1	+ 3.5	–	–
$Pt(PPh_3)_2(CF_3CF{=}CFCF_3)$	+ 2.5	+140.0	77.6	9.2
$Pd(PPh_3)_2(CF_3CF{=}CFCF_3)$	+ 4.0	+133.0	–	9.5

The higher shifts are associated with the fluorines bound to the olefinic carbons. The spectra also show a long range coupling between phosphorus and the substituent groups CF_3.

The "Dewar-benzene" platinum derivative,

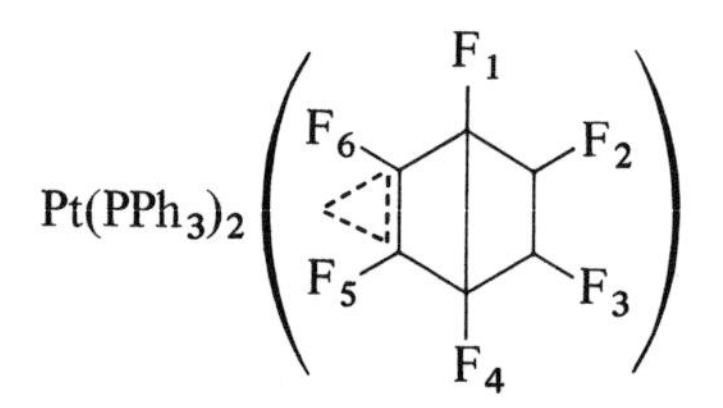

displays three bands of equal intensity increasing to high field at 48.0 (apparent singlet), 88.0 (complex multiplet) 99.8 (complex multiplet) p.p.m. relative to external CF_3COOH [308]. They were assigned to (F_2,F_3), (F_5,F_6) and to the bridgehead fluorine nuclei (F_1,F_4), respectively, on the basis of the chemical shifts. A $J_{Pt^{195}-F}$ = 220 Hz was also observed for the absorption at 88.0 p.p.m., as expected.

^1H n.m.r. spectra have also been used in order to support the coordination of *para*-benzoquinone with both the double bonds in $Pt(PPh_3)_2(C_6H_4O_2)$ [257]. This derivative in fact shows a complex, asymmetric triplet at τ 5.04, due to coupling of the olefinic hydrogen with two slightly different phosphorus atoms ($J_{P-H} \simeq 2$ Hz). Free *para*-benzoquinone gives a singlet at τ 3.30.

From the platinum satellites a J_{Pt-H} of 27 Hz was calculated. The corresponding 1,4-naphthoquinone derivative, $Pt(PPh_3)_2$(1,4-naphthoquinone) shows a doublet at τ 5.96 ($J_{P_{trans}-H}$ = 8 Hz) with platinum satellites (J_{Pt-H} = 48 Hz). Free 1,4-naphthoquinone gives a singlet at τ 3.10. The lower shift to high field of the resonance of the olefinic hydrogens upon complexation in the *para*-benzoquinone derivative, compared to the 1,4-naphthoquinone derivative, and the lower coupling with platinum, have been attributed to the fact that in the former case both the double bonds of the quinone are involved in the

TABLE 2.VIII

Molecular parameters for $M(PPh_3)_2$(alkene) complexes (M=Ni, Pt): (1)P, (2)P–M–C(1), C(2)

Complex	$d_{C(1)-C(2)}$(Å)	$d_{C(1)-C(2)}$(Å) (uncomplexed)	$d_{M-C(1)}$(Å)	$d_{M-C(2)}$(Å)	C(1)MC(2)°	$\theta°^a$
$Ni(PPh_3)_2(C_2H_4)$	1.46	1.344	1.99 (mean)	1.99 (mean)	43	5.0
$Pt(PPh_3)_2(C_2H_4)$	1.43	1.344	2.11 (mean)	2.11 (mean)	–	1.3
$Pt(PPh_3)_2(CNCH{=}CHCN)$	1.53(0.036)	–	2.05	2.16	42.6	5.2
$Pt(PPh_3)_2\{(CN)_2C{=}C(CN)_2\}$	1.49(5)	1.31	2.12(3)	2.10(3)	41.5(1.3)	8.3
$Pt(PPh_3)_2(Cl_2C{=}CCl_2)$	1.62(3)	–	2.03(3) (mean)	2.03(3) (mean)	47.1(1.0)	12.3(1.5)
$Pt(PPh_3)_2\{Cl_2C{=}C(CN)_2\}$	1.42(3)	–	2.00(2)[b]	2.10(2)[c]	40.6(9)	0.0
$Pt(PPh_3)_2(CH_3{-}C{=}C{-}CH_3)$ (C=C bridged by CH_2)	1.50	1.30	2.1(1)	2.11(1)	41.6	–

[a] Dihedral angle between P(1)MP(2) and C(1)MC(2); [b] Chlorine substituents; [c] Cyano substituents.
Data taken from refs. 338, 340, 344, 344b, 346, 346 *bis*, 347.

coordination, while in the latter case the quinone is coordinated as a mono-olefin substituted with electron-withdrawing substituents, such as the double bond in maleic anhydride [257]. The relevant $J_{P_{trans}-H}$ coupling constant observed in the latter case is also consistent with such an interpretation. Long-range coupling between phosphorus and the methyl groups of 1,4-tetra-methylbenzoquinone has been observed in the platinum-phosphine derivative of this quinone [257].

A detailed analysis of the ^1H n.m.r. spectra of the allene complexes $Pt(PPh_3)_2$(allene) (allene = $CH_2{=}C{=}CH_2$, PhCH=C=CHPh) has been reported [341].

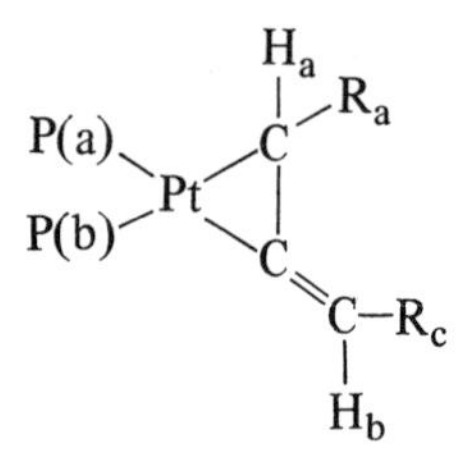

The data are reported on Table 2.IX. For $Pt(PPh_3)_2(CH_2{=}C{=}CH_2)$ the high field shift of H_a and the coupling with ^{195}Pt are typical of olefins bonded to platinum. Remarkable is the very long-range coupling between P_a and H_c (22.5 Hz). For $Pt(PPh_3)_2$(PhCH=C=CHPh), the resonance at $\tau = 6.25$ appears as a septet, due to the overlap of the splittings due to coupling of H_a with H_b, P_a and P_b.

TABLE 2.IX

^1H n.m.r. data ($CDCl_3$ as solvent) for $Pt(PPh_3)_2$(allene) complexes

Compound	τ	J(Hz)
$CH_2{=}C{=}CH_2$ (free)	5.30 (singlet)	–
H_a (complexed)	7.98 (triplet)[a]	J_{Pt-H_a} = 65
H_b (complexed)	5.46 (doublet)[b]	$J_{P_a-H_b}$ = 12.5
H_c (complexed)	4.00 (doublet)[b]	$J_{P_a-H_c}$ = 22.5
PhCH=C=CHPh (free)	3.44 (singlet)	–
H_a (complexed)	6.25 (triplet)[a]	$J_{P_a-H_a}$ = 5.5; J_{Pt-H_a} = 65; $J_{P_b-H_a}$ = 9.5
H_b (complexed)	4.18 (doublet)[b]	$J_{P_a-H_b}$ = 12.5; $J_{H_a-H_b}$ = 3.0

[a] Broad peak, with satellites due to coupling with ^{195}Pt.
[b] Coupling with ^{195}Pt not observed.
Data taken from ref. 341.

Also in this case a long-range coupling between H_b and P_a (12.5 Hz) is observed. Similar ^1H n.m.r. spectra have been obtained for the platinum derivatives of cyclic allenes [313], where couplings of platinum with protons such as H_b were also observed.

Among the electronic spectra, the most studied are those related to *para*-quinone derivatives [257]. For a series of $Pt(PPh_3)_2$ (*para*-quinone) complexes a very small perturbation of the $n \rightarrow \pi^*$ transition upon complexation has been found (a small shift to higher energies). This suggests that there is not only very little interaction between the carbonyl group and the metal, but also a very small shift to high energy of the quinone b_{2g} level (the group theoretical notation, D_{2h} symmetry, for the quinone orbitals is maintained for simplicity). The charge-transfer metal → quinone transition has been found at *ca.* 330–340 nm, and it appears at higher energies than in the related nickel(0)-quinone complexes, in accord with the fact that platinum *d* orbitals have a lower energy than nickel corresponding to a higher ionization potential [257].

(*e*) *Reactivity.* The reactions of the alkene derivatives of nickel, palladium and platinum with ligands such as phosphines have already been considered as a preparative method for the synthesis of ML_n complexes (M = Ni, Pd, Pt; L = substituted phosphine) (2.2.1(c)). Similarly, the exchange of a coordinated alkene with another alkene has already been discussed (see Section *a*). The exchange of a coordinated alkene with an alkyne represents a convenient method of synthesis of $M(PPh_3)_2$(alkyne) complexes, and will be discussed in the following paragraph. The rearrangement of a haloalkene derivative to the corresponding halo-vinyl complex, via the oxidative addition of the η-bonded haloalkene, has also been considered when discussing the oxidative addition reactions of halo derivatives to the ML_n complexes (2.2.3.3).

It is known that coordination of unsaturated hydrocarbons to the metals of the platinum group in a relatively high oxidation states yields derivatives susceptible to attack by nucleophiles [352]. On the other hand olefins coordinated to platinum in a formally zerovalent oxidation state may be attacked by a proton [333]:

$$Pt(PPh_3)_2(C_2F_4) + HX \rightarrow Pt(PPh_3)_2X(CF_2CF_2H) \quad (X = CF_3COO)$$

Similar reactions occur with $PtL_2(C_2F_4)$ ($L = AsPh_3, PEt_2Ph, PMePh_2, PBut_3$), and with hexafluoropropene, trifluoroethylene, chlorotrifluoroethylene, bromotrifluoroethylene and tetrachloroethylene complexes [334]. The ^{19}F n.m.r. spectra suggest that these complexes have a *trans* configuration of ligands around the platinum.

This isomerization contrasts with the action of CF_3COOH on related hexafluoro-2-butyne complexes (2.2.3.9(d)) which leads initially to *cis*-alkenyl derivatives. The same reactions studied with hydrochloric acid gave no reaction or simply *cis*-$Pt(PR_3)_2Cl_2$ complexes [334]. In view of the polarization of the platinum–alkene bond, which results in a drift of negative charge on the alkene, a direct protonation of the co-ordinated fluoro-olefin was considered to be more probable than an initial attack at platinum. However, this appears to contrast with the kinetic studies on the protonation reactions of a complex such as $Pt(PR_3)_2\{(CN)_2C{=}C(CN)_2\}$ (see later). The action of CF_3COOH upon $Pt(PPh_3)_2(C_2Cl_4)$ also leads to the alkyl complex, $Pt(PPh_3)_2(C_2Cl_4H)(OCOCF_3)$, but no reaction has been observed with the corresponding hexafluorocyclobutene complex [334].

In the particular case of a quinone derivative such as $Pt(PPh_3)_2$(*para*-benzoquinone), protonation with hydrochloric acid leads to *cis*-$Pt(PPh_3)_2Cl_2$ and hydroquinone [257]. An intermediate hydrido-compound might form by the oxidative addition of hydrogen chloride to the platinum atom, the reduction of the quinone moiety to hydroquinone taking place via the hydride hydrogens. This reduction is clearly facilitated by the π-back-donation to the quinone ligand, which should increase the electron density on the oxygen atoms. The reverse of protonation of the alkene bound to a metal:

$$Pt(PPh_3)_2H(X) + (CN)_2C{=}C(CN)_2 \rightarrow Pt(Ph_3)_2\{(CN)_2C{=}C(CN)_2\} + HX$$

(R = Ph, Et; X = Cl, Br, I, CN, etc.) has been carefully investigated [307]. This research is related to the important property of some complexes of metals in low oxidation states of catalysing the hydrogenation of alkenes in homogeneous phase, and under very mild conditions [353]. The insertion of the olefin into metal–hydrogen bonds has been postulated in many cases:

$$L_nM{-}H + \;\rangle C{=}C\langle \;\longrightarrow\; L_nM{-}\overset{|}{\underset{|}{C}}{-}\overset{|}{\underset{|}{C}}{-}H$$

With ethylene, *trans*-$Pt(PEt_3)_2H(Cl)$ has been shown to give $Pt(PEt_3)_2Cl(C_2H_5)$ [354], that is, the proposed insertion product. Presumably the reaction proceeds through coordination of the alkene η-bonded to the platinum complex and such an intermediate has also been isolated [307]:

$$Pt(PEt_3)_2H(CN) + (CN)_2C{=}C(CN)_2 \rightarrow Pt(PEt_3)_2H(CN)\{(CN)_2C{=}C(CN)_2\}$$

[ν(Pt–H), 2198; ν(CN), 2135; ν(CN), 2231 cm^{-1} of the coordinated alkene].

This type of derivative is considered to be the intermediate formed by a fast reaction between the alkene and the hydrido-platinum complexes, when the final product is $Pt(PR_3)_2\{(CN)_2C{=}C(CN)_2\}$ [307]. Two possible mechanisms have been proposed for the rate-determining step of the process:

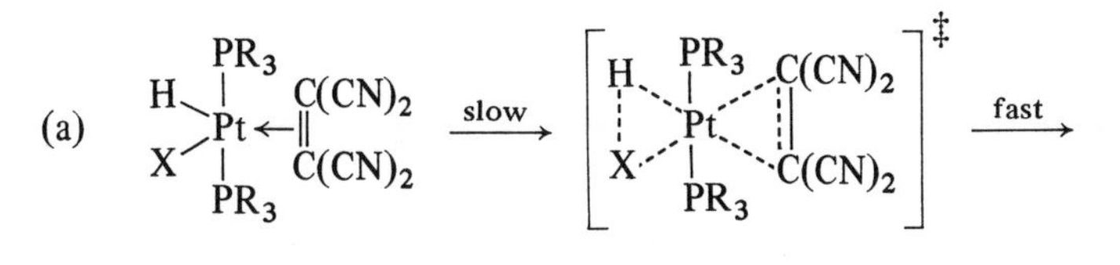

$$Pt(PR_3)_2\{(CN)_2C{=}C(CN)_2\} + HX$$

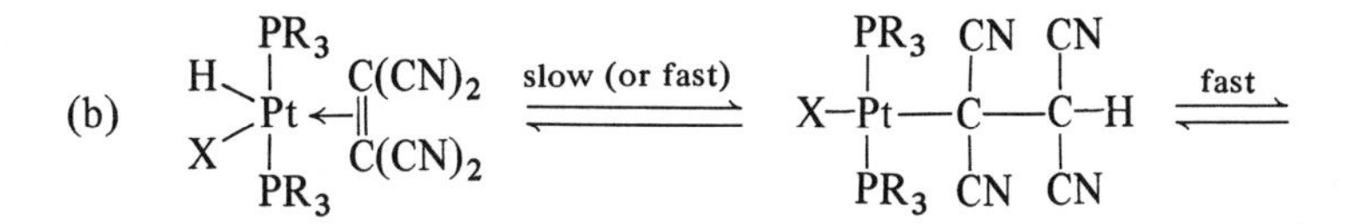

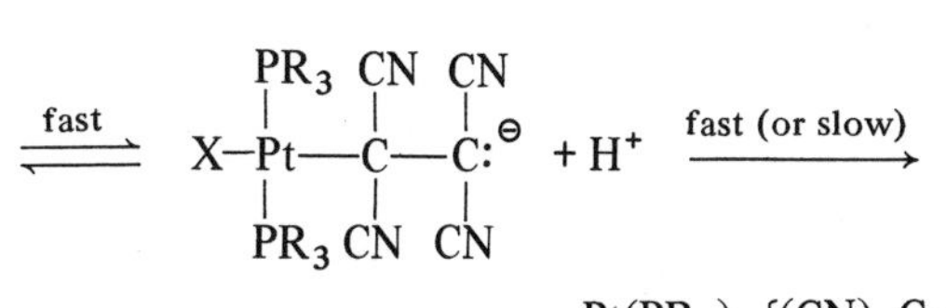

$$Pt(PR_3)_2\{(CN)_2C{=}C(CN)_2\} + X^-$$

Path (a) proceeds via formation of a three-centre transition state in which HX is eliminated to give the product. Path (b) is supported by the insertion reaction considered above; also, the acidic nature of protons attached to cyano carbons is well documented. However, when the alkene is tetracyanoethylene, as in those reactions studied kinetically, path (a) is preferred [307].

The kinetics of oxidative addition of alkyl halides to $Pt(PPh_3)_2(C_2H_4)$ where the ethylene molecule is displaced from the complex, have also been studied [53]:

$$Pt(PPh_3)_2(C_2H_4) + RX \rightarrow Pt(PPh_3)_2(R)X + C_2H_4$$

$$(X = I, Br;\ R = CH_3, C_6H_5CH_2)$$

$$Pt(PPh_3)_2(C_2H_4) + CH_2ICH_2I \rightarrow Pt(PPh_3)_2I_2 + 2\ C_2H_4$$

In each case the kinetics measured spectrophotometrically confirmed the rate law:

$$-d\ln[Pt(PPh_3)_2(C_2H_4)]_{total}/dt = k_2K[X]/K + [C_2H_4]$$

consistent with mechanism

$$Pt(PPh_3)_2(C_2H_4) \underset{\substack{\text{rapid}\\ \text{equilibrium}}}{\overset{K}{\rightleftharpoons}} Pt(PPh_3)_2 + C_2H_4$$

$$Pt(PPh_3)_2 + X \xrightarrow{K_2} \text{product} \qquad (X = CH_3I, \text{etc.})$$

$$[Pt(PPh_3)_2(C_2H_4)]_{total} = [Pt(PPh_3)_2(C_2H_4)] + [Pt(PPh_3)_2]$$

For K a value of $(3 \pm 1.5) \times 10^{-3}$ M has been found, which should indicate a notable dissociation into the active species $Pt(PPh_3)_2$. Support for the mechanism considered above has been derived from the kinetic investigation of the reaction [157]:

$$Pt(PPh_3)_2(C_2H_4) + SnPh_3Cl \rightarrow Pt(PPh_3)_2Cl(SnPh_3) + C_2H_4$$

However, this problem appears to be quite similar to the question of whether or not a species like $Pt(PPh_3)_3$ is effectively dissociated into $Pt(PPh_3)_2$ and PPh_3 (2.2.3) since as for $Pt(PPh_3)_3$, the dissociation of $M(PPh_3)_2(C_2H_4)$ complexes (M = Ni, Pt), has been recently shown to be questionable [55].

When R is a vinyl residue, as in RCl($R{=}CCl_2{=}CCl$), the final product is the chlorovinyl complex [355]:

$$Ni(PEt_3)_2(C_2H_4) + C_2Cl_4 \rightarrow Ni(PEt_3)_2Cl(CCl{=}Cl_2) + C_2H_4$$

presumably formed via a η-bonded $Ni(PEt_3)_2(C_2Cl_4)$ derivative. The reactions between $Ni(PEt_3)_2(C_2H_4)$, prepared *in situ* via the reduction of $Ni(acac)_2$ with $Al(OEt)Et_2$ in the presence of PEt_3, and C_6F_5Br, 1,3,5-trichlorobenzene, *ortho*-dichlorobenzene and *ortho*-chlorobromobenzene have been studied [356] as have those of $Ni(PPh_3)_2(C_2H_4)$ with the same reagents [356].

The products of the oxidative addition reactions, where ethylene is displaced, show the expected ν(Ni–X) at *ca.* 305–310 cm^{-1} (X = Br) and 360–380 cm^{-1} (X = Cl).

Some bands at around 415–430 cm^{-1} have been assigned to the nickel-phosphorus stretching vibrations, and a *trans* configuration has been proposed for these derivatives since in general only one ν(Ni–P) was observed [356]. However, as we have many times pointed out, such indications cannot be conclusive.

The reaction of $Ni(PEt_3)_2(C_2H_4)$ with 1,2,4-trichlorobenzene, followed by the decomposition of the products with hydrochloric acid, gave 7% *ortho*, 6% *meta* and 87% *para* dichlorobenzene, indicating that the attack occurs preferentially at the 2-position [356].

In contrast to the behaviour of $Ni(CH_2{=}CH{-}CN)_2$, the PPh_3 adduct $Ni(CH_2{=}CH{-}CN)_2(PPh_3)$ loses both its acrylonitrile ligands when reacted with allyl halides [356 *bis*]:

$$Ni(CH_2{=}CH{-}CN)_2(PPh_3) + C_3H_5X \xrightarrow[20^\circ]{benzene} (\pi\text{-}C_3H_5)Ni(X)(PPh_3) + 2CH_2{=}CH{-}CN$$

(X = Cl, Br)

In contrast to hexafluoro-2-butyne complexes of platinum (see later) no mercurated derivatives could be prepared from the reaction between $HgCl_2$ and $Pt(PPh_3)_2(C_2F_4)$, the product being only a mixture of *cis* and *trans*-$Pt(PPh_3)_2Cl_2$ [344]. Similarly, attempts to alkylate the alkene in this complex with *t*-butyl chloride leads to the same mixture.

$Pt(PPh_3)_2(C_2H_4)$ has been used as a convenient starting material for the synthesis of complexes with platinum-tin [212] or platinum-silicon [357] bonds:

$$Pt(PPh_3)_2(C_2H_4) \xrightarrow{HSnMe_3} \text{[a]nd } trans\text{-}Pt(PPh_3)_2(SnMe_3)_2$$

$$Pt(PPh_3)_2(C_2H_4) \xrightarrow{Me_3Sn{-}SnMe_3} trans\text{-}Pt(PPh_3)_2(SnMe_3)$$

$$Pt(PPh_3)_2(C_2H_4) \xrightarrow[\text{(in the presence of 1-hexene)}]{HSiMeCl_2} \text{[Pt(]}PPh_3)_2H(SiMeCl_2)$$

$$Pt(PPh_3)_2(C_2H_4) \xrightarrow{Me_3SnCl} Pt(PPh_3)_2Cl(SnMe_3)$$

$$Pt(PPh_3)_2(C_2H_4) \xrightarrow{HSiMeCl_2} Pt(PPh_3)_2(SiMeCl_2)_2$$

$$Pt(PPh_3)_2(C_2H_4) \xrightarrow{HSiCl_3} Pt(PPh_3)_2(SiCl_3)_2$$

A series of platinum-hydride derivatives containing platinum-silicon bonds has been obtained from the reaction of $Pt(PPh_3)_2(C_2H_4)$ with liquid R_3SiH compounds, without any solvent [182].

When an optically active silicon hydride such as (+)-HSi(Me)(Ph)(1-naphthyl) is used, cleavage of the product with $LiAlH_4$ gives back the original hydride with 97% overall retention of configuration [358]. The oxidative addition to $Pt(PPh_3)_2(C_2H_4)$ is also possible with the iron derivative π-$CpFe(CO)_2Cl$, leading to $Pt(PPh_3)_2Cl\{\pi\text{-}CpFe(CO)_2\}$ [212].

The first metal-pentafluorosulphur(VI) complex, $Pt(PPh_3)_2Cl(SF_5)$, has been synthesized by the reaction between $Pt(PPh_3)_2$ (*trans*-stilbene) and SF_5Cl in benzene [358 *bis*]. $Ni(PPh_3)_2(C_2H_4)$ has been reacted with BBr_3 and BPh_2Br [70], presumably with the aim of synthesizing 1:1 adducts, where the nickel complex should act as a donor base, in a way similar to that of *trans*-$IrClCO(PPh_3)_2$ in respect to BF_3 [359]. However, when the reaction was carried out in benzene, only $Ni(PPh_3)_2Br_2$ was obtained.

In ethyl ether with BPh_2Br, $\{Ni(PPh_3)_2Br\}_n$ and $\{Ni(PPh_3)_2BPh_2 \cdot \frac{1}{2}Et_2O\}_n$ with bridging BPh_2 groups were obtained, while with BBr_3, $Ni(PPh_3)_3Br$ was formed. The last nickel derivative was also synthesized from the oxidative addition of RBr (R = Et, *n*-Bu) on $Ni(PPh_3)_2(C_2H_4)$ in the presence of PPh_3 [70].

$M(PPh_3)_2(C_2H_4)$ complexes (M = Ni, Pt) thus show in general a reactivity similar to that of the corresponding $M(PR_3)_n$ (n = 3, 4) derivatives, with the advantage that the volatile ethylene may be removed from the reaction medium. From $Pt(PPh_3)_2(C_2H_4)$ and L = SO_2 [226], C_2H_2,O_2 [67] the $Pt(PPh_3)_2L$ adducts are easily obtained; with L=CO, $Pt(PPh_3)_2(CO)_2$ was isolated [67]. Some other reactions, as with benzyne precursors leading to cyclic azo-platinum derivatives [281], have already been considered (2.2.3.6). However, the reaction of $Pt(PPh_3)_2(C_2H_4)$ with diazonium salts probably involves a nucleophilic attack at the coordinated ethylene, although the nature of the products was not clarified [278]. This seems to be one of the few cases where the alkene bound in the complex is retained in the products.

Strangely enough, when diazonium salts are reacted with $Pt(PPh_3)_2(PhC{\equiv}CPh)$ cationic azo-complexes, with displacement of the alkyne, have been obtained and have been formulated as $[Pt(PPh_3)_2(N_2R)BF_4]$ [360]. Such derivatives react in mild conditions with hydrogen, with formation of ammonia and arylamines. The behaviour of these compounds seems to be rather different from that of the cationic azo-platinum complexes obtained from the reaction of $Pt(PPh_3)_3$ and diazonium salts (2.2.3.6). Bis(trifluoromethyl)diazomethane reacts with $Ni(PPh_3)_2(C_2H_4)$ or $Pt(PPh_3)_2$(*trans*-stilbene) with formation of:

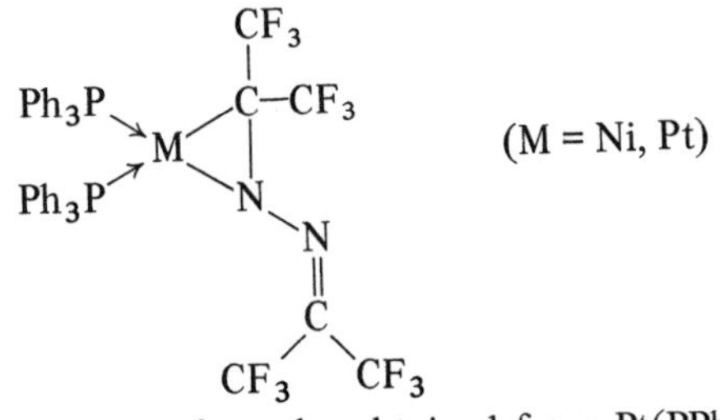

For M = Pt, the same compound can be obtained from $Pt(PPh_3)_4$ [277]. Also, $Pt(PPh_3)_2(C_2H_4)$ and sulphonyl azides in benzene gives $Pt(PPh_3)_2(N_4R_2)$ (R = *para*-$C_6H_4SO_2$), that is the same tetrazene complex obtained from $Pt(PPh_3)_4$ [284].

When the alkene is more firmly bound to platinum, as for $CH_2{=}CH{-}CN$ and

CH–CO
‖   O
CH–CO

, the reaction with organic azides is very slow or does not proceed at all [267].

Diazoalkane derivatives of nickel have been obtained from the reaction of $Ni(PPh_3)_2(C_2H_4)$ and R_2CN_2 (R = fluorenylidene) [361]. The product, of formula $Ni(PPh_3)_2(R_2CN_2)$, shows an absorption at around 1500 cm^{-1} in the i.r. spectrum due to the $>C{=}N{=}N$ group, and a side-on coordination of the diazoalkane to nickel has been proposed [361]. Particularly studied have been the reactions of such complexes with nitrogen and oxygen donor molecules. Besides the reactions above considered, $Ni(PPh_3)_2(C_2H_4)$ has been reacted with $(CF_3)_2CO$ [205], with formation of the expected $Ni(PPh_3)_2\{(CF_3)_2CO\}$, whose structure has been determined [255]. The nickel complex with $(CF_3)_2CNR$ (R = H, CH_3) gives the $Ni(PPh_3)_2\{(CF_3)_2CNR\}$ compounds [362]. Similar derivatives of formula $NiL_2\{(CF_3)_2CNMe\}$ [L = $PMePh_2$, L_2 = $C_2H_4(PPh_2)_2$] have been obtained by displacing 1,5-COD from Ni(1,5-COD)$\{(CF_3)_2CNMe\}$ with the ligands L [362].

The latter compound was isolated from the reaction between $Ni(1,5\text{-}COD)_2$ and $(CF_3)_2CNMe$. The platinum complexes, $Pt(PPh_3)_2\{(CF_3)_2CNR\}$ (R = H, Me) have been obtained similarly from the reactions of $Pt(PPh_3)_2$(*trans*-stilbene) and $(CF_3)_2CNR$ [362]. The platinum derivative (R = H) oxidatively adds organic acids such as CF_3COOH, with formation of $Pt(PPh_3)_2(CF_3CO_2)_2$-$\{(CF_3)_2CNR\}$, and a ring expansion occurs with hexafluoroacetone. In contrast the derivative with R = Me with hexafluoroacetone gives the already known substitution product $Pt(PPh_3)_2\{(CF_3)_2CO\}$ [254], and the nickel compound $Ni(PPh_3)_2\{(CF_3)_2CNH\}$ behaves analogously [360]. The reaction of $Pt(PPh_3)_2$(*trans*-stilbene) with CF_3CN has already been considered (2.2.3.6).

The ethylene molecule is easily displaced from $Pt(PPh_3)_2(C_2H_4)$ by reaction with *tert*-butyl isocyanide at −20°C, with formation of $Pt(PPh_3)_2(Bu^tNC)_2$ [363]. This derivative shows ν(NC) = 2030 cm^{-1}, 100 cm^{-1} lower than in the free isonitrile. By carbonylation in the solid state at 25 atmospheres $Pt(PPh_3)_2(Bu^tNC)(CO)$ is obtained [ν(NC) = 2110, ν(CO) = 1950 cm^{-1}]. Also, a series of oxidative addition reactions have been carried between $Pt(PPh_3)_2(Bu^tNC)_2$ and RX molecules (RX = I_2, CH_3I, CF_3I, $SnPh_3Cl$).

The reaction products, $Pt(PPh_3)_2(Bu^tNC)_2RX$, show ν(NC) at 2225-2190 cm^{-1} typical of isocyanides bonded to platinum(II), and are 1:1 electrolytes, the halogen X being the anion [363].

The reactions of $Ni(PPh_3)_2(C_2H_4)$ with organic peroxo compounds have also been studied [127]. Benzoyl peroxide gives $Ni(PhCO_2)_2$, while dialkylperoxides give the alkoxy derivatives $Ni(OR)_2$.

These reactions are related to the catalytic activity of low-valent metal complexes in the oxidation of alkenes, which in some cases proceed through the decomposition of organic hydroperoxides (2.2.3.5). The nickel–ethylene complex reacts also with dibenzyl and di-β-naphthyl disulphides, with formation of $Ni(SR)_2$ [127].

2.2.3.9. Reactions with Alkynes and Related Ligands

(*a*) *Methods of synthesis and collateral reactions.* The methods of synthesis used for the preparation of the alkene derivatives (2.2.3.8) can be in general employed for the synthesis of the alkyne derivatives.

The known $M(PR_3)_2$(alkyne) complexes (M = Ni, Pd, Pt) are reported in Tables 2.X, 2.XI and 2.XII.

A rather common reaction employed is the displacement of an alkene, particularly ethylene, previously bound in a zerovalent complex, by an alkyne

$$ML_2(C_2H_4) + RC{\equiv}CR' \rightarrow ML_2(RC{\equiv}CR') + C_2H_4$$

These reactions have also been kinetically studied, for M = Pt, L = PPh_3, R = Ph and R′ = H, and a prior dissociation of the starting complex has been suggested [53]:

$$Pt(PPh_3)_2(C_2H_4) \underset{k_{-1}}{\overset{k_1}{\rightleftharpoons}} Pt(PPh_3)_2 + C_2H_4$$

$$Pt(PPh_3)_2 + PhC{\equiv}CH \xrightarrow{k_2} Pt(PPh_3)_2(PhC{\equiv}CH)$$

The reaction rate should thus follow the law:

$$-d\ln[Pt(PPh_3)_2(C_2H_4)]/dt = \frac{k_1k_2[PhC{\equiv}CH]}{k_{-1}[C_2H_4] + k_2[PhC{\equiv}CH]}$$

for which the constants k_1 = 0.33 (s^{-1}), k_{-1} = 1.1 x 10^2 and k_2 = 2.8 x 10^2 (M^{-1} s^{-1}) have been determined. However we have already pointed out how the dissociation to give the reactive species $Pt(PPh_3)_2$ seems to be questionable (2.2.3.8e).

The use of hydrazine as reducing agent can sometimes give products other than those expected. For example during the attempted synthesis of $Pt(PPh_3)_2$ ($PhC{\equiv}CCOCH_3$) by reducing with hydrazine *cis*-$Pt(PPh_3)_2Cl_2$ in the presence of $PhC{\equiv}CCOCH_3$, a derivative of 7-methylbenzo-[3,4]-1,2-diazepine was obtained [373]:

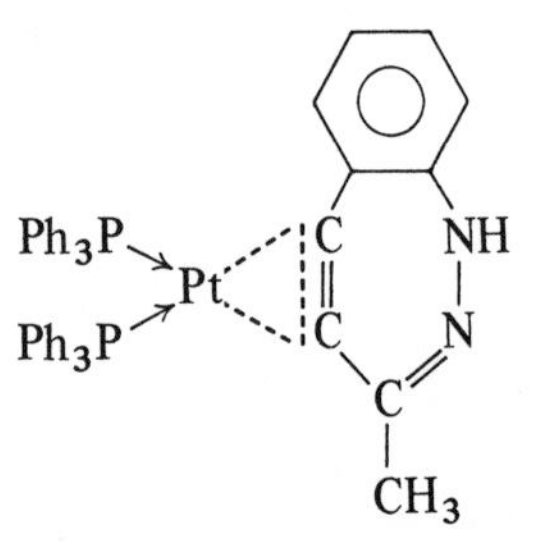

TABLE 2.X

Nickel(0) complexes with alkynes

Compound	Method of synthesis	Infrared absorptions (cm^{-1})	References
$Ni(PPh_3)_2(MeC{\equiv}CMe)$	$Ni(PPh_3)_2(C_2H_4) + MeC{\equiv}CMe$		103
$Ni(PPh_3)_2(PhC{\equiv}CMe)$	$Ni(PPh_3)_2(C_2H_4) + PhC{\equiv}CMe$	$\nu(C{\equiv}C)$, 1795	314, 364
$Ni(PPh_3)_2(PhC{\equiv}CPh)$	$Ni(PPh_3)_2(C_2H_4) + PhC{\equiv}CPh$	$\nu(C{\equiv}C)$, 1800	103, 314
$Ni(PPh_3)_2(CF_3C{\equiv}CCF_3)$	$Ni(PPh_3)_2(C_2H_4) + CF_3{\equiv}CCF_3$	$\nu(C{\equiv}C)$, 1790	314

TABLE 2.XI

Palladium(0) complexes with alkynes

Compound	Method of synthesis	Infrared absorptions (cm^{-1})	References
$Pd(PPh_3)_2(MeO_2CC{\equiv}CCO_2Me)$	$Pd(PPh_3)_4 + MeO_2CC{\equiv}CCO_2Me$	$\nu(C{\equiv}C)$, 1845–1830(sh)	314
	$Pd(dba)_2 + PPh_3 + MeO_2CC{\equiv}CCO_2Me$		365
	$Pd(PPh_3)_2\left(\begin{array}{l}CH{-}CO\\ \|\\ CH{-}CO\end{array}\!\!>O\right) + MeO_2CC{\equiv}CCO_2Me$		365
$Pd\{P(OPh)_3\}_2(MeO_2CC{\equiv}CCO_2Me)$	$Pd(dba)_2 + P(OPh)_3 + MeO_2CC{\equiv}CCO_2Me$	$\nu(C{\equiv}C)$, 1845	365
$Pd(PPh_3)_2(CNC{\equiv}CCN)$	$Pd(PPh_3)_2(CNCH{=}CHCN) + CNC{\equiv}CCN$	$\nu(C{\equiv}C)$, 1751 $\nu(C{\equiv}C)$, 2187	327
$Pd(PPh_3)_2(CF_3C{\equiv}CCF_3)$	$Pd(PPh_3)_4 + CF_3C{\equiv}CCF_3$	$\nu(C{\equiv}C)$, 1838–1811	314
$Pd(PMe_2Ph)_2(CF_3C{\equiv}CCF_3)$	$Pd(PMe_2Ph)_2Cl_2 + PMe_2Ph + NH_2{-}NH_2$ and then $CF_3C{\equiv}CCF_3$	$\nu(C{\equiv}C)$, 1837–1800	314
$Pd(PBut^{n}_{3})(CF_3C{\equiv}CCF_3)$	$Pd(PBut^{n}_{3})_2Cl_2 + PBut^{n}_{3} + NH_2{-}NH_2$ and then $CF_3C{\equiv}CCF_3$	$\nu(C{\equiv}C)$, 1837–1795	314

TABLE 2.XII

Platinum(0) complexes with alkynes

Compound	Method of synthesis	Infrared absorptions (cm^{-1})	References
$Pt(PPh_3)_2(CH{\equiv}CH)$	*cis*-$Pt(PPh_3)_2Cl_2 + NH_2{-}NH_2 + CH{\equiv}CH$		366
	$Pt(PPh_3)_2(p\text{-}NO_2C_6H_4CH{=}CHC_6H_4NO_2\text{-}p) + CH{\equiv}CH$		306
$Pt(PPh_3)_2(CF_3C{\equiv}CH)$	$Pt(PPh_3)_4 + CF_3C{\equiv}CH$	$\nu(C{\equiv}C)$, 1705	169, 368
$Pt(PPh_3)_2(MeC{\equiv}CH)$	*cis*-$Pt(PPh_3)_2Cl_2 + NH_2{-}NH_2 + MeC{\equiv}CH$	$\nu(C{\equiv}C)$, 1712	201
$Pt(PPh_3)_2(EtC{\equiv}CH)$	*cis*-$Pt(PPh_3)_2Cl_2 + NH_2{-}NH_2 + EtC{\equiv}CH$	$\nu(C{\equiv}C)$, 1705	201
$Pt(PPh_3)_2(OHCH_2C{\equiv}CH)$	*cis*-$Pt(PPh_3)_2Cl_2 + NH_2{-}NH_2 + OHCH_2C{\equiv}CH$	$\nu(C{\equiv}C)$, 1705	169
	$Pt(PPh_3)_3 + OHCH_2C{\equiv}CH$		367
$Pt(PPh_3)_2(MeOCH_2C{\equiv}CH)$	*cis*-$Pt(PPh_3)_2Cl_2 + NH_2{-}NH_2 + MeOCH_2C{\equiv}CH$	$\nu(C{\equiv}C)$, 1708	169
$Pt(PPh_3)_2(HO_2CC{\equiv}CH)$	$Pt(PPh_3)_4 + HO_2CC{\equiv}CH$	$\nu(C{\equiv}C)$, 1710	169
$Pt(PPh_3)_2(OHMe_2CC{\equiv}CH)$	*cis*-$Pt(PPh_3)_2Cl_2 + NH_2{-}NH_2$ (or $Pt(PPh_3)_4$) $+ OHMe_2CC{\equiv}CH$	$\nu(C{\equiv}C)$, 1684	169, 176
$Pt(PPh_3)_2(OHMeEtCC{\equiv}CH)$	*cis*-$Pt(PPh_3)_2Cl_2 + NH_2{-}NH_2 + HOMe_2CC{\equiv}CH$	$\nu(C{\equiv}C)$, 1680–1690	369
$Pt(PPh_3)_2(HOMe_2CC{\equiv}CH)$	*cis*-$Pt(PPh_3)_2Cl_2 + NH_2{-}NH_2$ (or $Pt(PPh_3)_4$) $+ OHMeEtCC{\equiv}CH$		176
$Pt(PPh_3)_2(HOPh_2CC{\equiv}CH)$	*cis*-$Pt(PPh_3)_2Cl_2 + NH_2{-}NH_2 + HOPh_2CC{\equiv}CH$		176
$Pt(PPh_3)_2$([cyclopentane ring]–C≡CH, OH)	*cis*-$Pt(PPh_3)_2Cl_2 + NH_2{-}NH_2$ (or $Pt(PPh_3)_4$) + [cyclopentane ring]–C≡CH, OH		176
$Pt(PPh_3)_2$([cyclohexane ring]–C≡CH, OH)	*cis*-$Pt(PPh_3)_2Cl_2 + NH_2{-}NH_2$ (or $Pt(PPh_3)_4$) + [cyclohexane ring]–C≡CH, OH		176
$Pt(PPh_3)_2$([cycloheptane ring]–C≡CH, OH)	*cis*-$Pt(PPh_3)_2Cl_2 + NH_2{-}NH_2$ (or $Pt(PPh_3)_4$) + [cycloheptane ring]–C≡CH		176
$Pt(PPh_3)_2(PhC{\equiv}CH)$	*cis*-$Pt(PPh_3)_2Cl_2 + NH_2{-}NH_2 + PhC{\equiv}CH$	$\nu(C{\equiv}C)$, 1684	169, 366 370, 371

	$Pt(PPh_3)_2(C_2H_4) + PhC{\equiv}CH$		53, 235
$Pt(PPh_3)_2(PhC{\equiv}CD)$	*cis*-$Pt(PPh_3)_2Cl_2 + NH_2{-}NH_2 + PhC{\equiv}CD$	$\nu(C{\equiv}C)$, 1642	201
$Pt(AsPh_3)_2(PhC{\equiv}CH)$	*cis*-$Pt(AsPh_3)_2Cl_2 + NH_2{-}NH_2 + PhC{\equiv}CH$	$\nu(C{\equiv}C)$, 1675	370
$Pt(PPh_3)_2(RC_6H_4C{\equiv}CH)$ (R = 4-NO_2, 3-NO_2, 4-Cl, 2-Cl, 4-Br, 4-F, 2-F, 4-Me, 4-MeO, 2-MeO, 3-MeO)	*cis*-$Pt(PPh_3)_2Cl_2 + NH_2{-}NH_2 + R{-}C_6H_4C{\equiv}CH$		371
$Pt\{P(C_6D_5)_3\}_2(RC_6H_4C{\equiv}CH)$ (R = H, 4-MeO, 3-MeO, 4-Me, 3-Me, 4-F, 4-Cl, 3-Cl, 4-NO_2, 3-NO_2)	*cis*-$Pt\{P(C_6D_5)_3\}_2Cl_2 + NH_2{-}NH_2 + RC_6H_4C{\equiv}CH$		372
$Pt(PPh_3)_2(CNC{\equiv}CCN)$	$Pt(PPh_3)_4$ or $Pt(PPh_3)_2(PhC{\equiv}CH)$ or $Pt(PPh_3)_2HCl + CNC{\equiv}CCN$		337
$Pt(AsPh_3)_2(CNC{\equiv}CCN)$	$Pt(AsPh_3)_4 + CNC{\equiv}CCN$		337
$Pt(PEt_3)_2(CNC{\equiv}CCN)$	$Pt(PEt_3)_2HCl + CNC{\equiv}CCN$		337
$Pt(PPh_3)_2(CF_3C{\equiv}CCF_3)$	*cis*-$Pt(PPh_3)_2Cl_2 + NH_2{-}NH_2$ (or $Pt(PPh_3)_4$) $+ CF_3C{\equiv}CCF_3$	$\nu(C{\equiv}C)$, 1775	197, 314, 370
$Pt(AsPh_3)_2(CF_3C{\equiv}CCF_3)$	$Pt(AsPh_3)_2(PhC{\equiv}CH) + CF_3C{\equiv}CCF_3$	$\nu(C{\equiv}C)$, 1775	370
$Pt(PMe_2Ph)_2CF_3C{\equiv}CCF_3)$	$K_2PtCl_4 + KOH + PMe_2Ph$, and then $CF_3C{\equiv}CCF_3$	$\nu(C{\equiv}C)$, 1767	314
$Pt(PBu^n_3)_2(CF_3C{\equiv}CCF_3)$	$K_2PtCl_4 + KOH + PBu^n_3$ and then $CF_3C{\equiv}CCF_3$	$\nu(C{\equiv}C)$, 1758	314
$Pt(PMePh_2)_2(CF_3C{\equiv}CCF_3)$	$Pt(PMePh_2)_4 + CF_3C{\equiv}CCF_3$	$\nu(C{\equiv}C)$, 1762	81
$Pt\{PMe_2(C_6F_5)\}_2(CF_3C{\equiv}CCF_3)$	$Pt\{PMe_2(C_6F_5)\}_4 + CF_3C{\equiv}CCF_3$	$\nu(C{\equiv}C)$, 1778	81
$Pt(PPh_3)_2$(cyclohexyne)	$Pt(PPh_3)_3 + NaHg_x$ + 1,2-dibromocyclohexene (Br, Br)	$\nu(C{\equiv}C)$, 1721	339
$Pt\{C_2H_4(PPh_2)_2\}$(cyclohexyne)	$Pt(PPh_3)_2$(cyclohexyne) $+ C_2H_4(PPh_2)_2$		339
$Pt(PPh_3)_2$(cycloheptyne)	$Pt(PPh_3)_3 + NaHg_x$ + 1,2-dibromocycloheptene (Br, Br)	$\nu(C{\equiv}C)$, 1770	339

TABLE 2.XII *continued*

Compound	Method of synthesis	Infrared absorptions (cm^{-1})	References
$Pt\{C_2H_4(PPh_2)_2\}$()	$Pt(PPh_3)_2$(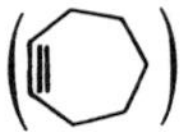) + $C_2H_4(PPh_2)_2$	$\nu(C{\equiv}C)$, 1761	339
$Pt(PPh_3)_2$(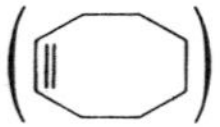)	$Pt(PPh_3)_4$ 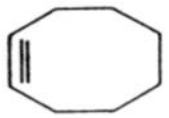 +		280
$Pt(PPh_3)_2(CH_3C{\equiv}CCH_3)$	*cis*-$Pt(PPh_3)_2Cl_2$ + $NH_2{-}NH_2$ + $CH_3C{\equiv}CCH_3$		373
$Pt(PPh_3)_2(EtC{\equiv}CEt)$	*cis*-$Pt(PPh_3)_2Cl_2$ + $NH_2{-}NH_2$ + $EtC{\equiv}CEt$	$\nu(C{\equiv}C)$, 1805	201
$Pt(PPh_3)_2(OHCH_2C{\equiv}CCH_2OH)$	*cis*-$Pt(PPh_3)_2Cl_2$ + $NH_2{-}NH_2$ + $OHCH_2C{\equiv}CCH_2OH$	$\nu(C{\equiv}C)$, 1780	169
$Pt(PPh_3)_2(MeOCH_2C{\equiv}CCH_2OMe)$	*cis*-$Pt(PPh_3)_2Cl_2$ + $NH_2{-}NH_2$ + $MeOCH_2C{\equiv}CCH_2OMe$	$\nu(C{\equiv}C)$, 1808	169
$Pt(PPh_3)_2(OHMe_2CC{\equiv}CCMe_2OH)$	*cis*-$Pt(PPh_3)_2Cl_2$ + $NH_2{-}NH_2$ + $OHMe_2CC{\equiv}CCMe_2OH$	$\nu(C{\equiv}C)$, 1745	169
$Pt(PPh_3)_2(OHMeEtCC{\equiv}CCEtMeOH)$	*cis*-$Pt(PPh_3)_2Cl_2$ + $NH_2{-}NH_2$ + $OHMeEtCC{\equiv}CCEtMeOH$		375
	$Pt(PPh_3)_2$(*trans*-$PhCH{=}CHPh$) + $OHMeEtCC{\equiv}CCEtMeOH$	$\nu(C{\equiv}C)$, 1745	261
$Pt(PPh_3)_2(MeO_2CC{\equiv}CCO_2Me)$	$Pt(PPh_3)_n$(n = 3,4) + $MeO_2CC{\equiv}CCO_2Me$	$\nu(C{\equiv}C)$, 1735 $\nu(C{=}O)$, 1685	301, 314
$Pt(PPh_3)_2(MeC{\equiv}CCMe_2OH)$	*cis*-$Pt(PPh_3)_2Cl_2$ + $NH_2{-}NH_2$ (or $Pt(PPh_3)_4$) + $MeC{\equiv}CCMe_2OH$	$\nu(C{\equiv}C)$, 1799	374
$Pt(PPh_3)_2(MeC{\equiv}CCPh_2OH)$	*cis*-$Pt(PPh_3)_2Cl_2$ + $NH_2{-}NH_2$ (or $Pt(PPh_3)_4$) + $MeC{\equiv}CCPh_2OH$	$\nu(C{\equiv}C)$, 1799	374
$Pt(PPh_3)_2$($MeC{\equiv}C$– HO) 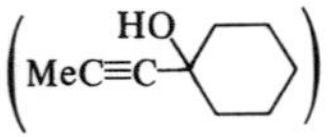	*cis*-$Pt(PPh_3)_2Cl_2$ + $NH_2{-}NH_2$ (or $Pt(PPh_3)_4$) 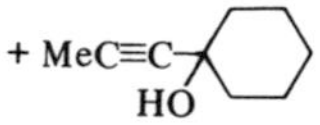 + + $MeC{\equiv}C$– HO	$\nu(C{\equiv}C)$, 1782	374
$Pt(PPh_3)_2$($MeC{\equiv}C$– HO CH_3 CH_3 CH_3 CH_3) 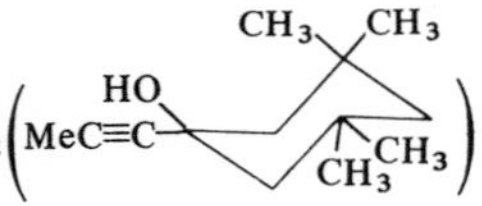	*cis*-$Pt(PPh_3)_2Cl_2$ + $NH_2{-}NH_2$ (or $Pt(PPh_3)_4$) 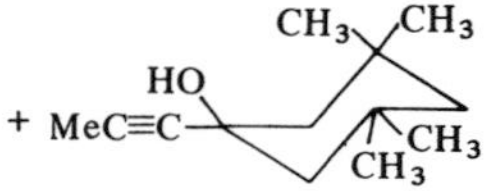 + + $MeC{\equiv}C$– HO CH_3 CH_3 CH_3 CH_3	$\nu(C{\equiv}C)$, 1789	374

$Pt(PPh_3)_2(MeC{\equiv}CC_2Me)$	*cis*-$Pt(PPh_3)_2Cl_2 + NH_2{-}NH_2 + MeC{\equiv}CC_2Me$	$\nu(C{\equiv}C)$, 2205 (uncomplexed) $\nu(C{\equiv}C)$, 1760	373
$Pt(PPh_3)_2(PhC{\equiv}CMe)$	*cis*-$Pt(PPh_3)_2Cl_2 + NH_2{-}NH_2 + PhC{\equiv}CMe$	$\nu(C{\equiv}C)$, 1756	201, 314, 364
$Pt(PPh_3)_2(PhC{\equiv}CCOMe)$	$Pt(PPh_3)_3 + PhC{\equiv}CCOMe$	$\nu(C{\equiv}C)$, 1725 $\nu(C{=}O)$, 1640	301
	$Pt(PPh_3)_2(MeC{\equiv}CMe) + PhC{\equiv}CCOMe$		373
$Pt(PPh_3)_2(PhC{\equiv}CEt)$	$Pt(PPh_3)_2(C_2O_4) + PhC{\equiv}CEt + h\nu$		133
$Pt(PPh_3)_2$ 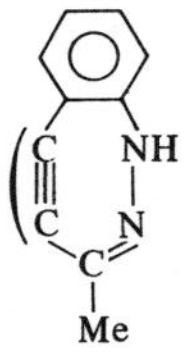	*cis*-$Pt(PPh_3)_2Cl_2 + NH_2{-}NH_2 + PhC{\equiv}CCOMe$	$\nu(C{\equiv}C)$, 1730 $\nu(C{=}N)$, 1655 $\nu(NH)$, 3380	373
$Pt(PPh_3)_2(PhC{\equiv}CPh)$	*cis*-$Pt(PPh_3)_2Cl_2 + NH_2{-}NH_2 + PhC{\equiv}CPh$	$\nu(C{\equiv}C)$, 1768–1740	314, 366 371
	$Pt(PPh_3)_2(C_2O_4) + PhC{\equiv}CPh + h\nu$		133
	$Pt(PPh_3)_2(CO_3) + PhC{\equiv}CPh + h\nu$		135
	$Pt(PPh_3)_2(OCOMe) + PhC{\equiv}CPh$		333
	$Pt(PPh_3)_2(C_2H_4) + PhC{\equiv}CPh$		157
	$Pt(PPh_3)_3 + PhC{\equiv}CPh$		157
$Pt(AsPh_3)_2(PhC{\equiv}CPh)$	$Pt(AsPh_3)_2(CO_3) + PhC{\equiv}CPh + h\nu$		135
$Pt(PPh_3)_2$(*p*-$MeC_6H_4C{\equiv}CC_6H_4Me$-*p*)	*cis*-$Pt(PPh_3)_2Cl_2 + NH_2{-}NH_2$ + *p*-$MeC_6H_4C{\equiv}CC_6H_4Me$-*p*-		201
$Pt(PPh_3)_2$(*p*-$NO_2C_6H_4C{\equiv}CC_6H_4NO_2$-*p*)	$Pt(PPh_3)_2\{$*o*-$C_6H_4(AsMe_2)_2\}$ + *p*-$NO_2C_6H_4C{\equiv}CC_6H_4NO_2$-*p*		366

This compound shows $\nu(C{\equiv}C)$ at 1730 cm^{-1}, a band at 1655 cm^{-1} which can be assigned to $\nu(C{=}N)$ and a weak band at 3380 cm^{-1} due to $\nu(NH)$. In the n.m.r. spectrum a single peak at τ 8.34 was assigned to the methyl resonance. The labilization of the *ortho* hydrogen of the phenyl seems to be analogous to that of ligands such as arylphosphines and arylazocompounds when bound to a metal, and possibly the ring closure proceeds through an intermediate hydrazone derivative.

Interestingly when hexafluorobut-2-yne, $CF_3C{\equiv}CCF_3$, was reacted with zerovalent nickel complexes such as $Ni(PPh_3)_2(C_2H_4)$, $Ni(AsMe_2Ph)_4$ or Ni(1,5-COD)$_2$ [376] derivatives of perfluorohexamethylbenzene were obtained

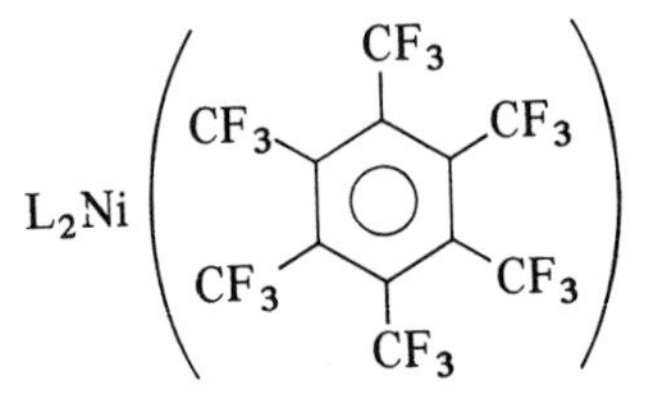

(L = PPh_3, $AsMe_2Ph$; L_2 = 1,5-COD). From the cyclo-octadiene derivative, similar compounds with L = $AsMe_2Ph$, $PMePh_2$, $P(OMe)_3$, $P(CH_2O)_3CMe$, or PPh_3, were obtained by displacing 1,5-COD with the ligands L [376].

From the reaction between Ni(1,5-COD)$_2$ and $CF_3C{\equiv}CCF_3$, a dimeric derivative was also isolated:

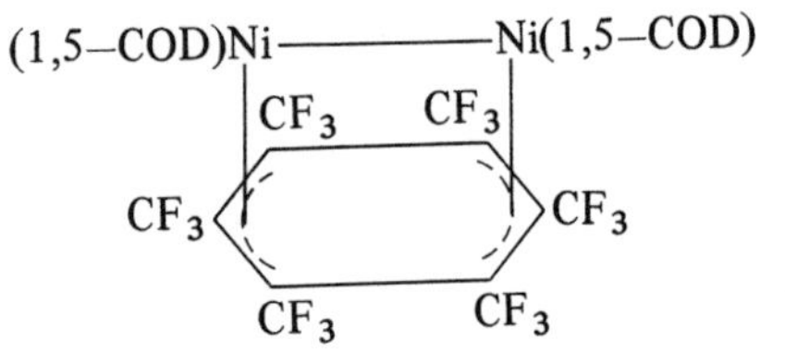

In the monomeric phosphine derivatives, the CF_3 signal in the ^{19}F n.m.r. spectrum appears as a triplet (J_{P-F} *ca.* 3–5 Hz), thus providing evidence for the symmetry of the bonded $C_6(CF_3)_6$ unit with respect to the two phosphorus atoms. This requires free rotation of the benzene molecule about the bond with nickel. However, this would result in pentacoordinated nickel(0), exceeding the rare gas configuration by two electrons. The possibility of a fluxional tetrahedral nickel(0), has also been considered where strong π-back-donation from nickel distorts the plane of the ring, which rotates by a 1,2-shift mechanism [376]:

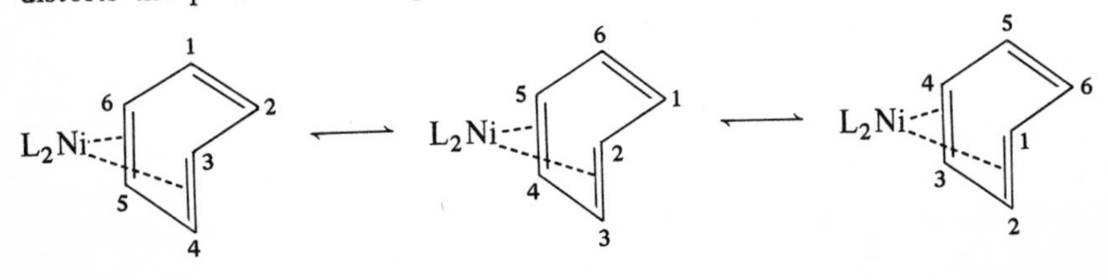

However, this tautomerism is very fast, since even at –90° only a slight broadening or a collapse to a single signal was observed for the resonance of the trifluoromethyl groups. A tetrakis(trifluoromethyl)nickel cyclopentadiene derivative has been proposed as the intermediate for the trimerization reaction of hexafluorobut-2-yne, in accord with the observation that from the reaction of metallocyclic derivatives of iridium(I) with alkynes, aromatic compounds were obtained [377]. This is in contrast to the so-called "π-complex, multicentre process", originally proposed for the polymerization and cyclization of alkynes, catalysed by zerovalent nickel complexes [1a]. According to this mechanism, three alkyne molecules successively coordinate with nickel and then the ring closure can take place.

In support of the former hypothesis, a palladiacyclopentadiene complex:

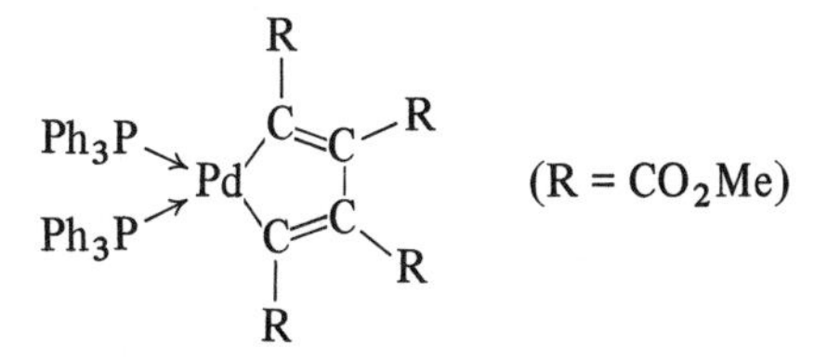

has been isolated from the reaction of $Pd(PPh_3)_2(MeO_2CC{\equiv}CCO_2Me)$ with excess alkyne [365].

From the reactions of zerovalent nickel complexes, having a chelating diphosphine as ligand, with alkynes such as 1-hexyne, Ph–C≡CH or 2-propyn-1-ol in benzene solution, coordination of the alkyne in solution was detected [ν(C≡C) $\simeq$ 1720–1750 cm^{-1}], but the complexes could not be isolated [101]. Polymerization of CH≡CH and PhC≡CH was also observed.

(*b*) *X-ray structures.* The structure of $Pt(PPh_3)_2(PhC{\equiv}CPh)$ has been determined [378]:

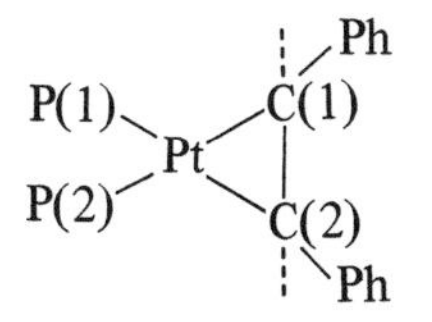

The phenyl substituents on the alkyne are bent at angles of 40° to the C–C direction and the carbon–carbon distance C(1)–C(2) is 1.32 ± 0.09 Å. This geometry is very similar to that of a *cis*-bent excited state of acetylene, which has a bond angle C(1)C(2)Ph of 142° and a carbon–carbon length of 1.38 Å. Thus this structure confirms that the wave function describing the metal–alkyne bond in these complexes will contain functions representing excited states of both metal and the alkyne (1.3.1.2). The coordination around platinum is distorted square planar, since platinum, the two phosphorus and one of the acetylenic carbons lie in a plane, which makes an angle θ of 14° with the line of the two acetylenic carbons. Other molecular parameters are: $d_{Pt-P(1)}$ = 2.28, $d_{Pt-P(2)}$ = 2.27 Å; C(1)PtC(2) = 39, P(1)PtP(2) = 102°.

From an X-ray study now in progress for $Pt(PPh_3)_2(CNC{\equiv}CCN)$, an almost planar structure [337] for the molecule, was the preliminary result. The C≡C distance in the complex is 1.40 Å, compared to 1.19 Å in free dicyanoacetylene.

The X-ray structures of small cyclic acetylenes coordinated to platinum in $Pt(PPh_3)_2$(alkyne) complexes have shown the usual distorted square planar coordination around platinum. In the cyclohexyne derivative [379]:

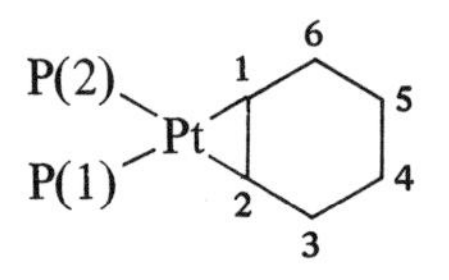

the planes defined by P(1)PtP(2) and C(1)PtC(2) form an angle θ = 3.2°. The C(1)–C(2) bond distance is 1.298 Å, while the remaining carbon–carbon distances in the ring are as expected, and only slightly shorter bonds were observed for C(1)–C(6) and C(2)–C(3) because of the different σ-orbital radii of carbon atoms approximately *sp* or sp^3 hybridized [379]. Steric strain is practically absent in the ring, and it seems almost exclusively relieved at the acetylenic carbons. Other molecular parameters are: $d_{Pt-C(1)}$ = 2.023, $d_{Pt-C(2)}$ = 2.055, $d_{Pt-P(1)}$ = 2.272, $d_{Pt-P(2)}$ = 2.269 Å; P(1)PtP(2) = 109.3, C(1)PtC(2) = 36.8, PtC(1)C(2) = 70, PtC(2)C(1) = 73, C(2)C(1)C(6) = 127, C(1)C(2)C(3) = 128°.

In the homologous cycloheptyne derivative [339]:

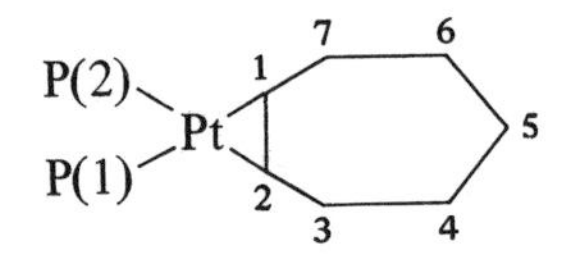

the dihedral angle θ was found to be 8.3°; the C(1)–C(2) distance of 1.294 Å is similar to that observed in the cyclohexyne derivative. The aliphatic carbon–carbon distances are normal, although significant deviations from the tetrahedral angles occur, presumably in an attempt to relieve ring strain. The most important molecular parameters are: $d_{Pt-P(1)}$ = 2.264, $d_{Pt-P(2)}$ = 2.272, $d_{Pt-C(1)}$ = 2.041, $d_{Pt-C(2)}$ = 2.068, $d_{C(1)-C(7)}$ = 1.48, $d_{C(7)-C(6)}$ = 1.52, $d_{C(6)-C(5)}$ = 1.47, $d_{C(5)-C(4)}$ = 1.50, $d_{C(4)-C(3)}$ = 1.57, $d_{C(3)-C(2)}$ = 1.50 Å; P(1)PtP(2) = 102.5, C(1)PtC(2) = 36.7, PtC(1)C(2) = 73, PtC(2)C(1) = 70, C(2)C(1)C(7) = 141, C(1)C(2)C(3) = 137°.

(*c*) *Spectral properties.* An infrared absorption at around 1700–1800 cm^{-1} is typical of the alkyne complexes we are considering here (Tables 2.X–XII), with a lowering of $\nu(C\equiv C)$ compared with the free alkyne of more than 400 cm^{-1}. This lowering was discussed when we considered the nature of the metal–alkyne interaction (1.3.1.2). This vibration becomes infrared active in the complexes, even when the alkyne is symmetrical, thus suggesting that upon complexation the alkynes distort from their linear geometry, as it has been determined for $Pt(PPh_3)_2(PhC\equiv CPh)$ (see Section (*b*)).

The entity of $\Delta\nu(C\equiv C)$ has been considered as an indication of the thermodynamic stability of the complexes [314, 327]. For the series of derivatives of the very strong π-acid hexafluorobut-2-yne (Table 2.XIII) values of

TABLE 2.XIII
$\nu(C\equiv C)$ and $\Delta\nu(C\equiv C)$ for $ML_2(CF_2C\equiv CCF_3)$ complexes

Compound	$\nu(C\equiv C)$	$\Delta\nu(C\equiv C)$
$CF_3C\equiv CCF_3$	2300 (Raman)	–
$Ni(PPh_3)_2(CF_3C\equiv CCF_3)$	1790	510
$Pd(PPh_3)_2(CF_3C\equiv CCF_3)$	1838–1811	475 (average shift)
$Pd(PMe_2Ph)_2(CF_3C\equiv CCF_3)$	1837–1800	482 (average shift)
$Pd(PBu^n_3)_2(CF_3C\equiv CCF_3)$	1837–1795	484 (average shift)
$Pt\{PMe_2(C_6F_5)\}_2(CF_3C\equiv CCF_3)$	1778	522
$Pt(PPh_3)_2(CF_3C\equiv CCF_3)$	1775	525
$Pt(AsPh_3)_2(CF_3C\equiv CCF_3)$	1775	525
$Pt(PMe_2Ph)_2(CF_3C\equiv CCF_3)$	1767	533
$Pt(PMePh_2)_2(CF_3C\equiv CCF_3)$	1762	538
$Pt(PBu^n_3)_2(CF_3C\equiv CCF_3)$	1758	542

See Tables 2.X, 2.XI and 2.XII for references.

$\Delta\nu(C\equiv C)$ are about 500 cm^{-1}, and they follow the order Pt $\simeq$ Ni $\gg$ Pd, which agrees with the prediction based on promotion energies (1.4) that palladium should be a poorer π-donor than nickel and platinum. With increasing basicity of the phosphine bound to the metal, a small increase on $\Delta\nu(C\equiv C)$ can also be observed.

For a series of platinum(II) derivatives of hexafluorobut-2-yne, $PtL_2(Cl)(CH_3)(CF_3C\equiv CCF_3)$ (L = PMe_3, $AsMe_3$, $SbMe_3$), $\Delta\nu(C\equiv C)$ was found to be about 435–475 cm^{-1} [380], which is *ca.* 100 cm^{-1} lower than in the zerovalent platinum complexes, where phosphines of basicity comparable to PMe_3 are present (Table 2.XIII).

According to the $\Delta\nu(C\equiv C)$ criterion, the alkyne derivatives could be ordered in a series of decreasing thermodynamic stability. However, this order does not always fit the kinetic results for the displacement reactions:

$$Pt(PPh_3)_2(\text{alkyne}) + \text{alkyne}^* \longrightarrow Pt(PPh_3)_2(\text{alkyne}^*) + \text{alkyne}$$

This is probably because this type of reaction does not simply proceed through a pre-dissociation of the type:

$$Pt(PPh_3)_2(\text{alkyne}) \xrightarrow{\text{slow}} Pt(PPh_3)_2 + \text{alkyne}$$

in which case the kinetic data should give direct information about the strength of the metal–alkyne interaction, but through an associative process (1.3.5).

For studying the steric and electronic influences of alkynes on the structures of platinum(0) alkyne complexes, a large series of hydroxyalkyne derivatives has been synthesized [176, 374]. Some types of hydrogen bonding must be present since ν(OH) is lowered by 45–260 cm^{-1} in the complexes [176]. The ^1H n.m.r. resonances of the hydroxy groups lie at higher τ in the complexes and this could indicate less association of these groups in the complexes than in free alkynes. However τ values can be affected by many different effects (1.3.2.1). For platinum(II) derivatives of hydroxyalkynes, $K[PtCl_3(\text{alkyne})]$, ν(OH) and τ(OH) are lower in the complexes, and this is consistent with hydrogen-bonding of OH to chlorine upon coordination [375].

Usually large phosphorus-hydrogen coupling constants have been determined in $M(PPh_3)_2(PhC{\equiv}CMe)$ (M = Ni, Pt) complexes [314, 364]. The platinum derivative shows the methyl resonance at τ 7.88, as a doublet with $J_{P_{trans}-H}$ = 6.2 Hz, and satellites due to J_{Pt-H} = 41.5 Hz. The resonances are further split by coupling with the second phosphorus atom ($J_{P_{cis}-H} \simeq 1.2$ Hz).

Similarly, the nickel derivative shows the methyl resonance at τ 7.86, with $J_{P_{trans}-H}$ = 4.5 Hz, further split by $J_{P_{cis}-H} \simeq 1$ Hz.

This implies that the rate of rotation of the alkyne about the metal–alkyne axis is very slow, on the n.m.r. time scale, and that this process has a very high activation energy [364]. By comparison, the ^1H n.m.r. spectrum of *cis*-$Pt(PPh_3)_2(CH_3)_2$ exhibits a quartet centred at τ 9.63, with satellites due to J_{Pt-H} = 69 Hz. Calculations of J_{P-H} *cis* and *trans* were not attempted, but they appeared to be of comparable magnitude. Similar couplings of fluorine with phosphorus and platinum were also observed for the CF_3 groups in $M(PPh_3)_2(CF_3C{\equiv}CCF_3)$ (M = Ni, Pd, Pt) complexes [314, 370]. The ^{19}F spectrum for M = Pt shows a doublet at 10.4 p.p.m. on the low field side of benzotrifluoride (J_{P-F} = 10.2 Hz) with satellites due to J_{Pt-H}= 64 Hz. It seems that in this case only the coupling with the phosphorus in *trans* position can be observed. Coupling to phosphorus or platinum is absent when the substituent on the alkyne is more remote, as in $M(PPh_3)_2(MeO_2CC{\equiv}CCO_2Me)$ (M = Pd, Pt) [314]. The position and couplings of the hydrogen resonances in monosubstituted alkyne complexes of platinum have been extensively studied [176, 372, 374] (see also 1.3.2). Resonances have been found at around τ 3.7, that is a region where alkenes absorb, for the derivatives of hydroxylalkynes of the type $Pt(PPh_3)_2(RC{\equiv}CH)$ (R = aliphatic substituent bearing a hydroxy group) [176, 374).

The signal is split in at least four resolvable peaks, with $J_{Pt-H} \simeq 60$ Hz, $J_{P_{trans}-H} \simeq 22$ Hz and $J_{P_{cis}-H} \simeq 2-10$ Hz. For a series of substituted phenylacetylenes of the type $Pt(PR_3)(XC_6H_4C{\equiv}CH)$ (R = C_6D_5), the resonances of the acetylenic hydrogens have been found at lower values, τ *ca.* 2.3–2.7 [372]. This explains why, when the ligand is triphenylphosphine, this signal is not detectable, since it lies under the phenyl resonances for the phosphine. The coupling constants have the usual values, similar to those reported above.

Some ^{31}P spectra for $Pt(PPh_3)_2$(alkyne) complexes have also been reported [201]. When the alkyne is symmetrical, e.g. $EtC{\equiv}CEt$, the two phosphorus atoms are equivalent and show a single resonance at *ca.* –27 to –30 p.p.m. relative to H_3PO_4, with $J_{Pt-P} \simeq 3400-3800$ Hz. However, when the alkyne is not symmetrical, e.g. $PhC{\equiv}CH$, the phosphorus nuclei are no longer equivalent, and two distinct signals with $J_{PP} \simeq 32-35$ Hz can be observed. Such values are consistent with the phosphines being mutually *cis* in these complexes.

Finally we briefly mention here that the mass spectra of $Pt(PPh_3)_2$(alkyne) (alkyne = cyclohexyne, cycloheptyne), show the molecular ions $\{Pt(PPh_3)_2(C_{n+2}H_{2n})\}^+$ (n = 4, 5) and the fragment ion $\{Pt(PPh_3)_2\}^+$ [339], and these are rare examples among the very few mass spectra of zerovalent complexes of nickel, palladium and platinum so far studied.

(*d*) *Reactivity.* Kinetic measurements of the exchange reactions:

$$Pt(PPh_3)_2(ac) + ac' \rightleftharpoons Pt(PPh_3)_2(ac') + ac$$

$$(ac = alkyne)$$

have already been discussed (1.3.5).

Protonation reactions of alkynes complexes of platinum have recently received considerable attention. The following mechanism seems to be the most probable [373]:

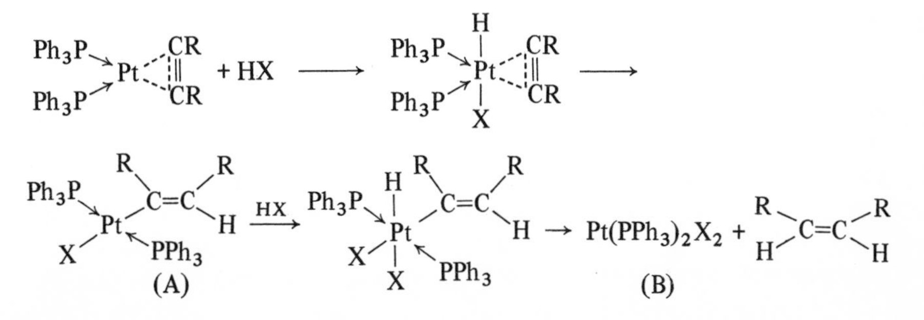

with a *cis* stereochemistry about the double bond in the vinyl intermediate (A), which can be isolated in some cases. This configuration has been elucidated by studying the protonation of $Pt(PPh_3)_2(p\text{-}CH_3C_6H_4C{\equiv}CH)$ with deuterated hydrochloric acid DCl, from which *trans*-$Pt(PPh_3)_2Cl(p\text{-}CH_3C_6H_5C{=}CHD)$ can be isolated [201]. This is in accord with the hydrocarbonylation of alkynes with $Ni(CO)_4$ and an aqueous acid, which involve an overall *cis* addition of hydrogen and of the carboxylic group to the triple bond in the Markownikoff sense, giving for example $CH_2{=}CPh(COOH)$ from $PhC{\equiv}CH$ [201].

By studying the protonation of $Pt(PPh_3)_2(CH_3C{\equiv}CCH_3)$ with different acids and in different solvents, *cis*-butene-2 was found as the major product, but also notable amounts of the *trans* isomer were obtained [373]:

	% *trans*-butene-2	
Acid	$CHCl_3$	aromatic solvent
HCl(g)	20	8
HCl(conc.)	26	9
CH_3COOH	22	14
CF_3COOH	42	14
$C_6H_2(NO_2)_3OH$	—	10
CH_3COSH	34	53
PhCOSH	30	44

In the presence of poisoning agents of hydrogenation catalysts, such as methyl sulphide, no appreciable variation in the products was observed. Unreacted butyne-2, butane or the isomerization product butene-1, were not detected. It has been assumed that isomerization, from the *cis* vinyl complex intermediate to the *trans* alkene, occurs because the *trans* isomer is more stable thermodynamically. A more complete isomerization to the *trans* form is observed with thioacids, since the sulphur ligands form more stable intermediates. The hydrogen transfer step probably occurs more slowly, thereby allowing a more complete isomerization.

Protonation of $Pt(PPh_3)_2(PhC{\equiv}CPh)$ with HCl in benzene gives *trans*-$Pt(PPh_3)_2Cl(CPh{=}CHPh)$ [201, 381], which upon further reaction with HCl in $CHCl_3$ is transformed to $Pt(PPh_3)_2Cl_2$ and *trans*-stilbene [381]. *Trans*-stilbene has also been obtained from $Pt(PPh_3)_2(PhC{\equiv}CPh)$ and CF_3COOH [333]. The observed formation of only *trans*-stilbene should thus imply that the platinum–carbon fission occurs either with a complete inversion of configuration, or that *cis*-stilbene is formed but is isomerized to the *trans*-isomer by the acidic medium. The protonation of $Pt(PPh_3)_2$(alkyne) complexes (alkyne = $PhC{\equiv}CMe$, $EtC{\equiv}CEt$, $MeC{\equiv}CH$, $EtC{\equiv}CH$, $PhC{\equiv}CH$ or $p\text{-}CH_3C_6H_4C{\equiv}CH$) with acids such as HCl or CF_3COOH have also been studied [201]. From $Pt(PPh_3)_2(PhC{\equiv}CMe)$ and CF_3COOH a *ca.* 1:1 mixture of the two isomers $Pt(PPh_3)_2X(CPh{=}CHMe)$ and $Pt(PPh_3)_2X(CMe{=}CHPh)$ (X = CF_3COO) has been obtained. The protonation of monosubstituted alkyne complexes is always in the Markownikoff sense, giving $Pt(PPh_3)_2X(CR{=}CH_2)$ ($X = Cl, CF_3COO$). The vinyl complexes show the expected $\nu(C{=}C)$ at around 1550–1570 cm^{-1}, while the chlorine derivatives also have $\nu(Pt{-}Cl)$ at around 247–279 cm^{-1}.

Similar vinyl complexes have been obtained by reacting $Pt(PPh_3)_2$(alkyne) complexes (alkyne = cyclohexyne, cycloheptyne) with CF_3COOH [339]. In this case the protonation can only give rise to a *cis* arrangement about the double bond:

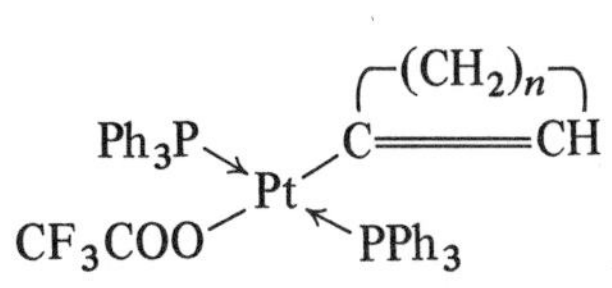

$Pt(PPh_3)_2(CF_3C{\equiv}CCF_3)$ and HCl [337, 381] or CF_3COOH [333] give the very stable vinyl complexes $Pt(PPh_3)_2X(CCF_3{=}CHCF_3)$ and a *cis* arrangement of the CF_3 groups have been demonstrated by the ^{19}F n.m.r. spectrum (J_{FF} = 11 Hz) [333]. A cyanoalkyne derivative such as $Pt(PPh_3)_2(CNC{\equiv}CCN)$ can be also protonated with HBr giving *trans*-$Pt(PPh_3)_2Br(CNC{=}CHCN)$, and the reverse reaction has also been shown to be possible [337]:

$$Pt(PEt_3)_2HCl + CNC{\equiv}CCN \xrightarrow{\text{benzene}} Pt(PEt_3)_2Cl(CNC{=}CHCN).$$

$Pt(PMePh_2)_2(CF_3C{\equiv}CCF_3)$ also reacts with HgX_2 ($X = Cl, Br$) in ethanol giving [259]:

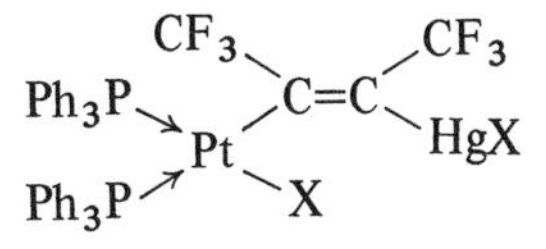

For X = Cl, $\nu(C{=}C)$ is at 1578 cm^{-1} and $\nu(Pt{-}Cl)$ is at 348–310 cm^{-1}. A *cis* configuration of the two CF_3 groups has been confirmed by the ^{19}F n.m.r. spectrum.

Alkyne derivatives of platinum have been shown to react slowly with chloroform, giving $Pt(PPh_3)_2Cl(CHCl_2)$ and the free alkyne [176, 314, 374]. $Pt(PPh_3)_2(PhC{\equiv}CR)$ (R = H, Ph) also react slowly with silicon hydrides R_3SiH, with formation of $Pt(PPh_3)_2H(SiR_3)$ [182]. The reaction is reversible and with excess alkyne the starting complex can be recovered.

Tetrachloro-*o*-benzoquinone readily undergoes a reaction with a benzene solution of $Pt(PPh_3)_2(PhC{\equiv}CPh)$ [261]:

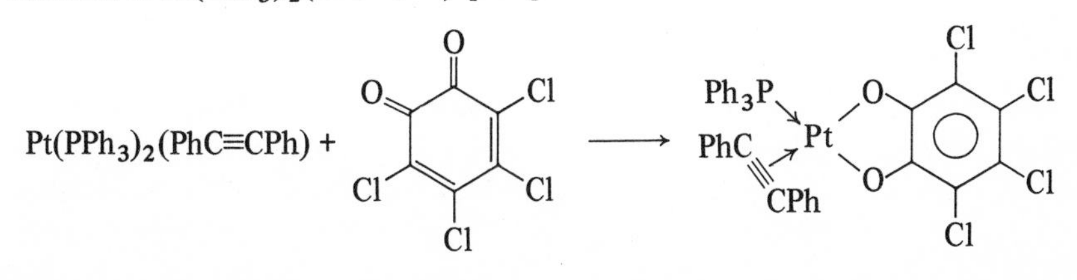

The presence of a weak band at 1961 cm^{-1} due to ν(C≡C) confirms the presence of an alkyne coordinated to platinum(II), and probably the alkyne is perpendicular to the coordination plane [261]. In contrast, treatment of the complexes $Pt(PPh_3)_2(PhC{\equiv}CH)$, $Pt(PPh_3)_2(EtMeOHCC{\equiv}CCOHMeEt)$, $Pt(PPh_3)_2(OHCH_2C{\equiv}CCH_2OH)$, and of the alkene derivatives $Pt(PPh_3)_2$(*trans*-PhCH=CHPh) and $Pt(PPh_3)_2(CH_2{=}CF_2)$, with the *ortho*-quinone leads to the displacement of the alkyne or of the alkene, and $Pt(PPh_3)_2$ (O O) is formed.

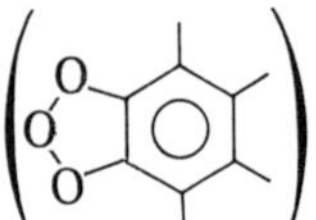

On the other hand, tetrachloro-*o*-benzoquinone has no action upon $Pt(PPh_3)_2(CF_3C{\equiv}CCF_3)$ and $Pt(PPh_3)_2(CF_2{=}CF_2)$: The factors which control the formation of a product of an oxidative reaction where the alkyne is still present, are thus very delicately balanced. Anyhow, the fact that these oxidative addition reactions are possible with alkyne (and alkene) derivatives of the type we are considering here, seems to be further evidence that such complexes are better considered as derivatives of a metal in a formally zero oxidation state, than complexes with rigid cyclopropene (or cyclopropane) structures. The lack of reactivity of the fluoroalkyne and fluoroalkene complexes reminds us of the inertness of complexes such as $Pt\{P(CF_3)F_2\}_4$ towards attack by hydrochloric acid (2.2.3.2), although $Pt(PPh_3)_2(CF_3C{\equiv}CCF_3)$ gives the vinyl complex with this acid (see above).

With excess bromine, $Pt(PPh_3)_2(CNC{\equiv}CCN)$ gives $Pt(PPh_3)_2Br_2$, while under more controlled conditions, that is by using pyridinium bromide perbromide as a mild brominating agent, $Pt(PPh_3)_2(BrCNC{=}CCNBr)$ is obtained [337].

2.2.3.10. Other Reactions

The strong basicity of the zerovalent complex $Pt(PPh_3)_3$, is nicely confirmed by its behaviour towards typical Lewis acids such as 1,3,5-trinitrobenzene and boron trichloride.

On considering the similarity between 1,3,5-trinitrobenzene (TNB) and tetracyanoethylene as acceptors in a variety of charge-transfer complexes, one expects that the former may also act as an acceptor towards $Pt(PPh_3)_3$ [382]. In fact by refluxing $Pt(PPh_3)_3$ and TNB in benzene, black crystals of $Pt(PPh_3)_2$(TNB) have been obtained. This compound shows ν(C=C), 1615, $\nu(NO_{2\,asym})$, 1520 and $\nu(NO_{2\,sym})$ 1340 cm^{-1}.

In the ^{1}H n.m.r. spectrum, besides the absorptions of the hydrogens of the phosphines, two broad lines were observed at τ 3.55 and 8.47. The broadening is probably due to coupling with ^{31}P and ^{195}Pt, and the position of the two resonances seems to indicate the presence of olefinic and aliphatic protons in the complex [382]. On this basis, the following structure has been proposed:

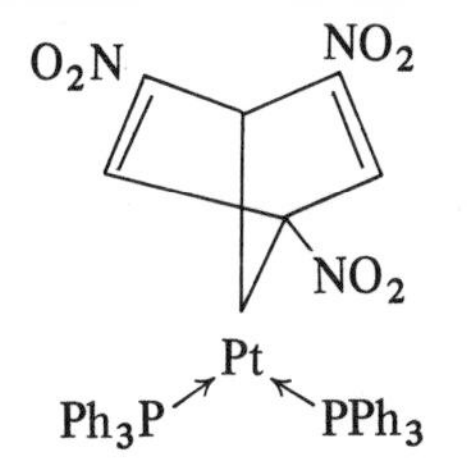

with two platinum–carbon σ-bonds. The alternative π-arene structure in fact violates the rare gas rule, and it does not explain the marked non-equivalence of the hydrogens in the ring. Although strong nucleophiles such as tetraphenylarsonium cyanide do not displace TNB from the complex and the expected Meisenheimer type cyanide ion-TNB adduct is not formed [382], a rigid structure corresponding to a tetrahedral platinum(0), where a strong π-back-donation from platinum distorts the plane of the ring, cannot be ruled out (see also 2.2.3.9a).

No complexes could be obtained by reacting $Pt(PPh_3)_3$ with *meta*-dinitrobenzene or with benzene itself. Picryl chloride gave a derivative of formula $Pt(PPh_3)_2Cl\{C_6H_2(NO_2)_3\}$ [382], which results from the oxidative addition of the aryl chloride to the platinum(0) complex. The Lewis acid–base adduct, $Pt(PPh_3)_2 . SiF_4$ has been obtained by treatment of $Pt(PPh_3)_3$ with SiF_4 in benzene [383].

From infrared studies, the silicon atom in the complex has been shown to be five-coordinate, and therefore a simple oxidative addition product such as $Pt(PPh_3)_2F(SiF_3)$ has been ruled out. The ^{19}F spectrum measured in acetone consists of a singlet at 137.3 p.p.m. to high field from CCl_3F, without any coupling to other nuclei.

Of the two possibilities which can explain such behaviour:

$$Pt(PPh_3)_2 . SiF_4 \rightleftharpoons [Pt(PPh_3)_2SiF_3]^+ + F^-$$

or

$$Pt(PPh_3)_2 . SiF_4 \rightleftharpoons Pt(PPh_3)_2 + SiF_4$$

the latter has been considered to be more likely.

On pyrolysis, or by reaction with iodine and DCl, $Pt(PPh_3)_2 . SiF_4$ gives silicon tetrafluoride. However, on treatment with ammonia, the expected Lewis salt $SiF_4 . 2NH_3$ is not formed, but rather a 1:1 adduct is obtained:

$$Pt(PPh_3)_2 . SiF_4 + NH_3 \rightarrow Pt(PPh_3)_2 . SiF_4 . NH_3$$

It has been considered that SiF_4 is acting as a diacid, the bases being ammonia and $Pt(PPh_3)_2$. In this latter case silicon must be considered sp^3d^2 hybridized, still having the $3d_{xz}$ and $3d_{yz}$ orbitals vacant and able to accept electron density from the filled $5d_{xz}$ and $5d_{yz}$ orbitals of the platinum atom with the possibility of a Si–Pt π-interaction in competition with the Pt–P π-interaction [383]. The platinum $4_{f_{7/2}}$ binding energy (73.6 eV) in $Pt(PPh_3)_2 \cdot SiF_4$ has a value close to that in $Pt(PPh_3)_2Cl_2$ (1.3.3), which implies a very strong Pt–Si multiple bond.

The strong Lewis acid BCl_3 can only interact with platinum via σ-bonds. Treatment of $Pt(PPh_3)_3$ with BCl_3 affords $Pt(PPh_3)_3 \cdot 2BCl_3$ [384]. The absence of i.r. absorptions from 400 to 200 cm^{-1} indicates that no Pt–Cl bonds are present in this complex, excluding an oxidative addition reaction having taken place.

Boron trichloride is in part regenerated by pyrolysis of $Pt(PPh_3)_3 \cdot 2BCl_3$, the adduct $PPh_3 \cdot BCl_3$ also being formed. Similarly the pyridine adduct $Py \cdot BCl_3$ is obtained when the complex is treated with excess pyridine, indicating that the latter is a stronger base than the zerovalent platinum complex. From infrared studies, the possibility that one molecule of BCl_3 is merely trapped in the crystal lattice as in a solvated complex, has been ruled out, and trigonal bipyramidal structure has been proposed for the bis-adduct:

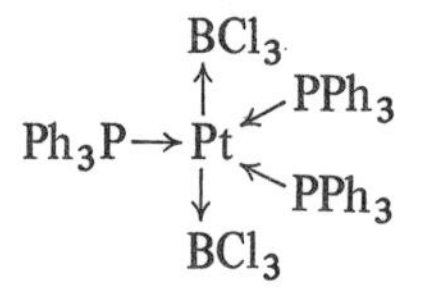

where the BCl_3 molecules are bonded to the platinum by electron pairs donated from the metal $p_z d_{z^2}$ hybrids.

When $Pt(PPh_3)_2 \cdot SiF_4$ is treated with boron trichloride SiF_4 is evolved and $Pt(PPh_3)_2 \cdot BCl_3$ is obtained [384]. This implies that BCl_3 is a stronger Lewis acid toward platinum than SiF_4, although in the latter case additional reinforcement of the metal-silicon bond should come from the Pt–Si π-interaction. Unfortunately binding energy data, which could indicate the extent of the electron transfer from the metal, have not been reported for the boron trichloride adducts.

When $Pt(PPh_3)_3$ is treated with excess aluminum trimethyl $[Al(CH_3)_3]_2$, the unstable $Pt(PPh_3)_2 \cdot 2Al(CH_3)_3$ is obtained [384]. The decomposition of this adduct does not afford free $[Al(CH_3)_3]_2$. This has been attributed to the fact that the aluminium-carbon bonds have lower energies (85 kcal mol^{-1}) than silicon-fluorine (165 kcal mol^{-1}) and boron-chlorine (128 kcal mol^{-1}) bonds, and thus the decomposition of the Pt–Al adduct probably involves an oxidative addition type reaction instead of a simple dissociation of the adduct to its components [384]. In general the drastic thermal decomposition of the adducts reported above also leads to some benzene among the products, generated by the degradation of triphenylphosphine.

The reaction between $Pt(PPh_3)_4$ or $Pt(PPh_3)_2$(*trans*-PhCH=CHPh) and $Fe_2(CO)_9$ gave several complexes, which could be separated by chromatography [80]. The air-stable, deep red $Fe_2Pt(CO)_9(PPh_3)$ was obtained, together with $Fe(CO)_4(PPh_3)$ and $Fe(CO)_3(PPh_3)_2$. The i.r. spectrum showed only terminal ν(CO) frequencies. Analogous complexes were obtained from the reaction between $Pt(AsPh_3)_4$ and $Fe_2(CO)_9$. Using other phosphorus ligands [L = $PMePh_2$, PMe_2Ph, $P(OPh)_3$, $PPh(OMe)_2$; $L_2 = C_2H_4(PPh_2)_2$, o-$C_6H_4(AsMe_2)_2$] [80], a second series of compounds of stoichiometry $Fe_2Pt(CO)_8L_2$ was obtained. No iron-platinum cluster compound could be isolated from the reaction between $Fe(CO)_5$ and $Pt(PMePh_2)_4$. Reaction of $Pt(PPh_3)_2$(*trans*-PhCH=CHPh) with $Fe_3(CO)_{12}$ afforded both type of complexes, with L = PPh_3, besides two products of composition $Pt_4(CO)_n(PPh_3)_3$ (n = 6, 8). Attempts to prepare analogous iron-palladium cluster compounds from reactions between $Fe_2(CO)_9$ and PdL_4 [L = PPh_3, $PMePh_2$, $P(OPh)_3$] were unsuccessful.

From the reaction between $Fe_2(CO)_9$ and $Pt\{P(OPh)_3\}_4$, $FePt_2(CO)_5\{P(OPh)_3\}_3$ and $Fe_2Pt(CO)_8\{P(OPh)_3\}_2$, have been obtained. The following structures have been assigned to these complexes [80]:

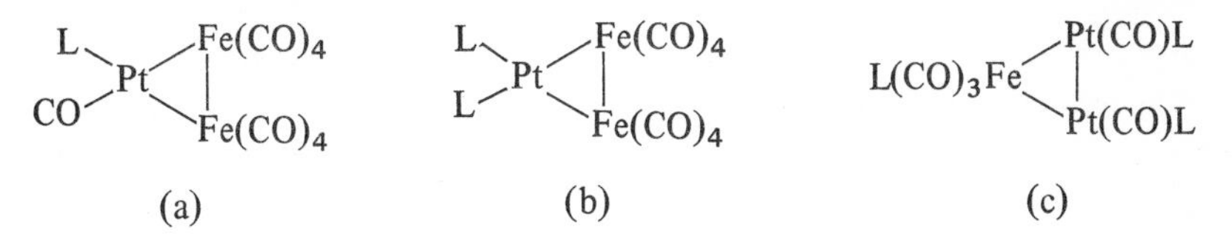

two of which (a and c) have been confirmed by X-ray studies [385, 386]. Ruthenium-platinum cluster compounds of stoichiometry $RuPt_2(CO)_5L_3$ were obtained from reactions between PtL_4 (L = $PMePh_2$, PMe_2Ph, $AsPh_3$) or $Pt(PPh_3)_2$ (*trans*-PhCH=CHPh) and $Ru_3(CO)_{12}$ [387]. They were obtained in mixtures with varying amounts of the phosphine-substituted triruthenium carbonyls, $Ru_3(CO)_{12-n}L_n$ (n = 1, 2 or 3 depending on L). Bridging and terminal CO groups were detected by i.r. spectroscopy in the ruthenium-platinum cluster compounds. In the reaction involving $Pt(PMe_2Ph)_4$, a second cluster derivative of formula $Ru_2Pt(CO)_7(PMe_2Ph)_3$ was also obtained. The reactions between PtL_4 [L = $PPh(OMe)_2$, $P(OPh)_3$] and $Ru_3(CO)_{12}$ were also studied, and cluster derivatives analogous to those reported above were obtained [387]. With $Pt\{C_2H_4(PPh_2)_2\}_2$, $Ru_3(CO)_{12}$ gives $Ru_2Pt(CO)_8\{C_2H_4(PPh_2)_2\}$, with an analogous structure to the iron derivatives (b) (see above).

Some reactions between $Os_3(CO)_{12}$ and platinum(0) phosphine complexes have also been studied [387], but no mixed cluster compounds have been obtained, probably as consequence of the greater stability of the Os_3 cluster in comparison to that of the Ru_3 unit. On the other hand, osmium-platinum complexes of stoichiometry $OsPt_2(CO)_5(PPh_3)_3$ and $[HOsPt(CO)_4(PPh_3)]_2$ have been obtained from the reaction between *cis*-$H_2Os(CO)_4$ and $Pt(PPh_3)_2$-

(C_2H_4) [387]. With $Pt(PMePh_2)_4$ and $H_2Os(CO)_4$, $OsPt_2(CO)_5(PMePh_2)_3$ and $Os_2Pt(CO)_7(PMePh_2)_3$ were isolated. The formation of these cluster complexes of Fe, Ru, Os with Pt could be explained by considering that a species like $Pt(PPh_3)_2$ has a chemical behaviour formally similar to that of carbenes [67, 387, 388], in that it can for instance insert into a metal-metal bond as does $SnCl_2$ [389]. However this similarity is more apparent than real, since carbenes behave mainly as electrophilic species, while low oxidation state complexes, and in particular $Pt(PPh_3)_2$, behave more as nucleophilic species [67].

A transfer reaction of azide ions occurs when $Pt(PPh_3)_n$ (n = 3, 4) and $AsPh_4[Au(N_3)_4]$ are dissolved in a tetrahydrofuran-benzene mixture [390]:

$$Pt(PPh_3)_3 + [Au^{III}(N_3)_4]^- \longrightarrow \left[(N_3)_2Au(\mu\text{-}N_3)_2Pt(PPh_3)_2 \right]^- \longrightarrow [Au^{I}(N_3)_2]^- + (N_3)_2Pt(PPh_3)_2$$

The reaction probably follows the bridge mechanism reported above, where N_3^- behaves formally as a pseudohalogen in an oxidative addition reaction.

The use of the commercially available $Ni(PF_3)_4$ as a source of trifluorophosphine in the syntheses of metal-PF_3 complexes has recently been reported [391]. Thermal reactions of $Ni(PF_3)_4$ and the pentamethylcyclopentadiene derivatives, $\{(CH_3)_5C_5MCl\}_2$ (M = Rh, Ir) give $\{(CH_3)_5C_5M(PF_3)_2\}$. In these reactions $Ni(PF_3)_4$ acts both as PF_3 donor and as a dehalogenating and reducing agent, as in the reaction [391]:

$$Mn(CO)_5Br \xrightarrow{Ni(PF_3)_4} Mn_2(CO)_{10-x}(PF_3)_x \qquad (x = 0,1,2)$$

In other cases, such as:

$$\pi\text{-}C_5H_5M(CO)_3CH_3 \xrightarrow{Ni(PF_3)_4} \pi\text{-}C_5H_5M(CO)_2(PF_3)CH_3$$
$$(M = Mo, W)$$

$Ni(PF_3)_4$ acts only as a source of trifluorophosphine, and in this respect it behaves like PtL_4 in some of its reactions with the carbonyls of the iron triad [80, 387].

Keten dimer reacts at room temperature with $Pt(PPh_3)_4$ giving the white adduct:

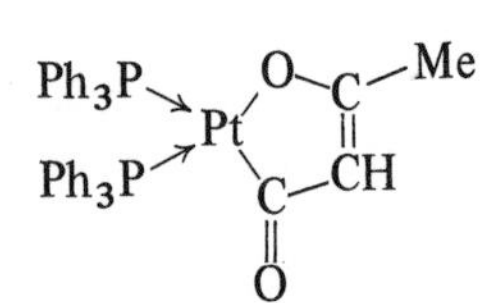

which shows ν(CO) at 1632 and ν(C=C) at 1659 cm^{-1} [392].

The reaction of organic radicals such as 1,2-diphenyl-2-picrylhydrazyl with $Pt(PPh_3)_4$ provides a new type of oxidative addition [262]:

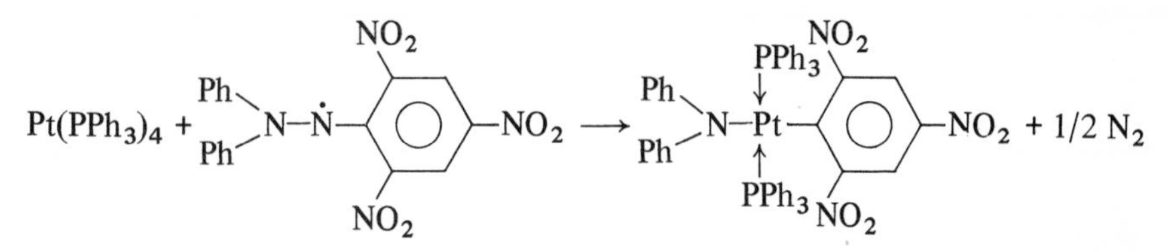

In contrast, no reaction was observed between the zerovalent platinum complex and diaryl and dialkylnitrogenoxide and tetra-arylhydrazine. We have already considered some oxidative addition reactions where carbon–carbon single bond cleavage occurs (2.2.3.3, 2.2.3.8(b)). Some other examples of this uncommon reaction are known. The triethylphosphine derivatives $Pt(PEt_3)_3$ and $Ni(PEt_3)_4$ give *trans*-$M(PEt_3)_2(Ph)(CN)$ (M = Ni, Pt) with benzonitrile at room temperature [132]. Treatment of $M(PPh_3)_4$ (M = Pd, Pt) with 1,1,2,2-tetracyano-cyclopropane gives [393]

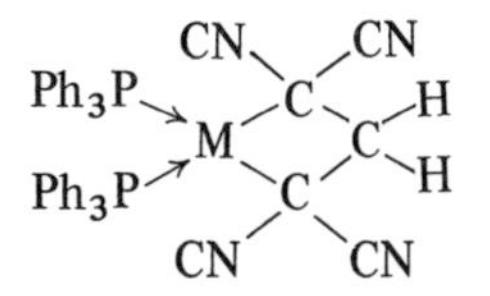

The platinum complex also reacts with tetracyanoethylene oxide with formation of the corresponding metallocyclic system containing metal–carbon σ-bonds. Such complexes are probably analogous to the intermediates thought to be involved in the isomerization of strained carbocycles [393].

2.2.4. Some Catalytic Properties

The extraordinary development of the chemistry of zerovalent derivatives of nickel, palladium and platinum in the last few years has been accompanied by many applications in the field of homogeneous catalysis. The catalytic properties of these compounds in oxidation reactions have already been discussed (2.3.3.5). We shall consider here some other recently reported aspects of the most important catalytic reactions.

The dimerization and oligomerization of butadiene by means of zerovalent nickel derivatives as catalysts have been widely studied in the very important work of Wilke and coworkers [1a, 116b, 300l]. The cyclization of acetylenic compounds has also received great attention [1a, 300d, 317] and it has frequently been postulated that the transition metal catalysed oligomerization of alkynes, 1,3-dienes, alkenes, as well as the dismutation of alkenes, are concerted processes [394]. This formulation however is in contrast to the stereochemistry of the isomers obtained from the cyclomerization of substituted butadienes, and a multistep mechanism seems now more reasonable (2.2.3.9) [394]. Some probable intermediates of the catalytic process such as $Ni(PCy_3)(butadiene)_2$ have been isolated [310, 394]. However they are better considered as derivatives of a dimer of butadiene bonded to nickel as a π-allyl [395] than a bisalkene complex.

The oligomerization of allenes has been recently studied, using zerovalent nickel complexes as catalysts. Bis(triphenylphosphine)nickel gives trimers to hexamers, waxes and polymers of allene at 75°C, but no dimers [57]. The mixed oligomerizations of allene with ethylene, and of allene with methyl acrylate have also been studied [57]. The allene trimer 1,2,4-trimethylene-cyclohexane has also been obtained from allene and $Ni\{P(O\text{-}o\text{-}C_6H_4C_6H_5)_3\}_2$ [105] and the nature of the complexes formed by the precursor of the cyclic trimer and nickel has been elucidated [105, 325, 326, 396]. The oligomerization of dimethylallene and 1,2-cyclononadiene proceeds at 60°C with the same nickel catalyst [326]. In the former case 2,5-dimethyl-3,4-dimethylene-1-hexene is obtained. $Ni(PPh_3)_4$ and $Ni(PPh_3)_2(C_2H_4)$ cyclopentamerize allene at high temperatures [396], while $Ni(1,5\text{-}COD)_2$ gives mainly a tetramer of allene when alkyl or aryl phosphines are present, or the trimer when phosphites are added to the reaction mixture [325]. The mixed allene-butadiene oligomerization with a nickel(0)-phosphite complex as catalyst has been studied [397].

The catalytic dimerization of malodinitrile proceeds with good yields in the presence of $M(PPh_3)_4$ complexes (M = Pd, Pt) which act as strong Lewis bases [399].

In contrast with the catalysts based on nickel(0) derivatives, palladium(0) complexes are less selective and give rise essentially to linear oligomerization of butadiene [398]. The reactions catalysed by palladium(0) complexes, when conducted in the presence of hydrogen donor molecules such as water, phenols, acids, malonates, nitro derivatives, amines, silanes, generally lead to butadiene dimers where the hydrogen donor molecule is incorporated into the products [398]:

$$CH_2{=}CH{-}CH{=}CH_2 + HX \xrightarrow[\text{catalyst}]{Pd^{(0)}} \begin{cases} XCH_2CH{=}CHCH_2CH_2CH_2CH{=}CH_2 + \\ CH_2{=}CH\underset{\displaystyle X}{\underset{|}{C}}HCH_2CH_2CH_2CH{=}CH_2 + \\ CH_2{=}CHCH{=}CHCH_2CH_2CH{=}CH_2 \end{cases}$$

(X = OH, OR, OPh, OCOR, $CH(COOR)_2$, $CHNO_2$, NHR, SiR_3)

The hydrosilation of butadiene with $HSiMe_3$ and $HSiEt_3$ gives 1-trimethyl or triethylsilylocta-2,6-diene in high yields in the presence of

$Pd(PPh_3)_2\left(\begin{array}{l}CH-CO\\ \| \qquad\quad O\\ CH-CO\end{array}\right)$ as catalyst [400]. $Pd(PPh_3)_2$ (p-benzoquinone: ring with =O at top and bottom) is also effective,

but not the $M(PPh_3)_4$ complexes (M = Pd, Pt). The reaction probably proceeds through a hydrido palladium intermediate, formed by the oxidative addition of the alkylsilanes to the palladium(0) complex, followed by the insertion of the diene into the Pd–H bond. Butadiene and water react in the presence of CO_2 and $Pd(PPh_3)_4$ as catalyst, yielding octa-2,7-dien-1-ol. The role of CO_2 is not clear; however $Pd(PPh_3)_3(CO_3)$ also acts as a catalyst for this reaction [401]. Tertiary phosphine palladium(0) complexes catalyse the reaction between butadiene and methanol or phenol to give 1-methoxyocta-2,7-diene and 1-phenoxyocta-2,7-diene, respectively [402, 403].

Similar reactions are observed with primary and secondary amines [404], carboxylic acids [404, 405] and methine compounds [406].

Some of the reactions above reported are also catalysed by nickel(0) catalysts. The mixture of $Ni(acac)_2$, phenyldi-isopropoxyphosphine and $NaBH_4$ is a catalyst for the telomerization of butadiene with alcohols, [407] and with primary and secondary amines [408], where butenyl substituted amines are obtained. The same catalyst is active in the reaction between butadiene and active methylene compounds such as benzylmethylketone and benzyl cyanide [409]. The various aspects of the catalytic process such as the real nature of the catalyst and the role of the hydrogen donor molecule have been discussed [398]. Important in this respect appears to be the observation that species such as $[\{C_2H_4(PPh_2)_2\}_2NiH]^+$ and $[\{C_2H_4(PPh_2)_2\}(PCy_3)PdH]^+$ also act as efficient catalysts for the reaction [410]:

$$\text{butadiene} \xrightarrow[\text{MeOH}]{60^\circ} n\text{-octa-1,3,7-triene} + \text{3-methoxyocta-1,7-diene} + \text{1-methoxyocta-2,7-diene}$$

Also, $[\{P(OEt)_3\}_4NiH]^+$ catalyses the coupling of butadiene and ethylene to give 1,4-hexadienes [411]. A kinetic study of the reaction between butadiene and the hydrido-nickel complex has shown that the rate limiting step is ligand dissociation to give $[L_3NiH]^+$, while the initial products are the *syn* and *anti* forms of the cationic π-crotyl nickel phosphite complexes [411]. The coupling between butadiene and ethylene is also catalysed by $Ni(1,5\text{-}COD)_2$ in the presence of various phosphines, and R_xAlCl_{3-x} compounds as cocatalysts [412]; the activity and the stereoselectivity are slightly affected by the nature of the added phosphine and the aluminium cocatalyst.

In the presence of catalytic amounts of $Pt(PPh_3)_2(C_2H_4)$ 1-hexene and $HSiMeCl_2$ are quantitatively transformed into $n\text{-}C_6H_{13}SiMeCl_2$ [357]. Other silanes can be used in the catalysed hydrosilation of 1-alkenes such as 1-hexene and $SiMe_3(CH{=}CH_2)$, but the addition of silanes to inner alkenes is not possible. 1,2-Addition of silanes is observed when the reaction is conducted with conjugated dienes [357]. The hydrosilation of styrene with $Ni\{(P(OAryl)_3\}_4$ catalysts gives equal amounts of the two isomers (α) and (β) [413]:

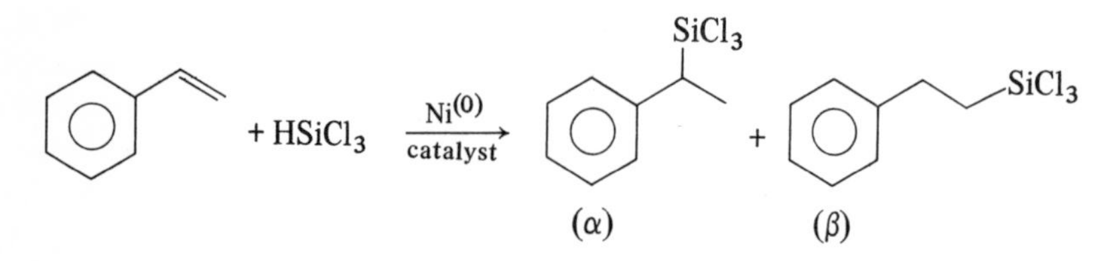

$Ni(PPh_3)_4$ does not promote the hydrosilation of terminal double bonds, while $Ni\{P(OPh)_3\}_4$ is not active for non-conjugated double bonds. The reaction between 1-octene and $HSiCl_3$ giving 1-trichlorosilyloctane in high yields is catalysed by $Pd(PPh_3)_4$, but any bivalent palladium compound is active when PR_3 derivatives (R = phenyl or alkyl) are added [414]. $Pt(PPh_3)_4$ selectively catalyses the addition of silanes to terminal olefinic double bonds, while hydrosilation of internal double bonds does not occur [415]. Interestingly, during these studies it has been observed that palladium metal is gradually transformed into the yellow $[Pd(PPh_3)_2]_x$, by heating with excess $HSiEt_3$ and PPh_3 [414]. The same zerovalent palladium complex can be obtained from $Pd(PPh_3)_4$ and $HSiEt_3$ at room temperature. It has also been observed that $Pt(PPh_3)_2(C_2H_4)$, when employed as catalyst in the hydrosilation of terminal double bonds, is rapidly transformed into the brown complex $[Pt(PPh_3) . C_6H_6]_x$ [357], probably tetrameric (2.2.1(e)).

The catalytic addition of hydrogen cyanide to double bonds with zerovalent nickel and palladium complexes as catalysts has received considerable attention, in view of the importance of these products for the synthesis of acids, esters, amides, amines and isocyanates [398, 416]. This reaction is particularly important for the synthesis of dinitriles when a conjugated diene is employed With butadiene, adiponitrile can be obtained catalytically, and this is a very important intermediate for the production of Nylon 66. The reaction proceeds through two steps.

In the first one 3-pentenenitrile is formed, which can be isomerized to 4-pentenenitrile. Further addition of HCN gives adiponitrile. An example of a catalyst able to isomerize the intermediate mononitrile is $Ni\{P(O{-}C_6H_4CH_3)_3\}_4$ in the presence of Lewis acids [417]. Addition of HCN to olefinic silanes are also catalysed by $Pt\{P(OPh)_3\}_4$ (yields 30-80%). In each case the (2-cyanoethyl)silane isomer is the major product [418].

The observed deactivation of the catalysts seems to be due to the formation of dicyanoderivatives such as $Pd\{P(OPh)_3\}_2(CN)_2$; this secondary reaction can be suppressed by using $Pd\{P(OPh)_3\}_4$ for example as catalyst in the presence of excess $P(OPh)_3$ [398, 416].

The carbonylation reactions of unsaturated organic compounds represent another important group of reactions catalysed by zerovalent complexes of nickel and palladium, but they involve mainly the carbonyl derivatives of these metals and are not considered here [see refs. 1a, 353, 398].

We briefly mention also some studies on the isomerization of alkenes catalysed by nickel(0) catalysts, which, however, proceeds only in the presence of acids such as HCl [310] or H_2SO_4 [419] and thus involve hydrido-nickel(II) derivatives. Very recently it has been reported that the isomerization of pent-1-ene to *cis*- and *trans*-pent-2-ene is catalysed by solutions of $Ni\{P(OEt)_3\}_4$ in benzene containing trifluoroacetic acid at 35°C [420]. The initial rate of the isomerization reaction does not decrease on addition of triethyl phosphite, but this reagent causes isomerization to stop after 40% of the pent-1-ene has reacted.

Finally we consider that only in very few cases has a zerovalent complex of nickel, palladium or platinum been found to be able to activate hydrogen in the catalytic hydrogenation of alkenes or alkynes [353, 398]. In fact it was observed that such complexes are in general inert towards molecular hydrogen (2.2.3.1) so that only when the basic triethylphosphine ligand is present has an activation of hydrogen been observed [132], with formation of $PtH_2(PEt_3)_3$ from the zerovalent complex and molecular hydrogen. It could be that complexes similar to $Pt(PEt_3)_3$ could act as catalysts in the hydrogenation reaction, but that hydrogen activation is not the only condition necessary in order to have a catalytic species [353, 398]. A catalytic reaction which could open a new chapter in the reactions catalysed by palladium(0) complexes has recently been discovered. It has been found that $Pd(PPh_3)_2(O{=}C_6H_4{=}O)$ is a catalyst for the formation of hydroquinone, ethers or anthraquinones from *para*-benzoquinone and alkenes [421].

2.3. COBALT, RHODIUM AND IRIDIUM

Cobalt(0) complexes having chelating diphosphines as ligands have been known for many years. It was reported that by reduction of a cobalt halide with sodium naphthalenide in the presence of $C_2H_4(PMe_2)_2$ in solution, the diamagnetic $Co\{C_2H_4(PMe_2)_2\}_2$ is obtained [422]. This was found to be isomorphous with the corresponding nickel(0) derivative, and thus a tetrahedral structure was suggested. However, when water is added or hydrogen is bubbled through the reaction mixture, the hydride $CoH\{C_2H_4(PMe_2)_2\}_2$ [ν(Co–H), 1855 cm^{-1}; τ_H, 18.7] is obtained [423]. Since this compound has the same colour and melting point as the derivative reported to be a zerovalent complex, some doubts about the real nature of the latter may arise. Analogously, the complex obtained by reduction of $Co\{C_2H_4(PPh_2)_2\}_2Br_2$ with $NaBH_4$ and at first formulated as $Co\{C_2H_4(PPh_2)_2\}_2$ [13] was shown later to be the hydrido complex $CoH\{C_2H_4(PPh_2)_2\}_2$ [424]. A zerovalent complex has been obtained according to the following reaction [425]:

$$5\ Co\{C_2H_4(PPh_2)_2\}_2Br_2 + 10\ KOH \xrightarrow[C_2H_5OH]{N_2} 4\ Co\{C_2H_4(PPh_2)_2\}_2$$

$$+ Co(OH)_2 + 2\{C_2H_4(PPh_2)_2\}O_2 + 10\ KBr + 4\ H_2O$$

This compound is paramagnetic, and very easily oxidized by air. It reacts with hydrogen giving the corresponding hydride, and it reduces $SnCl_2$ to metallic tin. On the other hand the chelating diphosphine o-$C_6H_4(PEt_2)_2$ dissolves finely divided cobalt at 200°C in a nitrogen atmosphere to form the violet $Co\{o\text{-}C_6H_4(PEt_2)_2\}_2$ [13], which strangely enough can also be obtained in a hydrogen atmosphere.

A monomeric, paramagnetic (1.70 B.M.), cobalt(0) complex having trimethylphosphine as the ligand has been reported [426]:

$$CoCl_2 + 4PMe_3 + 2\ NaHg_x \xrightarrow[20^\circ]{THF} Co(PMe_3)_4 + 2\ NaCl + 2x\ Hg$$

This dark brown complex shows an intense E.P.R. signal, with complex hyperfine structure caused by the cobalt nucleus with I = 7/2. Although it does not react with hydrogen, when the reduction is carried on in a hydrogen atmosphere $CoH(PMe_3)_4$ is also obtained, together with the zerovalent complex. $Co(PMe_3)_4$ reacts with NO giving $Co(NO)(PMe_3)_3$ and with azobenzene it gives the dark violet $Co(PMe_3)_2(PhN{=}NPh)$. The latter compound is still paramagnetic (2.02 B.M.).

By electroreduction of RhL_3Cl complexes (L = PPh_3, $PMePh_2$), the rhodium(0) derivatives RhL_4 have been obtained [427]. This reduction can be conducted on a preparative scale (1.9 g of the halide, with 70% yields). The triphenylphosphine derivative is diamagnetic and presumably dimeric. The methyldiphenylphosphine derivative is paramagnetic (less than one unpaired electron), and this could be due to a partial dissociation to the monomeric species.

Zerovalent complexes of cobalt, rhodium and iridium with phosphites as ligands are known. By refluxing the n-alkane suspensions of $MH(CO)(PPh_3)_3$ (M = Rh, Ir) in the presence of triarylphosphites, the white, diamagnetic, $[M\{P(OR)_3\}_4]_2$ have been obtained [428]. Extensive dissociation leaves some doubts on the true structure of these compounds. The same reaction with Rh and Ir is not extendable to alkylphosphites [428], but the diamagnetic, dissociated $[Co\{P(OEt)_3\}_4]_2$ has been reported [429].

The whole series of fluorophosphine complexes $M_2(PF_3)_8$ (M = Co, Rh, Ir) which does not have an equivalent in the carbonyl series, is known. We will not attempt to describe here the chemistry of these and of the fluorophosphine derivatives of the other metals, since they have been authoritatively reviewed [75, 430]. Among the most recent reports, we can remember the practical synthesis of $Co_2(PF_3)_8$ (50% yields) by condensation of vapours of cobalt metal with the phosphine at –196°C [21]. The bright yellow $Ir_2(PF_3)_8$ has been obtained, by irradiation of $HIr(PF_3)_4$ with u.v. light, with hydrogen evolution [431]. The ^{19}F spectrum shows a doublet at 5.85 p.p.m. (CCl_3F as external standard) with J_{P-F} = 1238 Hz, the usual value for terminal PF_3 groups. The mass spectrum shows the molecular peak, and successive loss of PF_3 groups; the rupture of the Ir–Ir bond followed by loss of PF_3 groups from the two halves of the molecule has also been observed. Another route for the synthesis of $M_2(PF_3)_8$ (M = Rh, Ir) has also been reported [432]:

$$K[Ir(PF_3)_4] + Ir(PF_3)_4Cl \rightarrow Ir_2(PF_3)_8 + KCl$$

For the rhodium complex, $[Rh(PF_3)_2Cl]_2$ and PF_3 have been used for the reaction with $K[Rh(PF_3)_4]$. Both complexes react with hydrogen at room temperature giving $HM(PF_3)_4$ (M = Rh, Ir) but iridium reacts at a much slower rate. The reaction with triphenylsilane leads to $M(PF_3)_4SiPh_3$ and $HM(PF_3)_4$ (M = Rh, Ir). Similar reactions occur with triphenylgermane; $Rh(PF_3)_4SiR_3$ complexes (R = Cl, OC_2H_5) have been synthesized analogously [432].

$Rh_2(PF_3)_8$ is similar to $Co_2(CO)_8$ in many chemical respects, and it reacts with a variety of alkynes (C_2H_2, PhC_2H, PhC_2Ph, MeC_2Me, EtC_2Me, MeC_2CO_2Me, *n*-$ButC_2H$, *t*-$ButC_2H$, or $CF_3C_2CF_3$) to give volatile, red or yellow solids or oils of formula $Rh_2(PF_3)_6$(alkyne) [433]. A X-ray structural analysis of $Rh_2(PF_3)_4(PPh_3)_2(PhC_2Ph) . Et_2O$, obtained by partially displacing PF_3 from $Rh_2(PF_3)_6(PhC_2Ph)$ by PPh_3 in ether, has shown that the two $Rh(PF_3)_2(PPh_3)$ moieties are symmetrically bridged by the diphenylacetylene [433]. The Rh–Rh distance [2.741(2) Å] indicates a metal–metal interaction, and the overall geometry of the molecule is similar to that of $Co_2(CO)_6(PhC_2Ph)$. The proton resonances of the $\equiv CH$ or $\equiv CCH_3$ groups in the $Rh_2(PF_3)_6$(alkyne) complexes are shifted downfield compared with the free ligands, and are split into symmetrical septets by coupling to six apparently equivalent ^{31}P nuclei, while complexes containing unsymmetrically substituted alkynes should have three non-equivalent PF_3 groups, two of which equivalent when the alkyne is symmetrically disubstituted [433]. The ^1H n.m.r. spectrum at low temperature shows this non-equivalence, and the stereochemical non-rigidity of these rhodium complexes is under investigation [433].

Among the most interesting dinitrogen complexes, the yellow cobalt compound obtained in the reduction of $Co(acac)_3$ with $AlEt_2(OEt)$ in the presence of PPh_3 and in a nitrogen atmosphere was considered initially to be $Co(PPh_3)_3(N_2)$ [$\nu(N_2)$, 2088 cm^{-1}] derived from $CO^{(0)}$ [434]. However, ^1H n.m.r. evidence [435] and an X-ray structural determination [436] have shown that this compound was not a zerovalent but instead the hydride $CoH(N_2)(PPh_3)_3$. This diamagnetic derivative was independently obtained from the interesting reaction [437]:

$$CoH_3(PPh_3)_3 + N_2 \rightleftharpoons CoH(N_2)(PPh_3)_3 + H_2$$

The chemical reactions of the yellow dinitrogen complex also showed that it should be considered as a hydrido derivative, and in a subsequent paper the first formulation was corrected [438]. Interestingly, if the reduction of $Co(acac)_3$ with $AlEt_2(OEt)$ in the presence of PPh_3 is conducted in an argon atmosphere, the dark brown $[Co(PPh_3)_2(C_2H_4)]_x$ (x possibly two) is formed [438]. An analogous reaction with ethylene abstraction from the aluminium alkyl has previously been observed in the synthesis of $Ni(PPh_3)_2(C_2H_4)$ (2.2.3.8). On the other hand, it has been reported that a paramagnetic zerovalent dinitrogen cobalt compound of formula $Co(PPh_3)_3(N_2)$ [$\nu(N_2)$, 2092.8 cm^{-1}] is the product of the reduction under nitrogen and a paramagnetic dihydrido complex of formula $H_2Co(PPh_3)_3$ is the product of the reduction under hydrogen when cobalt(II) instead of cobalt(III) acetylacetonate is employed [439].

The zerovalent cobalt dinitrogen complex has been shown to catalyse the isomerization of 1-hexene mainly to *cis*-hexene-2 [440]. The reaction of CoL_3Cl (L = PPh_3, PEt_2Ph) with sodium metal in tetrahydrofuran or toluene under nitrogen, gives the bridged dinitrogen complexes of cobalt(0), $L_3Co(N_2)CoL_3$ and cobalt(I), $Na[Co(N_2)L_3]$ [441] respectively.

An asymmetrically bridged dinitrogen complex has been obtained by the following reaction:

$$Na[Co(N_2)(PEt_2Ph)_3] + Co(PPh_3)_3Cl \rightarrow (PEt_2Ph)_3Co(N_2)Co(PPh_3)_3 + NaCl$$

These dinuclear dinitrogen complexes, which do not show any band assignable to $\nu(N_2)$ in their i.r. spectra, do not give any evidence for dissociation in solution to monomeric species [441].

2.4. IRON, RUTHENIUM AND OSMIUM

By reduction of *trans*-$Fe\{C_2H_4(PMe)_2\}_2Cl_2$ with sodium naphthalenide the zerovalent diamagnetic complex $Fe\{C_2H_4(PMe_2)_2\}_2$ has been obtained [422]. On the other hand the chelating diphosphine o-$C_6H_4(PEt_2)_2$ is able to dissolve finely divided cobalt and palladium but not iron to give the corresponding zerovalent complexes [13].

The reduction of *trans*-Ru$\{C_2H_4(PMe_2)_2\}_2Cl_2$ by alkali benzenide, naphthalenide, anthracenide and phenanthrenide has given a series of complexes which may be formulated, for example, as *cis*-RuH(2-$C_{10}H_7$)$\{C_2H_4(PMe_2)_2\}_2$ [442].

Similar derivatives of osmium have been obtained. However, most of their chemical reactions are consistent with the formulation Ru(arene)-$\{C_2H_4(PMe_2)_2\}_2$, that is as ruthenium(0) complexes. The naphthalene molecule is labile and a simple pyrolysis reaction gives free naphthalene and the diamagnetic Ru$\{C_2H_4(PMe_2)_2\}_2$. This compound, which might well be formulated as a ruthenium(0) derivative, shows ν(Ru—H) at 1791 cm^{-1}, with a shoulder at 1815 cm^{-1}. Also it has dual character, behaving in some respects as a zerovalent complex, but having the spectral properties of a hydride [442], so that a tautomeric equilibrium where the hydride form predominates in the solid state has been suggested:

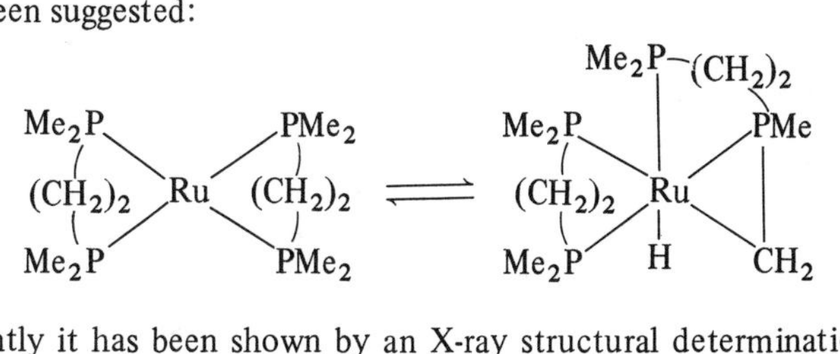

More recently it has been shown by an X-ray structural determination that the systems:

$$M^0\text{(naphthalene) }\{C_2H_4(PMe)_2\}_2 \rightleftharpoons \text{MH(naphtyl)}\{C_2H_4(PMe)_2\}_2$$
$$(M = Ru, Os)$$

which involves equilibria between metal(0) and metal(II) complexes, are shifted towards the hydridic forms in the solid state [443].

Attempts to reduce many other ruthenium(II) complexes having a chelating diphosphine as ligand to ruthenium(0) species by using sodium naphthalenide or metallic zinc in *N,N'*-dimethylformamide [a reducing system useful for the preparation of the zerovalent carbonyl complex $Ru(CO)_3(PPh_3)_2$] were generally unsuccessful [444]. The great tendency of ruthenium to abstract hydrogen from the ligands, giving hydrido-ruthenium(II) derivatives instead of zerovalent complexes, is also demonstrated by the formation of $Ru(PPh_2H)_4H_2$ when $Ru(PPh_2H)_4Cl_2$ is treated with $LiOCH_3$, that is with a reducing system usefully employed for the synthesis of zerovalent complexes of palladium and platinum [445].

The same reaction takes place when $Ru(PPh_3)_3Cl_2$ is treated with C_2H_5ONa and similar alkoxides, in the presence of phosphine: $Ru(PPh_3)_4H_2$ is obtained

[446]. In the absence of phosphine and in a nitrogen atmosphere, the known dinitrogen complex $Ru(PPh_3)_3(H_2)N_2$ was isolated, while in an argon atmosphere carbon monoxide abstraction from the alcohol was observed to form hydrido ruthenium(II) carbonyl derivatives [446]. Only in carbon monoxide did the reduction lead to the zerovalent $Ru(CO)_3(PPh_3)_2$ complex [446]. With *tert*-butoxide as reducing system, where abstraction of hydrogen and of carbon monoxide is less probable, the reaction gave uncharacterizable products. Some reductions of $Ru(PPh_3)_3Cl_2$ with sodium amalgam have also been attempted [446]. Although in this case some evidence has been found for the formation of zerovalent complexes, the products were too unstable to be fully characterized.

For the synthesis and chemical properties of the trifluorophosphine complexes $M(PF_3)_5$ (M = Fe, Ru, Os) we remind the reader of the most recent reviews in the field [75, 430]. We only mention the recent synthesis of $Fe(PF_3)_5$, although in low yields, by condensation of vapours of iron metal and PF_3 at −196°C [21].

From the ^{19}F n.m.r. spectra it has been recently shown that for these three fluorophosphine complexes the barrier to intramolecular rearrangement is less than 5 kcal mol^{-1} [447]. They have trigonal bipyramidal structures and the five PF_3 groups are equivalent in the ^{19}F n.m.r. spectrum:

	T°C	(p.p.m. from the upfield component of $CHClF_2$)
$Fe(PF_3)_5$	−81	74.04
$Ru(PF_3)_5$	−81	74.23
$Os(PF_3)_5$	−122	55.11

The recently developed technique which uses metal vapours for the synthesis of zerovalent metal complexes has also been used for obtaining $Fe(PF_3)(butadiene)_2$ [448, 449]; iron vapour and butadiene are condensed at −196°C, and PF_3 is then added. The same route has been applied to the synthesis of $Fe(PF_3)_2(toluene)$ [448]. A related butadiene complex, $Fe(PF_3)_3(butadiene)$, has been obtained by displacing carbon monoxide with PF_3 from $Fe(CO)_3(butadiene)$ [450]. In the ^{19}F spectrum at room temperature this compound shows a time-averaged equivalence, indicating some form of fluxional behaviour in the structure.

The ethylene-Fe° derivative $Fe\{C_2H_4(PPh_2)_2\}_2(C_2H_4)$ has been obtained by reducing $Fe(acac)_3$ with an aluminium alkyl in the presence of $C_2H_4(PPh_2)_2$ [451]. This derivative is transformed by u.v. irradiation into the hydride $HFe(C_6H_4PPhCH_2CH_2PPh_2)\{C_2H_4(PPh_2)_2\}$.

Some ruthenium–alkene complexes, considered to be zerovalent, have been obtained from the reactions of $Ru(PPh_3)_4H_2$ with alkenes. In this way, the compound reported as $Ru(PPh_3)_2\{(CN)_2C{=}C(CN)_2\}_3$ [452] and

$Ru(PPh_3)_3(RCH{=}CH_2)$ (R = H, Ph) are apparently formed [454]; $Ru(PPh_3)_3(C_2Cl_4)$ has been isolated by displacing styrene from $Ru(PPh_3)_3(PhCH{=}CH_2)$ with C_2Cl_4 [454].

The excess of the alkene acts as dehydrogenating agent, being transformed into the corresponding alkane. Acrylonitrile is transformed catalytically in these conditions to high polymers [452] (compare the reaction of the related $CoH(N_2)(PPh_3)_3$ with styrene which gives a paramagnetic binuclear cobalt complex derivative [453]).

A zerovalent iron-dinitrogen complex seems to be formed by reduction of $Fe(acac)_3$ with tri-*iso*-butyl aluminium, in the presence of PEt_3 and in a nitrogen atmosphere [455]. Although not isolated in a pure state it has the probable formula $Fe(PEt_3)_4(N_2)$ [$\nu(N_2)$ = 2038 cm^{-1}]. The same compound can be obtained from the reaction between $Fe(COT)_2$ and PEt_3 under nitrogen [455]. Some interesting ruthenium and osmium-nitrosyl complexes recently synthesized seem to be real zerovalent derivatives, although in the presence of nitrosyl ligands the formal oxidation state of the metal depends on the role assigned to the NO groups (1.3.1.3). By reduction of $Ru(NO)Cl_3L_2$ (L = PPh_3, $PMePh_2$, $PPh_2C_3H_7^i$, PPh_2Cy) with KOH in ethanol in the presence of excess L, $RuH(NO)L_3$ complexes have been obtained [456]. Analogous reductions of the osmium derivatives (L = PPh_3, $PMePh_2$) using 2-methoxyethanol as solvent have given the corresponding osmium complexes. These compounds show ν(M–H) at *ca.* 1900–2050 cm^{-1}, and ν(NO) at *ca.* 1600–1650 cm^{-1} and some of them have a marked catalytic activity in the isomerization of 1-hexene, while they are less active as hydrogenation catalysts [456]. An X-ray structural determination of $RuH(NO)(PPh_3)_3$ has shown a trigonal-bipyramidal structure with an axial linear nitrosyl group, generally considered to be the $[NO]^+$ group [457]. Also, for the related $[Ru(NO)\{C_2H_4(PPh_2)_2\}]^+$ complex, a linear coordination of the nitrosyl ligand was determined by X-ray [458].

On this basis these compounds have been considered to be ruthenium(0) derivatives. A ruthenium analogue of Vaska's iridium compound *trans*-$IrCl(CO)(PPh_3)_2$, namely $Ru(NO)(PPh_3)_2Cl$, has been obtained by reduction of $Ru(NO)(PPh_3)_2Cl_3$ with zinc in boiling benzene [459]. It shows a reactivity typical of zerovalent complexes, giving 1:1 adducts with O_2, CO and SO_2. Other reactions take place with CH_3I, HCl, Cl_2, Br_2, I_2, $HgCl_2$, $HgBr_2$ [460]. The carbonyl derivatives $M(NO)Cl(CO)(PPh_3)_2$ (M = Ru, Os) and similar complexes with an anion other than chlorine have also been independently isolated by a different route [460, 461]. $Ru(NO)Cl(CO)(PPh_3)_2$ with oxygen gives $Ru(O_2)Cl(NO)(PPh_3)_2$ [461] with carbon monoxide evolution, while the analogous osmium complex gives the carbonato derivative, $Os(CO_3)Cl(NO)(PPh_3)_2$ [460]. The ruthenium-dioxygen adduct having–NCS as the anionic ligand has been shown to be an efficient catalyst for the oxidation of triphenylphosphine to the corresponding oxide [462].

2.5. CHROMIUM, MOLYBDENUM, TUNGSTEN

$M\{C_2H_4(PMe_2)_2\}_3$ (M = Cr, Mo, W) complexes have been obtained by the usual reduction of a metal halide derivative with sodium naphthalenide [422]. For these metals $LiAlH_4$ can also be used as a reducing agent, while it cannot be used for instance with iron, otherwise hydrido derivatives instead of zerovalent complexes are formed [422]. Zerovalent molybdenum complexes, $Mo(L\text{–}L)_3$ [L–L = $C_2H_4(PMe_2)_2$, *o*-$C_6H_4(PEt_2)_2$, $C_2H_4(PPh_2)_2$] have also been obtained from the reactions of dibenzenemolybdenum with the ligands L–L in a sealed tube, under N_2, at 140°C for *ca.* 50 h [463]. Dibenzenechromium did not react under these conditions, neither did dibenzenemolybdenum with PPh_3. Dibenzenemolybdenum does react with PMe_3, but the product was not characterized [463]. More recently such reaction has been reinvestigated and it has been shown that dibenzenemolybdenum reacts at 90–120°C with a number of tricovalent phosphorus ligands with formation of $Mo(C_6H_6)L_3$ [464] [L = PMe_2Ph, $PMePh_2$, $P(OMe)_3$, $PPh_2(OMe)$, $P(OPh)_3$]. The compounds with L = PMe_2Ph and $PMePh_2$ can be reversibly protonated by mineral acids. The band at around τ 6 assigned to the benzene protons occurs as a 1 : 3 : 3 : 1 quartet, which must arise from coupling with three equivalent phosphorus nuclei [464].

Complete replacement of carbon monoxide in $Mo(CO)_6$ to give MoL_6 has been successful with L = PF_3 by either thermal or photochemical activation [for $M(PF_3)_6$ complexes (M = Cr, Mo, W) see refs. 75 and 430; for the synthesis of $Cr(PF_3)_6$ from chromium vapour and PF_3 see also ref. 21].

Very recently it has been shown that with photochemical activation it is also possible to obtain $M\{P(OC_3H_7\text{-}n)F_2\}_6$, $M\{P(OCH_3)_2F\}_6$ (M = Cr, Mo, W) and $Mo\{P(OCH_3)_3\}_6$ [465]. All these products are quite stable under nitrogen except for the last derivative which slowly decomposes. The ^{19}F n.m.r. spectra of these derivatives are quite complex; the fluorine resonance occurs at lower field in the compound than in the free ligand, with small differences in chemical shift between the chromium, molybdenum and tungsten compounds. Likewise the absolute magnitude of J_{P-F} is lowered by coordination. Steric effects probably explain why the ML_6 derivatives were not obtained with bulkier ligands. On the other hand, the failure to obtain $W\{P(OCH_3)_3\}_6$ cannot be attributed to steric effects since tungsten is larger than molybdenum. In this case it has been suggested that d_π-d_π back bonding is less efficient for the former as a consequence of the more diffuse nature of the metal *d* orbitals [465].

Very recently important results have been obtained in the field of zerovalent tungsten and molybdenum dinitrogen complexes. During these studies (see later) some zerovalent complexes of the type described above have also been isolated. Molybdenum dinitrogen complexes have been obtained by reducing $Mo(acac)_3$ with aluminium alkyls under nitrogen, in the presence of the appropriate ligand

[466]. In this way, $Mo(N_2)_2\{(CH_2)_n(PPh_2)_2\}_2$ (n = 1, 2, 3) and $Mo(N_2)$-$(PPh_3)_2 . C_6H_5CH_3$ (possibly polynuclear) have been obtained. From the reaction with $CH_2(PPh_2)_2$ as ligand, $Mo\{CH_2(PPh_2)_2\}_3$ has also been isolated.

The *trans*-stereochemistry of the $C_2H_4(PPh_2)_2$ derivative, which shows a strong $\nu(N_2)$ at 1970 cm^{-1} besides a very weak band at 2020 cm^{-1}, has been confirmed by an X-ray structural determination [467]. Similar complexes with $C_2H_4(PPh_2)_2$ and *cis*-$Ph_2PCH{=}CHPPh_2$ as ligands have been obtained either in one step by reducing $Mo(L{-}L)_2Cl_x$ (x = 2, 3) with sodium amalgam in a nitrogen atmosphere, or in two steps by first reducing $MoOCl_3(L{-}L)_2$ in tetrahydrofuran to $MoOCl_2(L{-}L)$(tetrahydrofuran) with $LiAlH_4$, and subsequently to *trans*-$Mo(N_2)_2(L{-}L)_2$ with zinc in the presence of an excess of the diphosphine [468]. The reduction of $MoCl_3(THF)_3$ in tetrahydrofuran solution with sodium amalgam or magnesium metal in the presence of a phosphine ligand under nitrogen, has also been applied to the synthesis of dinitrogen molybdenum complexes [469]. By this route $Mo(N_2)_2(L{-}L)_2$ [$L{-}L{=}C_2H_4(PPh_2)_2$, $Ph_2PCH{=}CHPPh_2$] complexes have been isolated. Also, $Mo(N_2)_2(PMe_2Ph)_4$ [$\nu(N_2)$ = 1951 and 2014 cm^{-1}] has been obtained, although not in a pure state, while when the reduction was carried under argon, $Mo(PMe_2Ph)_4$ was the product of the reaction. The last derivative is diamagnetic and monomeric, and was suspected to be a hydride, and not a zerovalent complex [469]. With PH_2Ph as ligand, the zerovalent $Mo(PH_2Ph)_6$ has been obtained even under nitrogen.

$Mo(N_2)_2L_4$ complexes [L = $PMePh_2$, PMe_2Ph; 2L = $C_2H_4(PPh_2)_2$] have also been synthesized by reducing the corresponding MoL_2Cl_4 derivatives with sodium amalgam under nitrogen [470]. For L = $PMePh_2$, $\nu(N_2)$ is at 1926 cm^{-1}, while for L = PMe_2Ph, $\nu(N_2)$ are at 2010 and 1937 cm^{-1} and thus a *cis* stereochemistry has been suggested for this last compound. The reduction of $MoCl_3(THF)_3$ with sodium amalgam has also been used to obtain *trans*-$Mo(N_2)_2(L{-}L)_2$ (L–L = 1-diphenylarsino-2-diphenylphosphinoethane or 1,2-diphenylarsinomethane) [471]. The synthesis of the analogous tungsten derivative has been achieved by reducing WL_2Cl_4 under nitrogen with sodium amalgam in the presence of excess ligand [472]: *cis*-$W(N_2)_2(PMe_2Ph)_4$ [$\nu(N_2)$ = 1931 and 1998 cm^{-1}] and *trans*-$W(N_2)_2\{C_2H_4(PPh_2)_2\}_2$ [$\nu(N_2)$ = 1953 cm^{-1}) were isolated. When the reduction was conducted under carbon monoxide, only complexes of the type *fac*-$W(CO)_3(PMe_2Ph)_3$ were obtained, even in the presence of excess phosphine.

Analogous to the cobalt hydride, $CoH_3(PPh_3)_3$, the molybdenum dihydride $C_6H_6Mo(PPh_3)_2H_2$ reacts reversibly with nitrogen [473] giving the dinuclear complex $\{C_6H_6Mo(PPh_3)_2\}_2N_2$ [$\nu(N_2)$ = 1910 cm^{-1}, Raman] which is reconverted to the starting complex by reaction with hydrogen.

In the presence of hydrogen $Mo(N_2)_2\{C_2H_4(PPh_2)_2\}_2$ is in equilibrium with $MoH_2\{C_2H_4(PPh_2)_2\}_2$ or $MoH_2\{C_2H_4(PPh_2)_2\}$-μ-$C_2H_4(PPh_2)_2$ depending on the solvent used [466]. With carbon monoxide, nitrogen is only very slowly displaced [466], but the reaction becomes much faster when u.v. irradiation is used and *cis*-$Mo(CO)_2\{C_2H_4(PPh_2)_2\}_2$ can be easily obtained [474]. The N_2 ligand in $M(N_2)_2\{C_2H_4(PPh_2)_2\}_2$ (M = Mo, W) complexes is able to act as a Lewis base towards Lewis acids such as $AlMe_3$ [475]; the adducts formed show $\nu(N_2)$ lowered to 1973 (M = Mo) or 1953 (M = W) cm^{-1}. Related molybdenum-dinitrogen complexes having PMe_2Ph and PEt_2Ph as ligands, similarly give adducts with $AlEt_3$ with a lowering of $\nu(N_2)$ of about 80–100 cm^{-1} compared with the parent complex [475 *bis*].

The reactivity of the dinitrogen complexes above reported has been studied with the aim of finding some way to reduce the coordinated nitrogen to ammonia on hydrazine.

The molybdenum dinitrogen complex when treated with iodine gave $[Mo(N_2)_2\{C_2H_4(PPh_2)_2\}_2]^+I_3^-$ and with HX(X = Cl, Br) the analogous $[Mo(N_2)_2\{C_2H_4(PPh_2)_2\}_2]^+X^-$, but no reaction of the coordinated dinitrogen was observed [476].

On the other hand Chatt and coworkers have reported that the coordinated dinitrogen in these molybdenum and tungsten complexes can be reduced with HX (X = Cl, Br) giving compounds which may be regarded as models for the first step of the sequence leading to ammonia [477]:

$$trans\text{-}M(N_2)_2\{C_2H_4(PPh_2)_2\}_2 + HX \rightarrow MX_2(N_2H_2)\{C_2H_4(PPh_2)_2\}_2$$

$$(M = Mo, W; X = Cl, Br).$$

The unusual N_2H_2 ligand can be attached to the metal in one of the following ways:

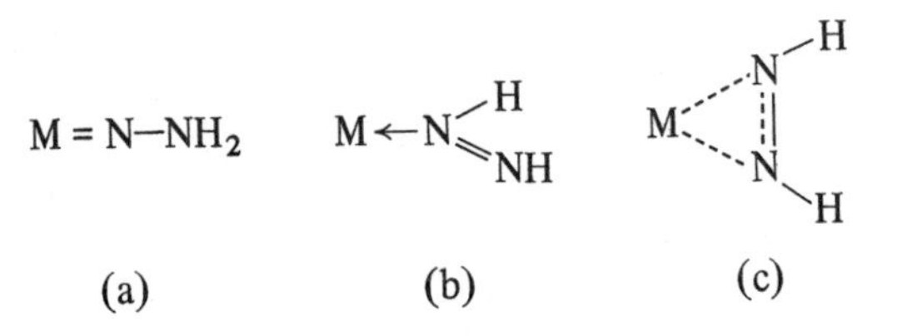

but nothing is known about the real structure of these complexes [477]. From the reaction between hydrochloric acid and the tungsten complex, $[WH\{C_2H_4(PPh_2)_2\}_2][HCl_2]$ has also been isolated. The tungsten dinitrogen complex by reaction with RCOCl compounds (R = Me, Et, Ph, *p*-$MeOC_6H_4$), was attacked on the coordinated dinitrogen [478]. Compounds such as $W\{C_2H_4(PPh_2)_2\}_2Cl_2(N_2HCOR)$ were isolated, from which HCl could be removed with NEt_3 to form the chelated aroyl or acylazo-*N,O*-derivatives $W\{C_2H_4(PPh_2)_2\}_2Cl(N{=}NCOR)$. In these compounds oxygen is involved in the bond with tungsten, since ν(C=O) is lowered from *ca.* 1700 cm^{-1} in the starting complexes, to *ca.* 1575–1500 cm^{-1} [478].

We briefly mention here the only zerovalent rhenium and vanadium phosphine derivatives so far reported, $Re_2(PF_3)_8$ [75, 430], and $V\{C_2H_4(PMe_2)_2\}_3$ [422]. The vanadium complex, obtained by the reduction of the haloderivative with sodium naphthalenide, is very unstable; it has a magnetic moment of 2.10 ± 0.25 B.M.

ISOCYANIDE COMPLEXES

2.6. Introduction

The isocyanide complexes of metals were reviewed in 1969 [298]. We will discuss here only the very recent aspects of the chemistry of zerovalent isocyanide derivatives. Most research, as for complexes having phosphines and related compounds as ligands, concerns the d^{10} metals, particularly nickel and palladium. This work has shown that the reactivity of the isocyanide complexes parallels that of the analogous phosphine derivatives, as anticipated on considering the ability of isocyanides to withdraw charge from the central metal atom, by either an inductive or π-electron mechanism [298].

2.7. Reactivity

Reaction between the binuclear [479] $Pd(Bu^tNC)_2$ and CF_2=CFX (X = Cl, Br) affords $Pd(Bu^tNC)_2(CF=CF_2)X$, with a single sharp i.r. band for ν(CN) at 2200 cm^{-1}, establishing a *trans* configuration for these derivatives [120]. Many acyl halides or haloformates add smoothly to $Pd(Bu^tNC)_2$, affording the corresponding *trans* acyl and alkoxycarbonyl compounds [187]. In contrast, $ClCO_2Ph$ gives to the chloride-bridged dimer, $Pd_2Cl_2(Bu^tNC)_4$, and diphenylcarbonate, with evolution of CO.

Methyl iodide reacts with $Pd(Bu^tNC)_2$ at 0°C affording *trans*-$Pd(Bu^tNC)_2I(CH_3)$. However even at 11°C, an insertion reaction takes place leading to an iminoacylpalladium(II) derivative, detected *in situ* by means of n.m.r. spectra:

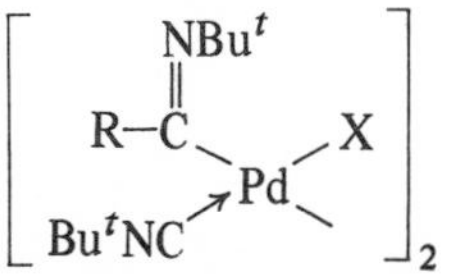

$$\left[\begin{array}{c} NBu^t \\ \| \\ R{-}C \searrow \quad \swarrow X \\ Pd \\ Bu^tNC \nearrow \quad \searrow \end{array} \right]_2 \qquad (X = I,\ R = Me)$$

The reaction between CH_3I and $Ni(Bu^tNC)_4$ leads to a similar σ-iminoacyl compound (A), derived from the successive insertion of Bu^tNC ligands into the Ni–Me bond [480]:

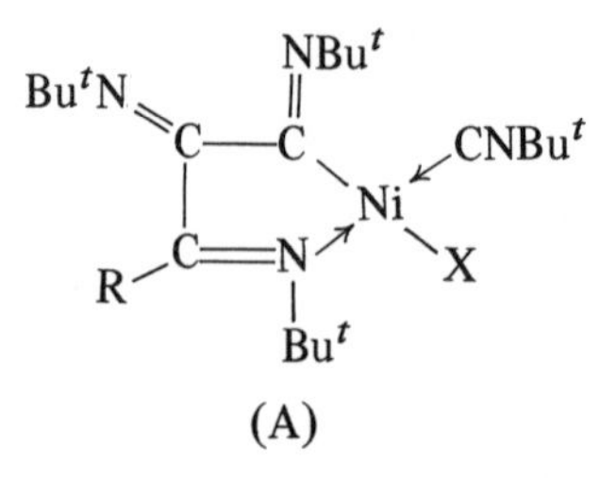

(A)

When benzoyl chloride was used instead of methyl iodide, two derivatives (B and C) were isolated [480]:

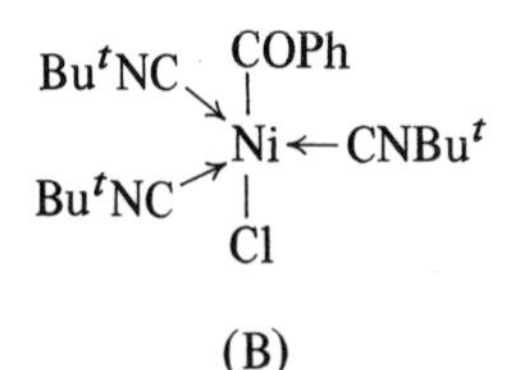

(B)

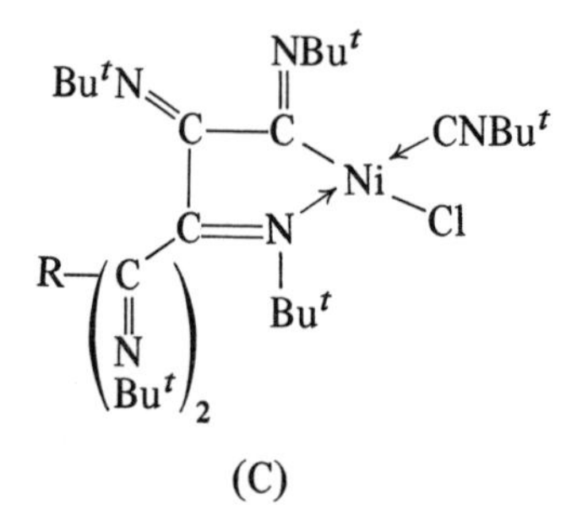

(C)

Compound (B) is presumably an intermediate in the formation of the products derived from the insertion reactions.

When an excess of Bu^tNC, or other isocyanides, was heated above room temperature in the presence of (A), the isocyanide was converted into a polyisocyanide, with a repeating unit (R–N) [480]. The above successive
‖
–C–

insertion reactions thus suggest a mechanism for the catalytic polymerization of isocyanides.

Similar reactions have been found to occur between $Ni(Bu^tNC)_4$ and acyl halides or haloformates and the results are summarized in the following scheme [187]:

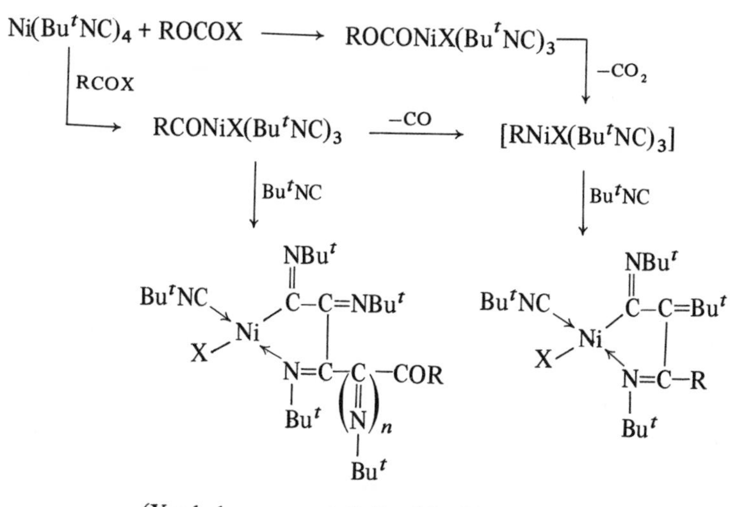

(X = halogen; n = 1,2; R = Me, $PhCH_2$)

A chlorobenzene solution of $Pd(Bu^tNC)_2$ reacts with $Pd(Bu^tNC)_2Cl_2$ giving $[Pd(Bu^tNC)_2Cl]_2 \cdot C_6H_5Cl$ [481]. This diamagnetic, dimeric, formally palladium(I) complex, presumably has a palladium–palladium bond. Similarly $[Pd(Bu^tNC)_2X]_2$ (X = Br, I) complexes have been isolated. The oxidative addition of organic halides RX, e.g. $PhCH_2I$ or XCH_2COOCH_3 (X = Cl, Br) to $Pd(Bu^tNC)_2$ led to the same derivatives, via the σ-alkyl isocyanide complexes *trans*-$Pd(Bu^tNC)X(R)$.

In view of the formal valence of one for the metal, the NC stretching frequency (*ca.* 2170 cm^{-1}) is reasonably low compared with that of the corresponding halogenopalladium(II) complexes.

Successful preparations of peroxo nickel and palladium isocyanide complexes have been obtained only with aliphatic isocyanides. By bubbling oxygen or air into a solution of $Ni(RNC)_4$ below −20°C the pale green peroxo complexes $Ni(RNC)_2(O_2)$ (R = Bu^t, Cy) [482] were obtained. Analogous reactions of oxygen with the cluster complex [479] $Ni(Bu^tNC)_2$ occur most readily, affording the same peroxo complex [482]. In a similar way the colourless peroxo palladium complex $Pd(Bu^tNC)_2(O_2)$ was obtained from $Pd(Bu^tNC)_2$. These derivatives are unstable in solution. In the solid state, the nickel complexes are fairly stable in air but they can explode even without shock when dried in amounts of more than 0.5 g. The magnetic susceptibility of

$Ni(Bu^tNC)_2(O_2)$ in the solid state is very low, a magnetic moment of 0.13 B.M. being found. The infrared spectrum of these peroxo complexes shows a moderately intense band at 800–900 cm^{-1}, and ^{18}O isotopic infrared studies indicated that the coordinated dioxygen has a side-on isosceles structure, except in the palladium complex, where a deviation from this structure was inferred from the spectrum [482, 483] (see also 1.3.1.3). For $M(Bu^tNC)_2(O_2)$ (M = Ni, Pd) compounds and for the related $Pt(PPh_3)_2(O)_2$ derivative the approximate force constants of O—O and M—O bonds were calculated.

The constants f(O—O) = 3.0–3.5 mdyn $Å^{-1}$ indicate a marked reduction of the coordinated dioxygen in comparison to free gaseous dioxygen (11.5 mdyn $Å^{-1}$), and are even lower than the force constant of the peroxide ion [f(O—O), 5.4 mdyn $Å^{-1}$] and close to that of O_2 in an excited state, $^3\Sigma_\mu^+$ [f(O—O), 3.2] or $^3\Sigma_\mu^-$ [f(O—O), 2.3] [483]. The M—O stretching force constants (2.1–2.4 mdyn $Å^{-1}$) show that f(M—O) is greater in the nickel than in the palladium complex, implying in the first case a significant contribution of π-back donation in the metal–oxygen bond (1.3.1.3 and 1.4). Chemical evidence supports this relative order of f(M—O). In fact, the catalytic oxygenation of isocyanides may be carried out at −20°C with $Pd(Bu^tNC)_2(O_2)$ as catalyst but only at room temperature with $Ni(Bu^tNC)_2(O_2)$ [482, 484]. Thus, M—O force constants seem to be indicative of the lability of the O_2 ligand, while the O—O stretches poorly correlate with such a property [483].

The NC stretching absorptions in these peroxo complexes occur at 2170–2200 cm^{-1} as two intense bands implying a *cis* geometry of the isocyanides [482]. In view of the high NC stretching frequency electron transfer to the O_2 ligand is such that a formal oxidation state of the metal approaches 2.

The NC stretching frequencies for palladium are higher than for nickel in accordance with the order of the promotion energies for the metals. From the results of the electronic spectra, an apparent hexacoordination in the solid state was proposed for these complexes:

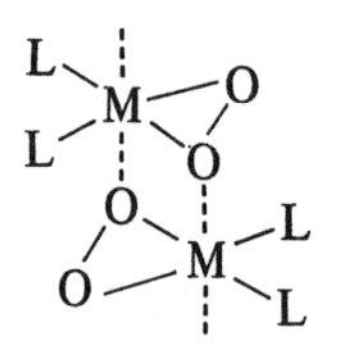

The oxygen is irreversibly bound, since heating of $Ni(Bu^tNC)_2(O_2)$ at 60°C *in vacuo* did not give appreciable dissociation.

Dissolution in a donor solvent destroys the oxygen bridging with formation of monomeric species, and an indication of the coordination of two molecules of CH_3CN to $Ni(Bu^tNC)_2(O_2)$ was obtained in cold chloroform solution in the presence of CH_3CN by means of n.m.r. spectroscopy.

At 30°C the oxygen atoms transfer to the isocyanide ligand. From the reaction between $Ni(Bu^tNC)_2(O_2)$ and an excess of Bu^tNC in toluene at ambient temperatures, Bu^tNCO and some $Ni(Bu^tNC)_4$ were isolated. When an excess of PPh_3 was added to the nickel peroxo complex, the oxygen atoms were transferred to the phosphine, affording Ph_3PO and $Ni(Bu^tNC)_2(PPh_3)_2$, while the reaction with CH_3NC (or CyNC) gave a mixture of Bu^tNCO and CH_3NCO (or CyNCO). Once isocyanate is formed within the coordination sphere, it will be readily replaced by isocyanides or other donor reagents available in the system. In the absence of these donor reagents, the main nickel species is a cluster complex of formula $[Ni(RNCO)]_x$, which, for R = Bu^t, can be isolated by treating $Ni(Bu^tNC)_2(O_2)$ with Bu^tNCO in toluene at ambient temperature. The reaction between $Ni(Bu^tNC)_2$ and Bu^tNCO gave a complex of probable stoichiometry $Ni(Bu^tNC)_2(Bu^tNCO)$. These isocyanate complexes gave $Ni(Bu^tNC)_4$ when reacted with excess Bu^tNC.

Coordination of a basic ligand L leading to a species such as $Ni(RNC)_2L(O_2)$ will enhance the π-electron transfer to oxygen; O–O bond cleavage and oxygen atom transfer to ligands is thus facilitated. On the other hand, treatment of $Ni(Bu^tNC)_2(O_2)$ with the strong π-acid $(CN)_2C{=}C(CN)_2$, which can compete with oxygen for π-bonding with the metal, gave $Ni(Bu^tNC)_2\{(CN)_2C{=}C(CN)_2\}$ while molecular oxygen is released. Analogously the reaction of the peroxo nickel complex with $BF_3 . Et_2O$ also releases the oxygen molecule [482]. Thus effective reduction of electron density of the metal leads to metal–oxygen bond cleavage.

The reactions of these dioxygen complexes $M(Bu^tNC)_2(O_2)$ (M = Ni, Pd) with a variety of reagents have recently been reported [485]. Carbon monoxide reacts with $Ni(Bu^tNC)_2(O_2)$ affording $Ni(Bu^tNC)_2(CO)_2$, CO_2 and a substantial amount of Bu^tNCO, while no carbonato complex could be isolated. This, together with the reactions with PPh_3 and RNC reported above, has been classified as "atom-transfer redox reactions" since they involve displacement of oxygenated ligands. "Atom-transfer oxidation reactions" are considered to be those where there is formation of a stable anion attached to the metal upon oxydation of a reactant such as N_2O_4, NO, CO or SO_2 [485]. Thus both the dioxygen nickel and palladium complexes react with N_2O_4 below −30°C affording the *trans* and *cis* dinitrato complexes $M(Bu^tNC)_2(NO_3)_2$, respectively. Nitric oxide gives $[Ni(Bu^tNC)_3(NO)]^+NO_3^-$ with the peroxo nickel complex in the presence of free Bu^tNC. This dark purple, diamagnetic complex, shows ν(NO) at 1825 cm^{-1} and ν(NC) at 2200 and 2210 cm^{-1}, and can be considered to be a derivative of nickel(0). The reaction with CO_2 takes place below 0°C giving the carbonato complex, $Ni(Bu^tNC)_2(CO_3)$, while no percarbonato derivatives were detected. Both the dioxygen complexes afford the sulphato derivatives $M(Bu^tNC)_2(SO_4)$ with SO_2 at −70°C. For comparison, the reaction between $Ni(Bu^tNC)_4$ and SO_2 has also been studied; $Ni(Bu^tNC)_3(SO_2)$ is formed, which on reaction with O_2 gives the same sulphato derivative as reported above. In contrast, the reaction between $Pd(Bu^tNC)_2(O_2)$ and SO_2 leads to the very stable $[Pd(Bu^tNC)_2(SO_2)]_3$, in which the SO_2 ligands are bidentate bridging two metals, as has been shown by a preliminary X-ray analysis [485]. Nitrosobenzene reacts with the nickel dioxygen complex in the presence of excess Bu^tNC, giving $Ni(Bu^tNC)_2(PhNO_2)$, a nitrobenzene complex, and Bu^tNCO. This complex is stable to air and the nitrobenzene ligand is not displaced by Bu^tNC or PhNO. The same complex can be obtained from $Ni(Bu^tNC)_4$ and $PhNO_2$, and it may be a derivative similar to $Pt(PPh_3)_2(TNB)$ (2.2.3.10). In order to examine the reactivity of coordinated nitroso groups a series of complexes $Ni(Bu^tNC)_2(p\text{-}XC_6H_4NO)$ ($X = H$, CH_3, Cl, etc.) has also been obtained presumably from the reactions of $Ni(Bu^tNC)_2$ with the nitroso compounds [485]. By reaction in *n*-hexane with oxygen $Ni(Bu^tNC)_2(O_2)$ and nitrosobenzene (for $X = H$) were obtained, while neither nitrobenzene nor the nitrobenzene complex was detected. However when the reaction was carried out in a solvent able to dissolve the nickel dioxygen adduct, the formation of nitrobenzene was observed. Similar displacement of the ligand leading to the dioxygen complex was observed for $Ni(Bu^tNC)_2(PhC_2Ph)$ and $Ni(Bu^tNC)_2(CS_2)$ [479].

"Metal-assisted electrophilic peroxidations" of organic moieties occur in the reactions of $Ni(Bu^tNC)_2(O_2)$ with benzoyl halogenides and trityl bromide or tetrafluoroborate:

$$Ni(Bu^tNC)_2(O_2) + 2\,PhCOX \longrightarrow Ni(Bu^tNC)_2X_2 + PhC(=O)OOC(=O)Ph \quad (X = Cl, Br)$$

$$Ni(Bu^tNC)_2(O_2) + 2\,Ph_3CX \longrightarrow Ni(Bu^tNC)_2X_2 + Ph_3COOCPh_3 \quad (X = Br, BF_4)$$

On the other hand numerous attempts to find some peroxidation of conjugated dienes or epoxidation of alkenes, typical reactions of excited states of oxygen, failed [485].

The reaction of hexafluoroacetone with $Ni(Bu^tNC)_2(O_2)$ gave the explosive:

$$(Bu^tNC)_2Ni\langle^{O-O}_{O-C(CF_3)_2}$$

which shows $\nu(O{-}O)$ at 900 cm^{-1} [486] and loses one oxygen atom in ethyl ether, giving

$$(Bu^tNC)_2Ni\langle^{O}_{O}\rangle C(CF_3)_2$$

When excess hexafluoroacetone was used, a second molecule of the ketone was coordinated by nickel as a donor ligand at the oxygen atom [$\nu(C{=}O)$, 1670 cm^{-1} in the complex; $\nu(C{=}O)$, 1810 cm^{-1} in the free ketone].

The reactions of excess hexafluoroacetone or hexafluoropropylidenimine, $(CF_3)_2X$ (X = CO, NH) with $Ni(Bu^tNC)_4$ [486] gave compounds similar to those already described with phosphines as ligands (2.2.3.5 and 6). Five membered rings are formed. In contrast, the reaction between the phenylisocyanide nickel derivative and hexafluoroacetone gave the 1 : 1 adduct $Ni(PhNC)_2\{(CF_3)_2CO\}$. 1 : 1 adducts of the type $Ni(Bu^tNC)_2\{(CF_3)_2X\}$ (X = CO, NH) have also been obtained by displacing cyclooctadiene from $Ni(1,5\text{-}COD)\{(CF_3)_2X\}$ with Bu^tNC [486]. These compounds undergo insertion reactions with $(CF_3)_2NH$ (X = CO) and $(CF_3)_2CO$ (X = NH). The complex $Ni(PhNC)_2\{(CF_3)_2CO\}$ also gave an insertion product with $(CF_3)_2NH$. The structure of $(Bu^tNC)_2NiC(CF_3)_2NHC(CF_3)_2O$ has been determined by X-ray and has been already discussed (2.2.3.6).

The reactions of $Pd(Bu^tNC)_2$ with $(CF_3)_2X$ (X = CO, NH), leading to five-membered rings, have also been studied [120]. With $(CF_3)_2NMe$, the unstable 1 : 1 adduct was isolated. The reactions between bis(trifluoromethyl) diazomethane and $Ni(Bu^tNC)_4$ or $Pd(RNC)_2$ (R = Bu^t, Cy) afforded derivatives [277] which seem to be isostructural with the analogous azine derivative obtained from $Pt(PPh_3)_4$, whose structure has been determined (2.2.3.6).

A series of diazoalkane nickel(0) complexes have been obtained from the reaction between $Ni(Bu^tNC)_2$ and R_2CN_2 (R = fluorenylidene, Ph_2C, $(NC)_2C$) [361]. They have the stoichiometry $Ni(Bu^tNC)_2(R_2CN_2)$ with a side-on coordination involving the C=N=N system (see also 2.2.3.8e). The NC stretching vibrations are higher than that in the starting complexes, indicating electron withdrawal from metal to the diazoalkane ligand. These metal compounds are very sensitive to air, and the corresponding palladium complexes are even more so [361].

Nickel and palladium isocyanide complexes having unsaturated organic molecules such as alkenes and related derivatives as ligands have been recently reported. An equimolar mixture of $Ni(Bu^tNC)_2$ and azobenzene in ether affords the orange $Ni(Bu^tNC)_2(PhN{=}NPh)$, diamagnetic and extremely air-sensitive [487, 488]. From i.r. and n.m.r. evidence a side-on (olefin-like) coordination without *ortho*-metallation was proposed. Its chemical reactivity also supported this view, since reactions with oxygen or tetracyanoethylene, L, gave the corresponding $Ni(Bu^tNC)_2L$ complexes and azobenzene [488]. The crystal structure of the azobenzene complex has been determined, and it has confirmed this type of bonding [489]. The coordination about the nickel atoms is trigonal if the azobenzene is regarded as a monodentate ligand:

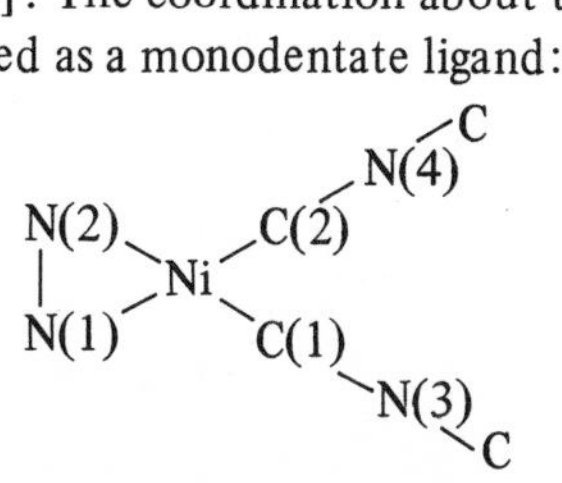

The dihedral angle between the N(1)NiN(2) and C(1)NiC(2) planes is only 1.2(3)°, and thus the inner coordination sphere is essentially planar. The average Ni–C and C≡N distances are 1.841(5) and 1.156(6) Å respectively, and each C≡N–C fragment is essentially linear. The N(1)–N(2) distance is 1.385(5) Å, significantly longer than the corresponding distance in free azobenzene (the corresponding bond distance in *cis*-azobenzene is 1.23 Å, while the two different molecules in the asymmetric unit of the *trans* form have N–N distances of 1.243(3) and 1.172(3) Å).

A comparison with structural data on other azo compounds and with N–N single bond lengths, indicates a lengthening of the N–N distance to that of a single bond [489]. The nitrogen atoms of azobenzene are equidistant from the metal [1.898(4) Å], with a Ni–N distance somewhat shorter than the sum of covalent radii. The average N–N–C (phenyl) angle is 111.8(3) Å, little different from those for free *trans*-azobenzene. There is a large dihedral angle of 49.0(3) Å between the planes of the two phenyl rings, probably to relieve steric strain resulting from non-bonded interactions of the azobenzene nitrogen atoms with *ortho* carbon atoms of the phenyl rings. The approximately tetrahedral geometry assumed by the bonded nitrogen atoms suggests that the lone-pair electrons occupy orbitals directed away from the metal [489]. Activated alkenes such as tetracyanoethylene, maleic anhydride, or fumaronitrile (TCNE, MA, or FN respectively) react with $Ni(Bu^tNC)_4$ affording $Ni(Bu^tNC)_3$(alkene) derivatives [488]. Weaker π-acids such as ethylene, diphenylacetylene or azobenzene failed to replace the isocyanide ligands. $M(Bu^tNC)_2(L)$ complexes (M = Ni, Pd) (L = TCNE, FN, MA, dimethyl maleate) can be easily obtained from $M(Bu^tNC)_2$ and the ligands L. Similar complexes can also be obtained from $Ni(Bu^tNC)_2(CS_2)$ or $Ni(Bu^tNC)_2(PhC_2Ph)$ and the alkene. With excess MA and $Ni(Bu^tNC)_2$, $Ni(Bu^tNC)_2(MA)_2$ was isolated, but other strong or moderate d_π-acceptors failed to give bis-alkene complexes. A complex of probable stoichiometry $Ni(Bu^tNC)_2(\text{allene})_2$ was the product of the reaction between $Ni(Bu^tNC)_2$ and allene in cold *n*-hexane or benzene, but extensive dissociation of this derivative prevented its isolation in a pure form. The $Ni(Bu^tNC)_3(L)$ complexes are moderately air sensitive in solution, with the exception of the TCNE derivative which is quite inert. They show the usual lowering of the stretching frequencies of the electron-withdrawing groups present on the alkenes and the high-field shift of the resonance of the olefinic protons.

Generally the ν(NC) frequencies of the isocyanide in $Ni(Bu^tNC)_3L$ complexes are lower than those of the corresponding $Ni(Bu^tNC)_2(L)$ derivatives, while the olefinic protons of the bis-isocyanide compounds are generally more effectively shielded.

The TCNE derivative shows a rapid isocyanide ligand exchange detected by n.m.r. However the dissociation of the isocyanide ligand must be very small since the molecular weights measured in benzene correspond to the theoretical value.

The $M(Bu^tNC)_2L$ complexes are very sensitive to air, particularly in solution, and it has been observed that the higher the NC stretching frequencies of the isocyanide, the greater the stability of the metal–alkene bond, which shows the importance of the π-bond in the metal–alkene interaction. A correlation between ν(NC) of the isocyanide and the electron affinities of the alkene in the sense that ν(NC) increases with increasing the electron affinity of the alkene has been observed for the $M(Bu^tNC)_2L$ derivatives. The PhN=NPh (but not the PhC_2Ph, see later) derivatives also obeys these correlations. The $M(Bu^tNC)_2L_2$ complexes have a thermal stability which depends to a great extent upon the ligand.

A comparison of the i.r. data of $Ni(Bu^tNC)_2(CO)_2$, obtained from $Ni(CO)_4$ and Bu^tNC, with those for $Ni(Bu^tNC)_4$, $Ni(PPh_3)_2(CO)_2$ and $Ni(CO)_4$, has suggested the following order of d_π acceptor strength: $PPh_3 < Bu^tNC < CO$ [488]. The compound $Pd(PhNC)_2(TCNE)$ obtained from the reaction of $Pd(PhNC)_2$ and TCNE [329], has NC stretching modes lower than the corresponding Bu^tNC derivative, as expected. PPh_3 and $AsPh_3$ displace one isocyanide, giving the Pd(PhNC)(L)(TCNE) complexes (L = PPh_3, $AsPh_3$).

Fluoroalkene derivatives of nickel and palladium have also been reported. Both 1,1-dicyano-2,2-bis(trifluoromethyl)ethylene and hexafluoropropene give stable 1 : 1 adducts with $Pd(Bu^tNC)_2$ [120], in contrast with tetrafluoroethylene. On the other hand, $Ni(Bu^tNC)_4$ reacts with C_2F_4, but with formation of the octafluoronickelacyclopentane derivative, $Ni(Bu^tNC)_2(CF_2)_4$ [486]. Similarly, hexafluorobutadiene and $Ni(Bu^tNC)_4$ give the nickelacyclopent-3-ene derivative,

$Ni(Bu^tNC)_2\left(\begin{matrix} CF_2-CF \\ \| \\ CF_2-CF \end{matrix}\right)$ while simple alkene derivatives have been obtained with $(CF_3)_2C{=}C(CN)_2$ or $CF_3CF{=}CF_2$. The X-ray structural determination of $Ni(Bu^tNC)_2\{(CN)_2C{=}C(CN)_2\}$ has shown the usual trigonal coordination about the nickel atom [490].

The central C–C bond length of 1.476(5) Å is only little shorter than that of 1.52(3) Å found in $Pt(PPh_3)_2(TCNE)$. However, a large deviation from planarity of the $ML_2(C_2)$ portion of this trigonal molecule was observed in this derivative (23.9°), and although such distortions could be partially electronic in nature [491], they could also arise as a consequence of packing energy in the solid state [490].

Reactions of tetrakis(aryl isocyanide)nickel complexes with diphenylacetylene gave the diiminocyclobutenes

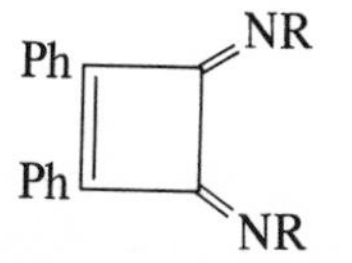

a reaction reminiscent of the reaction of metal carbonyls with alkynes. In order to elucidate the nature of the intermediate products, compounds of the series $Ni(RNC)_2(PhC{\equiv}CPh)$ (R = aryl substituent) were synthesized by adding PhC_2Ph to a mixture of $Ni(1,5\text{-}COD)_2$ and RNC [492]. Heating of these complexes gave the corresponding diiminocyclobutenes. However pyrolysis of the compound where R = 2,6-$Me_2C_6H_4$ gave mainly the cyclopentadienone derivative,

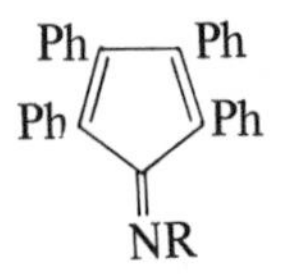

The complexes $M(Bu^tNC)_2(RC_2R)$ (M = Ni, Pd; R = CO_2Me) have been obtained by reacting $M(Bu^tNC)_2$ first with diphenylacetylene, and then displacing this alkyne with dimethyl acetylenedicarboxylate [488]. The nickel derivatives are active catalysts for polymerization of alkynes, including dimethyl acetylenedicarboxylate [479]. In general the palladium derivatives are more labile than the corresponding nickel derivatives.

The diphenylacetylene-nickel complex, $Ni(Bu^tNC)_2(PhC_2Ph)$ has an essentially planar structure [dihedral angle θ of 2.6(7)° between the NiCC (of acetylene) and NiCC (of Bu^tNC) planes], with trigonal coordination about the nickel atom [493]. The phenyl rings of the coordinated alkyne are bent away from the metal by about 31° from the C–C bond axis, with a *cis* geometry of the alkyne. The average C–C distance of 1.284(16) Å is longer than the bond length of 1.19(2) Å reported for free PhC_2Ph in the solid state, but shorter than the same distance in $Pt(PPh_3)_2(PhC_2Ph)$ [1.32(9) Å]. A general discussion concerning the molecular parameters of coordinated alkynes (and alkenes), the type of metal–alkyne interaction, and the reactivity of the metal–alkyne complexes has been reported, but more systematic structural studies on this type of complexes are needed in order to have a large number of experimental data [493].

OTHER COMPLEXES

There are a variety of other complexes of metals which could be classified as zerovalent derivatives.

However, complexes such as the cyano and alkynyl derivatives of nickel, palladium and platinum of formula $K_4[M(CN)_4]$ (M = Ni, Pd) or $K_4[M(C{\equiv}CR)_4]$ (M = Ni, Pd, Pt) have not received great attention since their discovery [1a]. Also the nitrosyl derivatives $M(NO)(\pi\text{-}C_5H_5)$ (M = Ni, Pd and Pt) [1a] have for the most part been studied from the point of view of their spectroscopic properties. Some ammonia and ethylendiamine derivatives such as $Pt(NH_3)_4$ and $Pt(en)_2$ are rather peculiar in view of the absence of stabilization through the π-back-donation process with these ligands, and their real zerovalent nature has not yet been confirmed [1a]. Another class of derivative which could be included among the zerovalent complexes is that of the compounds having dithiolate derivatives as ligands. However the non-innocent nature of the ligands, which may be considered to be

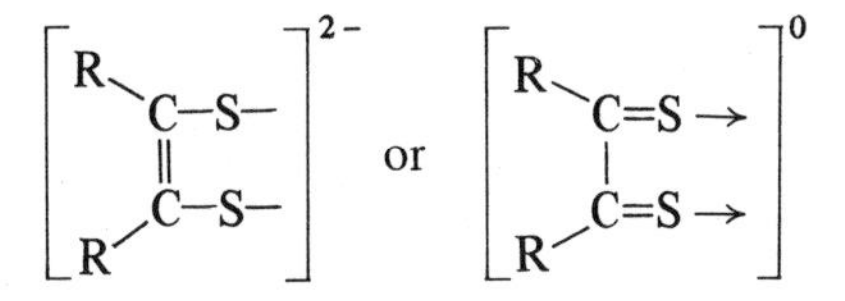

makes the oxidation state of these compounds uncertain. For $[Ni(S_2C_2R_2)_n]^z$ (n = 2, 3; z = 0, −1, −2 or −3 if n = 3; R = organic group) recent measurements of the ionization energies of the $Ni_{2p_{3/2}}$ electrons have shown that they are close to that of elemental Ni°, and lower than those of nickel(I) and nickel(II) complexes [494].

A similar problem exists for the derivatives of 2,2-dipyridine and 1,10-phenanthroline. A large number of formally zerovalent complexes having such heterocyclic derivatives as ligands is known. The chemistry of these complexes has received more attention in the last few years. However, a recent careful infrared study, conducted by using the metal isotope technique, of these compounds in oxidation states which appear to be −1 or +1 and +2 if the organic ligand is considered as a neutral species, has shown that in more reduced complexes, an increasing fraction of the electron population once considered to be on the metal actually resides in the ligand orbitals, so that the electron density around the metal remains fairly constant through the series [495]. In fact the ligand spectra of complexes such as $[Cr(bipy)_3]^\circ$ are similar to that of Li(bipy) in which the ligand is negatively charged. Other physical measurements, particularly with photoelectron spectroscopy and related techniques, are required to give a more definite answer to the question of the real oxidation state of the metals present in these derivatives.

REFERENCES

1. (a) L. Malatesta, R. Ugo and S. Cenini, *Adv. Chem. Ser., Amer. Chem. Soc.*, **62**, 318 (1966) and references therein.
(b) G. Booth, *Adv. Inorg. Chem. Radiochem.*, **6**, 1 (1964).
(c) P. M. Maitlis, "The Organic Chemistry of Palladium", Vol. 1. "Metal Complexes", Academic Press, New York (1971).
2. L. Malatesta and M. Angoletta, *J. Chem. Soc.*, 1186 (1957).
3. L. Malatesta and C. Cariello, *J. Chem. Soc.*, 2323 (1958).

4. L. Malatesta and R. Ugo, *J. Chem. Soc.*, 2080 (1963).
5. R. Ugo, *Coord. Chem. Rev.*, **3**, 319 (1968).
6. C. K. Jørgensen, *Inorg. Chem.*, **3**, 1201 (1964).
7. F. R. Hartley, *Organometallic Chem. Rev.* (A), **6**, 119 (1970).
8. L. Mond, C. Langer and F. Quinke, *J. Chem. Soc.*, **57**, 749 (1890).
9. L. Pauling, "The nature of the chemical bond", Cornell University Press, Ithaca, New York (1939).
10. L. D. Quin, *J. Amer. Chem. Soc.*, **79**, 3681 (1957).
11. L. Maier, *Angew. Chem.*, **71**, 574 (1959).
12. J. Chatt and F. A. Hart, *J. Chem. Soc.*, 1378 (1960).
13. J. Chatt, F. A. Hart and D. T. Rosevear, *J. Chem. Soc.*, 5504 (1961).
14. J. Chatt and A. A. Williams, *J. Chem. Soc.*, 3061 (1951).
15. G. Wilkinson, *J. Amer. Chem. Soc.*, **73**, 5501 (1951).
16. G. B. Street and A. B. Burg, *Inorg. Nuclear Chem. Letters*, **1**, 47 (1965).
17. C. W. Garland, R. C. Lord and P. F. Troians, *J. Phys. Chem.*, **69**, 1195 (1965).
18. T. Kruck and K. Baur, *Chem. Ber.*, **98**, 3070 (1965).
19. T. Kruck, K. Baur and W. Lang, *Chem. Ber.*, **101**, 138 (1968).
20. (a) J. F. Nixon and M. D. Sexton, *Inorg. Nuclear Chem. Letters*, **4**, 275 (1968).
    (b) *ibid.*, *J. Chem. Soc.* (*A*), 1089 (1969).
21. P. L. Timms, *Chem. Comm.*, 1033 (1969).
22. S. Carrà and R. Ugo, *Inorg. Chim. Acta Rev.*, **1**, 49 (1967).
23. J. W. Irvine, Jr. and G. Wilkinson, *Science*, **113**, 742 (1951).
24. L. Malatesta and A. Sacco, *Ann. Chim.* (*Italy*), **44**, 134 (1954).
25. G. Wilkinson, *Z. Naturforsch.*, **9b**, 446 (1954).
26. M. Bigorgne and A. Zelwer, *Bull. Soc. Chim. France*, **1986 (1960)**.
27. A. Loutellier and M. Bigorgne, *Bull. Soc. Chim. France*, 3186 (1965).
28. R. J. Clark and E. O. Brimm, *Inorg. Chem.*, **4**, 651 (1965).
29. R. Schmutzler, *Chem. Ber.*, **96**, 2435 (1963).
30. R. Schmutzler, *Inorg. Chem.*, **3**, 415 (1964).
31. R. Schmutzler, *Chem. Ber.*, **98**, 552 (1965).
32. A. B. Burg and G. B. Street, *Inorg. Chem.*, **5**, 1532 (1966).
33. J. F. Nixon and M. D. Sexton, *J. Chem. Soc.* (*A*), 321 (1970).
34. C. B. Lindahl and W. L. Jolly, *Inorg. Chem.*, **3**, 1634 (1964).
35. J. R. Leto and M. F. Leto, *J. Amer. Chem. Soc.*, **83**, 2944 (1961).
36. R. F. Clark and C. D. Storrs, *Chem. Abstr.*, **57**, 16662c (1962).
37. (a) J. G. Verkade, R. E. McCarley, D. G. Hendricker and R. W. King, *Inorg. Chem.*, **4**, 228 (1965).
    (b) see also: K. J. Coskran, T. J. Huttemann and J. G. Verkade, *Advan. Chem. Ser.* (*Amer. Chem. Soc.*), **63**, 590 (1963).
38. D. G. Hendricker, R. E. McCarley, R. W. King and J. G. Verkade, *Inorg. Chem.*, **5**, 639 (1966).
38*bis*. P. Cassoux, J. M. Savariault, and J. F. Labarre, *Bull. Soc. Chim. France*, 741 (1969).
39. G. S. Reddy and R. Schmutzler, *Inorg. Chem.*, **6**, 823 (1967).
40. M. Beherens and A. Müller, *Z. Anorg. Chem.*, **341**, 124 (1965).
41. M. Beherens, A. Müller and M. Preiss, *Proc. 2nd Intern. Symp. Organomet. Chem.*, 72 (Aug. 30–Sept. 3, 1965).
42. R. J. Clark, P. I. Hoberman and E. O. Brimm, *J. Inorg. Nuclear Chem.*, **27**, 2109 (1965).

42*bis*. T. Kruck, K. Baur, K. Glinka and M. Stadler, *Z. Naturforsch.*, **23b**, 1147 (1968).
43. J. J. Levison and S. D. Robinson, *Inorg. Synth.*, **13**, 105 (1972).
44. R. L. Keiter and J. G. Verkade, *Inorg. Chem.*, **9**, 404 (1970).
45. K. S. Wheelock, J. H. Nelson and H. B. Jonassen, *Inorg. Chim. Acta*, **4**, 399 (1970).
46. S. Takahashi, K. Sonogashira and N. Hagihara, *J. Chem. Soc. Japan*, **87**, 610 (1966); *Chem. Abstr.*, **65**, 14485 (1966).
47. (a) C. A. Tolman, *J. Amer. Chem. Soc.*, **92**, 2956 (1970).
47. (b) C. A. Tolman, *Chem. Soc. Rev.*, **1**, 337 (1972).
48. N. von Kutepow, H. Seibt and F. Meier, *U.S. Patent*, 3,346,608 (1967).
49. J. Browning, C. S. Cundy, M. Green and F. G. A. Stone, *J. Chem. Soc. (A)*, 20 (1969).
50. (a) C. S. Cundy, M. Green and F. G. A. Stone, *J. Chem. Soc. (A)*, 1647 (1970).
(b) C. S. Cundy, PhD. Thesis, Bristol (1969).
51. A. Pidcock and G. G. Roberts, *J. Chem. Soc. (A)*, 2922 (1970).
52. M. Trabelsi, A. Loutellier and M. Bigorgne, *J. Organometallic Chem.*, **40**, C45 (1972).
53. J. P. Birk, J. Halpern and A. L. Pickard, *J. Amer. Chem. Soc.*, **90**, 4491 (1968).
54. W. C. Seidel and C. A. Tolman, *Inorg. Chem.*, **9**, 2354 (1970).
55. C. A. Tolman, W. C. Seidel and D. H. Gerlach, *J. Amer. Chem. Soc.*, **94**, 2669 (1972).
56. S. Takahashi and N. Hagihara, *Nippon Kagaku Zasshi*, **88**, 1306 (1967); *Chem. Abstr.*, **69**, 27514g (1968).
57. R. J. De Pasquale, *J. Organometallic Chem.*, **32**, 381 (1971).
58. R. D. W. Kemmitt and R. D. Moore, *J. Chem. Soc. (A)*, 2472 (1971).
59. Y. Takahashi, T. Ito, S. Sakai and Y. Ishii, *J. Chem. Soc. (D)*, 1065 (1970).
60. K. Moseley and P. M. Maitlis, *J. Chem. Soc. (D)*, 982 (1971).
61. F. Seel, K. Ballreich and R. Schmutzler, *Chem. Ber.*, **94**, 1173 (1961).
62. T. Kruck and M. Höfler, *Angew. Chem. Internat. Edn.*, **6**, 563 (1967).
63. T. Kruck, M. Höfler, K. Baur, P. Junkes and K. Glinka, *Chem. Ber.*, **101**, 3827 (1968).
63*bis*. D. F. Bachman, E. D. Stevens, T. A. Lane and J. T. Yoke, *Inorg. Chem.*, **11**, 109 (1972).
64. T. Kruck, M. Höfler, H. Jung, and H. Blume, *Angew. Chem. Internat. Edn.*, **8**, 522 (1969).
65. O. Glemser, E. Niecke and A. Müller, *Angew Chem. Internat. Edn.*, **5**, 583 (1966).
66. R. Ugo, F. Cariati and G. La Monica, *Inorg. Synth.*, **11**, 105 (1968).
67. R. Ugo, G. La Monica, F. Cariati, S. Cenini and F. Conti, *Inorg. Chim. Acta.*, **4**, 390 (1970).
68. R. Van der Linde and R. O. de Jongh, *J. Chem. Soc. (D)*, 563 (1971).
69. (a) K. Jonas and G. Wilke, *Angew. Chem. Internat. Edn.*, **8**, 519 (1969).
(b) P. W. Jolly, K. Jonas, C. Krüger and Y. H. Tsay, *J. Organometallic Chem.*, **33**, 109 (1971).
70. C. S. Cundy and H. Nöth, *J. Organometallic Chem.*, **30**, 135 (1971).
71. L. W. Gosser and C. A. Tolman, *Inorg. Chem.*, **9**, 2350 (1970).
72. K. A. Jensen, B. Nygaard, G. Elisson and P. M. Nielsen, *Acta Chem. Scand.*, **19**, 768 (1965).

73. J. Chatt, F. A. Hart and H. R. Watson, *J. Chem. Soc.*, 2537 (1962).
74. T. Kruck and K. Baur, *Angew. Chem. Internat. Edn.*, **4**, 521 (1965).
75. T. Kruck, *Angew. Chem. Internat. Edn.*, **6**, 53 (1967).
76. J. Chatt and H. R. Watson, *Nature,* **189**, 1003 (1961).
77. J. Chatt and G. A. Rowe, *Nature,* **191**, 1191 (1961).
78. D. T. Rosevear and F. G. A. Stone, *J. Chem. Soc.* (*A*), 164 (1968).
79. M. Bressan, G. Favero, B. Corain and A. Turco, *Inorg. Nuclear Chem. Letters,* **7**, 203 (1971).
80. M. I. Bruce, G. Shaw and F. G. A. Stone, *J. C. S. Dalton,* 1082 (1972).
81. H. C. Clark and K. Itoh, *Inorg. Chem.*, **10**, 1707 (1971).
82. R. B. King and P. N. Kapoor, *Inorg. Chem.*, **11**, 1524 (1972).
83. J. J. Levison and S. D. Robinson, *J. Chem. Soc.* (*A*), 96 (1970).
84. L. G. Cannell, *U.S. Patent,* 3,102,899; *Chem. Abstr.*, **60**, 1362e (1964).
85. A. D. Allen and C. D. Cook, *Proc. Chem. Soc.*, 218 (1962).
86. J. Chatt and B. L. Shaw, *J. Chem. Soc.*, 5075 (1962).
87. J. A. Chopoorian, J. Lewis and R. S. Nyholm, *Nature,* **190**, 528 (1961).
88. G. C. Dobinson, R. Mason, G. B. Robertson, R. Ugo, F. Conti, D. Morelli, S. Cenini and F. Bonati, *Chem. Comm.*, 739 (1967).
89. R. Ugo and S. Cenini, unpublished results.
90. Pi-Chang Kong and D. M. Roundhill, *Inorg. Chem.*, **11**, 749 (1972).
91. D. R. Coulson, *Inorg. Synth.*, **13**, 121 (1972).
92. M. Maier, F. Basolo and R. G. Pearson, *Inorg. Chem.*, **8**, 795 (1969).
93. M. Maier and F. Basolo, *Inorg. Synth.*, **13**, 112 (1972).
94. H. F. Klein and H. Schmidbaur, *Angew. Chem. Internat. Edn.*, **9**, 903 (1970).
95. P. Roffia, F. Conti and G. Gregorio, *Chimica e Industria,* **54**, 317 (1972).
96. P. Roffia, F. Conti and G. Gregorio, *Chimica e Industria,* **53**, 361 (1971).
97. D. A. White and G. W. Parshall, *Inorg. Chem.*, **9**, 2358 (1970).
98. G. Wilke, E. W. Müller and M. Kröner, *Angew. Chem.*, **73**, 33 (1961).
99. B. Bogdanovic, M. Kröner and G. Wilke, *Ann.*, **699**, 1 (1966).
100. B. Corain, P. Rigo and M. Bressan, *Chimica e Industria,* **51**, 386 (1969).
101. B. Corain, M. Bressan and P. Rigo, *J. Organometallic Chem.*, **28**, 133 (1971).
102. G. R. Van Hecke and W. D. Horrocks, Jr., *Inorg. Chem.*, **5**, 1968 (1966).
103. G. Wilke and G. Hermann, *Angew Chem. Internat. Edn.*, **1**, 549 (1962).
104. M. A. McCall and H. W. Coover, *British Patent,* 1,146,074 (1969).
105. M. Englert, P. W. Jolly and G. Wilke, *Angew. Chem. Internat. Edn.*, **10**, 77 (1971).
106. R. F. Clark, C. D. Storrs and C. G. McAlister, *Chem. Abstr.*, **59**, 11342 (1963).
107. R. A. Schunn, *Inorg. Synth.*, **13**, 124 (1972).
108. K. Kudo, M. Hidai and Y. Uchida, unpublished results quoted in: K. Kudo, M. Hidai, T. Murayama and Y. Uchida, *J. Chem. Soc.* (*D*). 1701 (1970).
109. R. S. Vinal and L. T. Reynolds, *Inorg. Chem.*, **3**, 1062 (1964).
110. M. Meier, F. Basolo and R. G. Pearson, *Inorg. Chem.*, **8**, 795 (1969).
111. W. C. Drinkard, D. R. Eaton, J. P. Jesson and R. V. Lindsey, Jr., *Inorg. Chem.*, **9**, 392 (1970).
112. T. J. Huttemann, Jr., B. M. Foxman, C. R. Sperati and J. G. Verkade, *Inorg. Chem.*, **4**, 950 (1965).
113. A. A. Orio, B. B. Chastain and H. B. Gray, *Inorg. Chim. Acta,* **3**, 8 (1969).

113*bis*. C. W. Weston, G. W. Bailey, J. H. Nelson and H. B. Jonassen, *J. Inorg. Nuclear Chem.*, **34**, 1752 (1972).
114. G. F. Svatos and E. E. Flagg, *Inorg. Chem.*, **4**, 422 (1965).
115. T. Kruck and K. Baur, *Z. Anorg. Chem.*, **364**, 192 (1969).
116. (a) G. Wilke and B. Bogdanovic, *Angew. Chem.* **73**, 756 (1961).
116. (b) G. Wilke *et al.*, *Angew Chem. Internat. Edn.*, **2**, 105 (1963).
117. J. K. Becconsall, B. E. Job and S. O'Brien, *J. Chem. Soc.* (*A*) 423 (1967).
118. E. O. Fischer and H. Werner, *Chem. Ber.*, **95**, 703 (1962).
119. B. F. G. Johnson, T. Keating, J. Lewis, M. S. Subramanian and D. A. White, *J. Chem. Soc.* (*A*), 1793 (1969).
120. H. D. Empsall, M. Green, S. K. Shakshooki and F. G. A. Stone, *J. Chem. Soc.* (*A*), 3472 (1971).
121. H. D. Empsall, M. Green and F. G. A. Stone, *J. C. S. Dalton*, 96 (1972).
122. J. R. Olechowski, C. G. McAlister and R. F. Clark, *Inorg. Chem.*, **4**, 246 (1965).
123. J. R. Olechowski, C. G. McAlister and R. F. Clark, *Inorg. Synth.*, **9**, 181 (1967).
124. B. Bogdanović, H. Bönnemann and G. Wilke, *Angew. Chem. Internat. Edn.*, **5**, 582 (1966).
125. J. F. Nixon, *J. Chem. Soc.* (*A*), 1136 (1967).
126. J. F. Nixon, *Chem. Comm.*, 34 (1966).
127. H. Schott and G. Wilke, *Angew. Chem. Internat. Edn.*, **8**, 877 (1969).
128. G. Wilke and H. Schott, *Angew. Chem. Internat. Edn.*, **5**, 583 (1966).
129. W. Kuran and A. Musco, *J. Organanometallic Chem.*, **40**, C47 (1972).
130. H. C. Volger and K. Vrieze, *J. Organometallic Chem.*, **9**, 527 (1967).
131. H. T. Dodd and J. F. Nixon, *J. Organometallic Chem.*, **32**, C67 (1971).
132. D. H. Gerlach, A. R. Kane, G. W. Parshall, J. P. Jesson and E. L. Muetterties, *J. Amer. Chem. Soc.*, **93**, 3543 (1971).
133. D. M. Blake and C. J. Nyman, *J. Amer. Chem. Soc.*, **92**, 5359 (1970).
134. D. M. Blake and C. J. Nyman, *Chem. Comm.*, 483 (1969).
135. (a) D. M. Blake and R. Mersecchi, *J. Chem. Soc.* (*D*), 1045 (1971).
(b) D. M. Blake and L. M. Leung. *Inorg. Chem.*, **11**, 2879 (1972).
136. R. D. Gillard, R. Ugo, F. Cariati, S. Cenini and F. Bonati, *Chem. Comm.*, 869 (1966).
137. P. Fitton and J. E. McKeon, *Chem. Comm.*, 4 (1968).
138. J. A. Osborn, *Chem. Comm.*, 1231 (1968).
139. B. F. G. Johnson, J. Lewis and D. A. White, *J. Chem. Soc.* (*A*), 2699 (1971).
140. W. C. Smith, *Inorg. Synth.*, **6**, 201 (1960).
141. P. Heimbach, *Angew. Chem. Internat. Edn.*, **3**, 648 (1964).
142. D. Titus, A. A. Orio and H. B. Gray, *Inorg. Synth.*, **13**, 117 (1972).
143. V. G. Albano, P. L. Bellon and V. Scatturin, *Chem. Comm.*, 507 (1966).
144. V. G. Albano, P. L. Bellon and M. Sansoni, *Chem. Comm.*, 899 (1969); see also *errata, ibid.*, 188 (1970).
145. V. G. Albano, P. L. Bellon and M. Manassero, *J. Organometallic Chem.*, **35**, 423 (1972).
146. V. G. Albano, G. M. Basso Ricci and P. L. Bellon, *Inorg. Chem.*, **8**, 2109 (1969).
147. J. H. Darling and J. S. Ogden, *Inorg. Chem.*, **11**, 666 (1972).
148. J. C. Marriott, J. A. Salthouse, M. J. Ware and J. M. Freeman, *J. Chem. Soc.*, (*D*), 595 (1970).

149. Y. Morino, K. Kuchitsu and T. Moritani, *Inorg. Chem.*, **8**, 867 (1969).
150. R. L. Kuczkowski and D. R. Lide, *J. Chem. Phys.*, **46**, 357 (1967).
151. R. W. Kiser, M. A. Krassoi and R. J. Clark, *J. Amer. Chem. Soc.*, **89**, 3653 (1967).
152. R. E. Sullivan and R. W. Kiser, *Chem. Comm.*, 1425 (1968).
153. S. Pignataro and S. Cenini, unpublished results.
154. (a) L. A. Woodward and J. R. Hall, *Nature*, **191**, 831 (1958).
(b) *ibid.*, *Spectrochim. Acta*, **16**, 654 (1960).
155. L. H. Jones, *J. Chem. Phys.*, **28**, 1215 (1958).
156. H. Stammreich, K. Kawal, O. Sala and P. Krumholtz, *J. Chem. Phys.*, **35**, 2168 (1961).
157. J. P. Birk, J. Halpern and A. L. Pickard, *Inorg. Chem.*, **7**, 2672 (1968).
158. D. R. Eaton and S. R. Stuart, *J. Amer. Chem. Soc.*, **90**, 4170 (1968).
159. C. J. Nyman, C. E. Wymore and G. Wilkinson, *Chem. Comm.*, 407 (1967).
160. F. Cariati, R. Mason, G. B. Robertson and R. Ugo, *Chem. Comm.*, 408 (1967).
161. F. Cariati, R. Ugo and F. Bonati, *Inorg. Chem.*, **5**, 1128 (1966).
162. R. D. W. Kemmitt, R. D. Peacock and J. Stocks, *J. Chem. Soc.* (*A*), 846 (1971).
163. J. T. Dumler and D. M. Roundhill, *J. Organometallic Chem.*, **30**, C35 (1971).
164. R. A. Schunn, *Inorg. Chem.*, **9**, 394 (1970).
165. C. A. Tolman, *J. Amer. Chem. Soc.*, **92**, 4217 (1970).
166. C. A. Tolman, *Inorg. Chem.*, **11**, 3128 (1972).
167. K. Thomas, J. T. Dumler, B. W. Renoe, C. J. Nyman and D. M. Roundhill, *Inorg. Chem.*, **11**, 1795 (1972).
168. D. M. Roundhill, P. B. Tripathy and B. W. Renoe, *Inorg. Chem.*, **10**, 727 (1971).
169. P. B. Tripathy and D. M. Roundhill, *J. Organometallic Chem.*, **24**, 247 (1970).
170. M. J. Church and M. J. Mays, *J. Chem. Soc.* (*A*) 3074 (1968).
171. R. Ugo, G. La Monica, S. Cenini, A. Segre and F. Conti, *J. Chem. Soc.* (*A*), 522 (1971).
172. A. E. Keskinen and C. V. Senoff, *J. Organometallic Chem.*, **37**, 201 (1972).
173. D. M. Roundhill, *Inorg. Chem.*, **9**, 254 (1970).
174. J. H. Nelson, D. L. Schmitt, R. A. Henry, D. W. Moore and H. B. Jonassen, *Inorg. Chem.*, **9**, 2678 (1970).
175. I. Harvie and R. D. W. Kemmitt, *J. Chem. Soc.* (*D*), 198 (1970).
176. J. H. Nelson, H. B. Jonassen, and D. M. Roundhill, *Inorg. Chem.*, **8**, 2591 (1969).
177. L. S. Meriwether, M. F. Leto, E. C. Colthup and G. W. Kennerly, *J. Org. Chem.*, **27**, 3930 (1962).
178. A. Furlani, I. Collamati and G. Sartori, *J. Organometallic Chem.* **17**, 463 (1969).
179. W. Beck, K. Shorpp and F. Kern, *Angew. Chem. Internat. Edn.*, **10**, 66 (1971).
180. J. Chatt, C. Eaborn and P. N. Kapoor, *J. Chem. Soc.* (*A*), 881 (1970).
181. W. Fink and A. Wenger, *Helv. Chim. Acta*, **54**, 2186 (1971).
182. C. Eaborn, A. Pidcock and B. Ratcliff, *J. Organometallic Chem.*, **43**, C5 (1972).

183. C. D. Cook and G. S. Jauhal, *Canad. J. Chem.*, **45**, 301 (1967).
184. S. Cenini and S. Maiorana, unpublished results.
185. M. Keubler, personal communication.
186. P. Fitton, M. P. Johnson and J. E. McKeon, *Chem. Comm.*, 6 (1968).
187. S. Otsuka, M. Naruto, T. Yoshida and A. Nakamura, *J.C.S. Chem. Comm.*, 396 (1972).
188. B. Corain, M. Martelli, *Inorg. Nuclear Chem. Letters*, **8**, 39 (1972).
189. C. R. Green and R. J. Angelici, *Inorg. Chem.*, **11**, 2095 (1972).
190. D. Commereuc, I. Douek and G. Wilkinson, *J. Chem. Soc.* (*A*), 1771 (1970).
191. A. J. Mukhedkar, M. Green and F. G. A. Stone, *J. Chem. Soc.* (*A*) 947 (1970).
192. M. Hidai, T. Kashiwagi, T. Ikeuchi and Y. Uchida, *J. Organometallic Chem.*, **30**, 279 (1971).
193. P. Fitton and E. A. Rick, *J. Organometallic Chem.*, **28**, 287 (1971).
194. D. R. Coulson, *Chem. Comm.*, 1530 (1968).
195. J. Chatt and B. L. Shaw, *J. Chem. Soc.*, 705 (1959).
196. C. O'Connor, *J. Inorg. Nuclear Chem.*, **32**, 2299 (1970).
197. W. J. Bland and R. D. W. Kemmitt, *J. Chem. Soc.* (*A*), 1278 (1968).
197*bis*. F. Faraone, L. Silvestro, S. Sergi and R. Pietropaolo, *J. Organometallic Chem.*, **46**, 379 (1972).
198. M. C. Baird and G. Wilkinson, *J. Chem. Soc.* (*A*), 865 (1967).
199. (a) W. J. Bland, J. Burgess and R. D. W. Kemmitt, *J. Organometallic Chem.*, **14**, 201 (1968).
(b) *ibid.*, **15**, 217 (1968).
200. W. J. Bland, J. Burgess and R. D. W. Kemmitt, *J. Organometallic Chem.*, **18**, 199 (1969).
201. B. E. Mann, B. L. Shaw and N. I. Tucker, *J. Chem. Soc.* (*A*), 2667 (1971).
202. J. Lewis, B. F. G. Johnson, K. A. Taylor and J. D. Jones, *J. Organometallic Chem.*, **32**, C62 (1971).
203. M. Green, R. B. L. Osborn, A. J. Rest and F. G. A. Stone, *J. Chem. Soc.* (*A*), 2525 (1968).
204. A. J. Mukhedkar, M. Green and F. G. A. Stone, *J. Chem. Soc.* (*A*), 3023 (1969).
205. J. Ashley-Smith, M. Green and F. G. A. Stone, *J. Chem. Soc.* (*A*), 3019 (1969).
206. J. Browning, M. Green and F. G. A. Stone, *J. Chem. Soc.* (*A*), 453 (1971).
206*bis*. A. Greco, M. Green, S. K. Shakshooki and F. G. A. Stone, *J. Chem. Soc.* (*D*), 1374 (1970).
207. J. P. Collman, J. N. Cawse and J. W. Kang, *Inorg. Chem.*, 2574 (1969).
208. W. H. Baddley and L. M. Venanzi, *Inorg. Chem.*, **5**, 33 (1966).
209. W. H. Baddley, C. Panattoni, G. Bandoli, D. A. Clemente and U. Belluco, *J. Amer. Chem. Soc.*, **93**, 5590 (1971).
210. J. L. Burmeister and L. M. Edwards, *J. Chem. Soc.* (*A*), 1663 (1971).
211. A. J. Layton, R. S. Nyholm, G. A. Pneumaticakis and M. L. Tobe, *Chem. and Ind.*, 465 (1967).
212. M. Akhtar and H. C. Clark, *J. Organometallic Chem.*, **22**, 233 (1970).
213. B. Cetinkaya, M. F. Lappert, J. McMeeking and D. Palmer, *J. Organometallic Chem.*, **34**, C37 (1972).
214. R. Ugo, G. La Monica and S. Cenini, unpublished results.

215. P. E. Garrou and G. E. Hartwell, *J.C.S. Chem. Comm.*, 881 (1972).
216. B. J. Argento, P. Fitton, J. E. McKeon and E. A. Rick, *Chem. Comm.*, 1427 (1969).
217. M. Bressan, G. Favero, B. Corain and A. Turco, *Inorg. Nuclear Chem. Letters*, **7**, 203 (1971).
218. R. Zanella, R. Ros and M. Graziani, Fifth National Meeting on Inorganic Chemistry, Taormina, September (1972).
219. J. Chatt and D. M. P. Mingos, *J. Chem. Soc.* (*A*), 1243 (1970).
220. A. C. Skapski and P. G. H. Troughton, *Chem. Comm.*, 170 (1969).
221. C. D. Cook, unpublished results.
222. W. D. Bonds and J. A. Ibers, *J. Amer. Chem. Soc.*, **94**, 3413 (1972).
223. J. J. Levison and S. D. Robinson, *Chem. Comm.*, 198 (1967); *ibid.*, *J.C.S. Dalton*, 2013 (1972).
224. M. C. Baird, G. Hartwell, Jr., R. Mason, A. I. M. Rae and G. Wilkinson, *Chem. Comm.*, 92 (1967).
225. S. J. La Placa and J. A. Ibers, *Inorg. Chem.*, **5**, 405 (1966).
226. C. D. Cook and G. S. Jauhal, *J. Amer. Chem. Soc.*, **89**, 3066 (1967).
227. T. Kashiwagi, N. Yasuoka, T. Ueki, N. Kasai, M. Kakudo, S. Takahashi and N. Hagihara, *Bull. Chem. Soc., Japan*, **41**, 296 (1968).
228. R. Mason and A. I. M. Rae, *J. Chem. Soc.* (*A*) 1767 (1970).
229. T. Kashiwagi, N. Yasuoka, T. Ueki, N. Kasai and M. Kakudo, *Bull. Chem. Soc. Japan*, **40**, 1998 (1967).
230. A. P. Ginsberg and W. E. Silverthorn, *Chem. Comm.*, 823 (1969).
231. M. Green, R. B. L. Osborn and F. G. A. Stone, *J. Chem. Soc.* (*A*), 944 (1970).
232. R. Ugo, *Engelhard Industries Technical Bulletin*, **11**, 45 (1970).
233. G. Wilke, H. Schott and P. Heimbach, *Angew Chem. Internat. Edn.*, **6**, 92 (1967).
234. K. Sonogashira and N. Hagihara, *Mem. Inst. Sci. Ind. Res. Osaka University*, **22**, 165 (1965).
235. C. D. Cook and G. S. Jahual, *Inorg. Nuclear Chem. Letters*, **3**, 31 (1967).
236. C. J. Nyman, C. E. Wymore and G. Wilkinson, *J. Chem. Soc.* (*A*) 561 (1968).
236*bis*. V. J. Choj and C. J. O'Connor, *Coord. Chem. Rev.*, **9**, 145 (1972).
237. R. Ugo and S. Cenini, First International Symposium on "New Aspects of the Chemistry of Metal Carbonyls and Derivatives" Venice, September (1968).
238. J. P. Collmann, M. Kubota and J. W. Hosking, *J. Amer. Chem. Soc.*, **89**, 4809 (1967).
239. J. J. Levison and S. D. Robinson, *J. Chem. Soc.* (*A*), 762 (1971).
240. P. J. Hayward, D. M. Blake, G. Wilkinson and C. J. Nyman, *J. Amer. Chem. Soc.*, **92**, 5873 (1970).
241. W. Horn, E. Weissberger and J. P. Collman, *Inorg. Chem.*, **9**, 2367 (1970).
242. J. A. McGinnety, R. J. Doedens and J. A. Ibers, *Inorg. Chem.*, **6**, 2243 (1967).
243. T. Kashiwagi, N. Yasuoka, N. Kasai and M. Kakudo, *Chem. Comm.*, 743 (1969).
244. C. D. Cook, Pei-Tak Cheng and S. C. Nyburg, *J. Amer. Chem. Soc.*, **91**, 2123 (1969).

245. J. Halpern and A. L. Pickard, *Inorg. Chem.*, **9**, 2798 (1970).
246. R. Ugo, F. Conti, S. Cenini, R. Mason and G. B. Robertson, Chem. Comm., 1498 (1968).
246*bis*. I. S. Kommikov, Yu. D. Koreshkov, T. S. Lobeeva and M. E. Volpin, *Invest. Akad. Nauk. SSSR, Ser. Khim.*, **5**, 1181 (1972).
247. P. J. Hayward and C. J. Nymann, *J. Amer. Chem. Soc.*, **93**, 617 (1971).
248. P. J. Hayward, S. J. Saftich and C. J. Nyman, *Inorg. Chem.*, **10**, 1311 (1971).
249. S. Cenini and R. Ugo, unpublished results.
250. A. Fusi, R. Ugo, F. Fox, A. Pasini and S. Cenini, *J. Organometallic Chem.*, **26**, 417 (1971), and references therein.
251. S. Cenini, A. Fusi and G. Capparella, *J. Inorg. Nuclear Chem.*, **33**, 3576 (1971).
252. E. W. Stern, *J. Chem. Soc.* (*D*), 736 (1970).
253. R. A. Sheldon, *J. Chem. Soc.* (*D*), 788 (1971).
254. B. Clarke, M. Green, R. B. L. Osborn and F. G. A. Stone, *J. Chem. Soc.* (*A*), 167 (1968).
255. R. Countryman and B. R. Penfold, *J. Chem. Soc.* (*D*), 1598 (1971).
256. J. S. Valentine and D. Valentine, Jr., *J. Amer. Chem. Soc.*, **92**, 5795 (1970).
257. S. Cenini, R. Ugo and G. La Monica, *J. Chem. Soc.* (*A*) 416 (1971), and references therein.
258. Y. S. Sohn and A. L. Balch, *J. Amer. Chem. Soc.*, **93**, 1290 (1971).
259. D. M. Barlex, R. D. W. Kemmitt and G. W. Littlecott, *J. Chem. Soc.* (*D*), 199 (1971).
260. R. Mason, Personal Communication.
261. D. M. Barlex, R. D. W. Kemmitt and G. W. Littlecott, *J. Organometallic Chem.*, **43**, 225 (1972).
262. W. Beck, K. Schorpp and K. H. Stetter, *Naturforsch.*, **26b**, 684 (1971).
263. G. A. Pneumaticakis, *Chem. Comm.*, 275 (1968).
264. S. Cenini, R. Ugo, G. La Monica and S. D. Robinson, *Inorg. Chim. Acta*, **6**, 182 (1972).
265. R. D. Feltham, *Inorg. Chem.*, **3**, 119 (1964).
266. R. H. Reimann and E. Singleton, *J. Organometallic Chem.*, **32**, C44 (1971).
267. S. Cenini, M. Pizzotti and F. Porta, unpublished results.
268. R. D. Feltham, *Inorg. Chem.*, **3**, 116 (1964).
268*bis*. B. F. G. Johnson, S. Bhaduri and N. G. Connelly, *J. Organometallic Chem.*, **40**, C36 (1972).
269. J. H. Enemark, *Inorg. Chem.*, **10**, 1952 (1971).
270. W. J. Bland, R. D. W. Kemmitt, I. W. Nowell and D. R. Russell, *Chem. Comm.*, 1065 (1968).
271. C. A. Tolman, *Inorg. Chem.*, **10**, 1540 (1971).
272. M. Green, R. B. L. Osborn and F. G. A. Stone, *J. Chem. Soc.* (*A*), 3083 (1968).
273. H. F. Klein and J. F. Nixon, *J. Chem. Soc.* (*D*), 42 (1971).
274. S. Cenini and R. Ugo, unpublished results.
275. G. W. Parshall, *J. Amer. Chem. Soc.*, **89**, 1822 (1967).
276. D. J. Cardin, B. Cetinkaya, M. F. Lappert, Lj. Manojlović-Muir and K. W. Muir, *J. Chem. Soc.* (*D*), 400 (1971).

277. J. Clemens, R. E. Davis, M. Green, J. D. Oliver and F. G. A. Stone, *J. Chem. Soc. (D)*, 1095 (1971).
278. S. Cenini, R. Ugo and G. La Monica, *J. Chem. Soc. (A)*, 3441 (1971).
279. M. E. Volpin, V. B. Shur, R. V. Kudryavstev and L. A. Peodayko, *Chem. Comm.*, 1038 (1968).
280. T. L. Gilchrist, F. J. Graveling and C. W. Rees, *Chem. Comm.*, 821 (1968).
281. C. D. Cook and G. S. Jauhal, *J. Amer. Chem. Soc.*, **90**, 1464 (1968).
282. S. Patai, "The Chemistry of the Azido Group", Interscience (1971).
283. M. J. McGlinchey and F. G. A. Stone, *J. Chem. Soc. (D)*, 1265 (1970).
284. W. Beck, M. Bauder, G. La Monica, S. Cenini and R. Ugo, *J. Chem. Soc. (A)*, 113 (1971).
285. G. La Monica, S. Cenini, P. Sandrini and F. Zingales, *J. Organometallic Chem.*, **50**, 287 (1973).
286. (a) S. Cenini, W. Beck, *et al.*, to be published.
(b) S. Cenini and G. La Monica, Fifth National Meeting on Inorganic Chemistry, Taormina, September (1972).
287. J. Ashley-Smith, M. Green and F. G. A. Stone, *J. C. S. Dalton*, 1805 (1972).
288. B. Hessett, J. H. Morris and P. F. Perkins, *Inorg. Nuclear Chem. Letters*, 1149 (1971).
289. S. D. Robinson and M. F. Uttley, *J. Chem. Soc. (D)*, 1315 (1971).
290. S. D. Robinson and M. F. Uttley, *J.C.S. Chem. Comm.*, 184 (1972).
291. F. Cariati and R. Ugo, *Chimica e Industria*, **48**, 1288 (1966).
292. P. Chini and G. Longoni, *J. Chem. Soc. (A)*, 1542 (1970).
293. S. Cenini, unpublished results.
294. K. Kudo, M. Hidai and Y. Uchida, *J. Organometallic Chem.*, **33**, 393 (1971).
295. B. Corain, M. Bressan and G. Favero, *Inorg. Nuclear Chem. Letters*, **7**, 197 (1971).
296. J. R. Olechowski, *J. Organometallic Chem.*, **32**, 269 (1971).
297. R. D. Johnston, F. Basolo and R. G. Pearson, *Inorg. Chem.*, **10**, 247 (1971).
298. L. Malatesta and F. Bonati, "Isocyanide Complexes of Metals", Wiley-Interscience, London (1969).
299. M. Meier, F. Basolo and R. G. Pearson, *Inorg. Chem.*, **8**, 795 (1969).
300. (a) R. G. Guy and B. L. Shaw, *Adv. Inorg. Chem. Radiochem.*, **4**, 77 (1962).
(b) M. A. Bennett, *Chem. Rev.*, **62**, 611 (1962).
(c) P. M. Treichel and F. G. A. Stone, *Adv. Organometallic Chem.*, **1**, 143 (1964).
(d) G. N. Schrauzer, *Adv. Organometallic Chem.*, **2**, 1 (1964).
(e) R. Jones, *Chem. Rev.*, **68**, 785 (1968).
(f) W. H. Baddley, *Inorg. Chim. Acta Rev.*, **2**, 7 (1968).
(g) F. A. Bowden and A. B. P. Lever, *Organometallic Chem. Rev.*, **3**, 227 (1968).
(h) U. Belluco, B. Crociani, R. Pietropaolo and P. Uguagliati, *Inorg. Chim. Acta Rev.*, **3**, 19 (1969).
(i) F. R. Hartley, *Chem. Rev.*, **69**, 799 (1969).
(l) P. Heimbach, P. W. Jolly and G. Wilke, *Adv. Organometallic Chem.*, **8**, 29 (1970).

(m) J. H. Nelson and H. B. Jonassen, *Coord. Chem. Rev.*, **6**, 27 (1971).
(n) M. Herberhold, "Metal π-Complexes", Vol. II, Elsevier, (1972).
301. S. Cenini, R. Ugo and G. La Monica, *J. Chem. Soc.* (*A*), 409 (1971).
302. G. W. Parshall and F. N. Jones, *J. Amer. Chem. Soc.*, **87**, 5356 (1965).
303. G. La Monica, S. Cenini, G. Navazio and P. Sandrini, *J. Organometallic Chem.*, **31**, 89 (1971).
304. T. Ito, Y. Takahashi and Y. Ishii, *J.C.S. Chem. Comm.*, 629 (1972).
305. G. N. Schrauzer, *Chem. Ber.*, **94**, 642 (1961).
306. J. Chatt, B. L. Shaw and A. A. Williams, *J. Chem. Soc.*, 3269 (1962).
307. P. Uguagliati and W. H. Baddley, *J. Amer. Chem. Soc.*, **90**, 5446 (1968).
308. B. L. Booth, R. N. Haszeldine and N. I. Tucker, *J. Organometallic Chem.*, **11**, P5 (1968).
309. D. J. Cook, M. Green, N. Mayne and F. G. A. Stone, *J. Chem. Soc.* (*A*), 1771 (1968).
310. P. W. Jolly, I. Tkatchenko and G. Wilke, *Angew. Chem. Internat. Edn.*, **10**, 328 (1971).
311. H. Bönnemann, *Angew. Chem. Internat. Edn.*, **9**, 736 (1970).
312. B. Bogdanović and M. Velić, *Angew. Chem. Internat. Edn.*, **6**, 803 (1967).
313. J. P. Visser and J. E. Ramakers, *J.C.S. Chem. Comm.*, 178 (1972).
314. E. O. Greaves, C. J. L. Lock and P. M. Maitlis, *Canad. J. Chem.*, **46**, 3879 (1968).
315. P. W. Jolly and K. Jonas, *Angew. Chem. Internat. Edn.*, **7**, 731 (1968).
316. R. G. Miller, P. A. Pinke, R. D. Stauffer and H. J. Golden, *J. Organometallic Chem.*, **29**, C42 (1971).
317. W. Reppe, N. Von Kutepow and A. Magin, *Angew. Chem. Internat. Edn.*, **8**, 727 (1969).
317*bis*. H. Müller, D. Wittenberg, H. Seibt and E. Sharf, *Angew. Chem. Internat. Edn.*, **4**, 327 (1965).
318. G. Herrmann, Dissertation TH Aachen (1963).
319. M. Dubini and F. Montino, *Chimica e Industria*, **49**, 1283 (1967).
320. G. N. Schrauzer, S. Eichler and D. A. Brown, *Chem. Ber.*, **95**, 2755 (1962).
321. G. N. Schrauzer, *J. Amer. Chem. Soc.*, **82**, 1008 (1960).
322. G. N. Schrauzer, *J. Amer. Chem. Soc.*, **81**, 5310 (1959).
323. G. Wilke, *Angew. Chem.*, **72**, 581 (1960).
324. K. Jonas, P. Heimbach and G. Wilke, *Angew. Chem. Internat. Edn.*, **7**, 949 (1968).
325. S. Otsuka, A. Nakamura, T. Yamagata and K. Tani, *J. Amer. Chem. Soc.*, **94**, 1037 (1972).
326. M. Englert, P. W. Jolly and G. Wilke, *Angew. Chem. Internat. Edn.*, **11**, 136 (1972).
327. G. L. McClure and W. H. Baddley, *J. Organometallic Chem.*, **27**, 155 (1971).
328. D. M. Roundhill and G. Wilkinson, *J. Chem. Soc.* (*A*), 506 (1968).
329. T. Boschi, P. Uguagliati and B. Crociani, *J. Organometallic Chem.*, **30**, 283 (1971).
330. J. P. Visser, C. G. Leliveld and D. N. Reinhoudt, *J.C.S. Chem. Comm.*, 178 (1972).
331. K. Suzuki and H. Okuda, *J. Chem. Soc., Japan*, **45**, 1938 (1972).
332. W. J. Bland and R. D. W. Kemmitt, *J. Chem. Soc.* (*A*), 2062 (1969).

333. D. M. Barlex, R. D. W. Kemmitt and G. W. Littlecott, *Chem. Comm.*, 613 (1969).
334. R. D. W. Kemmitt, B. Y. Kimura, G. W. Littlecott and R. D. Moore, *J. Organometallic Chem.*, **44**, 403 (1972).
335. H. C. Clark, K. R. Dixon and W. J. Jacobs, *J. Amer. Chem. Soc.*, **90**, 2259 (1968).
336. J. Ashley-Smith, M. Green and D. C. Wood, *J. Amer. Chem. Soc.* (*A*) 1847 (1970).
337. G. L. McClure and W. H. Baddley, *J. Organometallic Chem.*, **25**, 261 (1970).
338. A. McAdam, J. N. Francis and J. A. Ibers, *J. Organometallic Chem.*, **29**, 149 (1971).
339. M. A. Bennett, G. B. Robertson, P. O. Whimp and T. Yoshida, *J. Amer. Chem. Soc.*, **93**, 3797 (1971).
340. J. P. Visser, A. J. Schipperijn, J. Lukas, D. Bright and J. J. De Boer, *J. Chem. Soc.* (*D*), 1266 (1971).
341. S. Otsuka, A. Nakamura and Z. Tani, *J. Organometallic Chem.*, **14**, P30 (1968).
342. W. Wong, S. J. Singer, W. D. Pitts, S. F. Watkins and W. H. Baddley, *J.C.S. Chem. Comm.*, 672 (1972)
343. J. P. Visser and J. E. Ramakers-Blom, *J. Organometallic Chem.*, **44**, C63 (1972).
344 (a) C. D. Cook, C. H. Koo, S. C. Nyburg and M. T. Shiomi, *Chem. Comm.*, 426 (1967).
(b) P. T. Cheng, C. D. Cook, S. C. Nyburg and K. Y. Wan, *Inorg. Chem.*, **10**, 2210 (1971).
(c) P. T. Cheng, C. D. Cook, C. H. Koo, S. C. Nyburg and M. T. Shiomi, *Acta Cryst.*, **27B**, 1904 (1971).
345. W. Dreissig and H. Dietrich, *Acta Cryst.* **B24**, 108 (1968).
346. (a) C. Panattoni, G. Bombieri, U. Belluco and W. H. Baddley, *J. Amer. Chem. Soc.*, **90**, 798 (1968).
(b) G. Bombieri, E. Forsellini, C. Panattoni, R. Graziani and G. Bandoli, *J. Chem. Soc.* (*A*), 1313 (1970).
346*bis*. C. Panattoni, R. Graziani, G. Bandoli, D. A. Clemente and U. Belluco, *J. Chem. Soc.*, (*B*), 371 (1970).
347. J. N. Francis, A. McAdam and J. A. Ibers, *J. Organometallic Chem.*, **29**, 131 (1971).
348. M. Kadonaga, N. Yasuoka and N. Kasai, *J. Chem. Soc.* (*D*), 1597 (1971) and references therein.
349. T. G. Hewitt, K. Hanzenhofer and J. J. De Boer, *J. Organometallic Chem.*, **18**, P19 (1969).
350. C. Krüger and Y. H. Tsay, *J. Organometallic Chem.*, **34**, 387 (1972).
351. H. P. Fritz and G. N. Schrauzer, *Chem. Ber.*, **94**, 650 (1961).
352. A Panunzi, A. De Renzi and G. Paiaro, *J. Amer. Chem. Soc.*, **92**, 3488 (1970).
353. R. Ugo, Editor, "Aspects of Homogeneous Catalysis", Vol. 1, Carlo Manfredi, Milano (1970).
354. J. Chatt and B. L. Shaw, *J. Chem. Soc.*, 5075 (1962).
355. R. G. Miller, D. R. Fahey and D. P. Kuhlman, *J. Amer. Chem. Soc.*, **90**, 6248 (1968).

356. D. R. Fahey, *J. Amer. Chem. Soc.*, **92**, 402 (1970).
356*bis*. M. Dubini, and F. Montino, *J. Organometallic Chem.*, **6**, 188 (1966).
357. K. Yamamoto, T. Hayashi and M. Kumada, *J. Organometallic Chem.*, **28**, C37 (1972).
358. C. Eaborn, D. J. Tune and D. R. M. Walton, *J.C.S. Chem. Comm.*, 1223 (1972).
358*bis*. R. D. W. Kemmitt, R. D. Peacock and J. Stocks, *Chem. Comm.*, 554 (1969).
359. (a) R. N. Scott, D. F. Shriver and L. Vaska, *J. Amer. Chem. Soc.*, **90**, 1079 (1968).
(b) R. N. Scott, D. F. Shriver, and D. D. Lehman, *Inorg. Chim. Acta*, **4**, 73 (1970).
360. L. Toniolo, G. De Luca and C. Panattoni, Fifth National Meeting on Inorganic Chemistry, Taormina, 25–29 September (1972).
361. S. Otsuka, A. Nakamura, T. Koyama and Y. Tatsuno, *J.C.S. Chem. Comm.*, 1105 (1972).
362. J. Ashley-Smith, M. Green and F. G. A. Stone, *J. Chem. Soc.* (*A*) 3161 (1970).
363. G. A. Larkin, R. Mason and M. G. H. Wallbridge, *J. Chem. Soc.* (*D*), 1054 (1971).
364. E. O. Greaves, R. Bruce and P. M. Maitlis, *Chem. Comm.*, 860 (1967).
365. T. Ito, S. Hasegawa, Y. Takahashi and Y. Ishii, *J.C.S. Chem. Comm.*, 629 (1972).
366. J. Chatt, G. A. Rowe and A. A. Williams, *Proc. Chem. Soc.*, 208 (1957).
367. S. Cenini and R. Ugo, unpublished results.
368. D. A. Harbourne and F. G. A. Stone, *J. Chem. Soc.* (*A*), 1765 (1968).
369. A. Furlani, P. Bicev, M. V. Russo and P. Carusi, *J. Organometallic Chem.*, **29**, 321 (1971).
370. J. L. Boston, S. O. Grim and G. Wilkinson, *J. Chem. Soc.*, 3468 (1963).
371. (a) A. D. Allen and C. D. Cook, *Canad. J. Chem.*, **42**, 1063 (1964).
(b) *ibid.*, **41**, 1084 (1963).
372. C. D. Cook and K. Y. Wan, *Inorg. Chem.*, **10**, 2696 (1971).
373. P. B. Tripathy, B. W. Renoe, K. Adzamli and D. M. Roundhill, *J. Amer. Chem. Soc.*, **93**, 4406 (1971).
374. J. H. Nelson, J. J. R. Reed and H. B. Jonassen, *J. Organometallic Chem.*, **29**, 163 (1971).
375. J. Chatt, R. G. Guy, L. A. Duncanson and D. T. Thompson, *J. Chem. Soc.*, 5170 (1963).
376. J. Browning, C. S. Cundy, M. Green and F. G. A. Stone, *J. Chem. Soc.* (*A*), 448 (1971).
377. J. P. Collman, J. W. Kang, W. F. Little and M. F. Sullivan, *Inorg. Chem.*, **7**, 1298 (1968).
378. J. O. Glanville, J. M. Stewart and S. O. Grim, *J. Organometallic Chem.*, **7**, P9 (1967).
379. G. B. Robertson and P. O. Whimp, *J. Organometallic Chem.*, **32**, C69 (1971).
380. H. C. Clark and R. J. Puddephatt, *Inorg. Chem.*, **10**, 18 (1971).
381. P. B. Tripathy and D. M. Roundhill, *J. Amer. Chem. Soc.*, **92**, 3825 (1970).
382. M. C. Baird, *J. Organometallic Chem.*, **16**, P16 (1969).

383. T. R. Durkin and E. P. Schram, *Inorg. Chem.*, **11**, 1048 (1972).
384. T. R. Durkin and E. P. Schram, *Inorg. Chem.*, **11**, 1054 (1972).
385. R. Mason, J. Zubieta, A. T. T. Hsieh, J. Knight, and M. J. Mays, *J.C.S. Chem. Comm.*, 200 (1972).
386. V. G. Albano, G. Ciani, M. I. Bruce, G. Shaw and F. G. A. Stone, *J. Organometallic Chem.*, **42**, C99 (1972).
387. M. I. Bruce, G. Shaw and F. G. A. Stone, *J.C.S. Dalton*, 1781 (1972).
388. J. Halpern, *Chem. Eng. News.*, **44**, 68 (1964).
389. F. Bonati, S. Cenini, D. Morelli and R. Ugo, *J. Chem. Soc.* (*A*), 1052 (1966).
390. W. Beck and K. Shorpp, *Angew. Chem. Internat. Edn.*, **9**, 735 (1970).
391. R. B. King and A. Efraty, *J. Amer. Chem. Soc.*, **94**, 3768 (1972).
392. T. Kobayashi, Y. Takahashi, S. Sakai and Y. Ishii, *Chem. Comm.*, 1373 (1968).
393. M. Lenarda, R. Ros, M. Graziani and U. Belluco, *J. Organometallic Chem.*, **46**, C29 (1972).
394. P. W. Jolly, I. Tkatchenko and G. Wilke, *Angew Chem. Internat. Edn.*, **10**, 329 (1971) and references therein.
395. J. M. Brown, B. T. Golding and M. J. Smith, *J. Chem. Soc.* (*D*)., 1240 (1971).
396. S. Otsuka, A. Nakamura, S. Ueda and K. Tani, *J. Chem. Soc.* (*D*)., 863 (1971).
397. P. Heimbach, H. Selbeck, and E. Troxler, *Angew. Chem. Internat. Edn.*, 659 (1971).
398. (a) P. M. Maitlis, "The Organic Chemistry of Palladium", Vol. II, Catalytic Reactions, Academic Press (1971).
(b) F. Conti, Recenti aspetti della chimica e proprietà catalitiche dei complessi fosfinici zerovalenti di palladio e platino, *Chimica e Industria*, in press, and references therein.
399. K. Schorpp, P. Kreutzer and W. Beck, *J. Organometallic Chem.*, **37**, 397 (1972).
400. S. Takahashi, T. Shibano and N. Hagihara, *Chem. Comm.*, 161 (1969).
401. K. E. Atkins, W. E. Walker and R. M. Manyik, *J. Chem. Soc.* (*D*)., 330 (1971).
402. E. J. Smutny, *J. Amer. Chem. Soc.*, **89**, 6793 (1967).
403. S. Takahashi, T. Shibano and N. Hagihara, *Tetrahedron Letters*, 245 (1967).
404. S. Takahashi, R. Shibano and N. Hagihara, *Bull. Chem. Soc. Japan*, **41**, 454 (1968).
405. W. E. Walker, R. M. Manyik, K. E. Atkins and M. L. Farmer, *Tetrahedron Letters*, 3817 (1970).
406. G. Hata, K. Takahashi and A. Miyake, *Chem. and Ind.*, 1836 (1969); *ibid. J. Org. Chem.*, **36**, 2116 (1971).
407. J. C. Shields and W. E. Walker, *J. Chem. Soc.* (*D*)., 193 (1971).
408. R. Baker, D. E. Halliday and T. N. Smith, *J. Chem. Soc.* (*D*)., 1583 (1971).
409. R. Baker, D. E. Halliday and T. N. Smith, *J. Organometallic Chem.*, **35**, C61 (1972).
410. M. L. H. Green and H. Munakata, *J. Chem. Soc.* (*D*), 549 (1971).
411. C. A. Tolman, *J. Amer. Chem. Soc.*, **92**, 6777 (1970).
412. A. C. L. Su and J. W. Collette, *J. Organometallic Chem.*, **36**, 177 (1972).
413. E. W. Bennett and P. J. Orenski, *J. Organometallic Chem.*, **28**, 137 (1971).
414. M. Hara, K. Ohno and J. Tsuji, *J. Chem. Soc.* (*D*)., 247 (1971).
415. W. Fink, *Helv. Chim. Acta.*, **54**, 1304 (1971).

416. E. S. Brown, "Aspects of Homogeneous Catalysis" (R. Ugo, ed.), *2*, (1973) in press.
417. W. C. Drinkard, Jr. and R. J. Kassal, *French Patent*, 1,529,134.
418. E. S. Brown, E. A. Rick and F. D. Mendicino, *J. Organometallic Chem.*, **38**, 37 (1972).
419. C. A. Tolman, *J. Amer. Chem. Soc.*, **94**, 2994 (1972).
420. D. Bingham, D. E. Webster and P. B. Wells, *J.C.S. Dalton* 1928 (1972).
421. P. Roffia, F. Conti, G. Gregorio, G. F. Pregaglia and R. Ugo, "Catalysis by palladium salts. III and IV", *J. Organometallic Chem.*, in press.
422. J. Chatt and H. R. Watson, *J. Chem. Soc.*, 2545 (1962).
423. R. A. Schunn, *Inorg. Chem.*, **9**, 2567 (1970).
424. A. Sacco and R. Ugo, *J. Chem. Soc.*, 3274 (1964).
425. A. Sacco and M. Rossi, *Chem. Comm.*, 602 (1965).
426. H. F. Klein, *Angew. Chem. Internat. Edn.*, **10**, 343 (1971).
427. D. C. Olson and W. Keim, *Inorg. Chem.*, 2028 (1969).
428. S. D. Robinson, *Chem. Comm.*, 521 (1968).
429. M. E. Volpin and I. S. Kolomnikov, *Izvest. Akad. Nauk. S.S.S.R., Ser. Khim.*, 2041 (1966).
430. J. F. Nixon, *Adv. Inorg. Chem. Radiochem.*, **13**, 364 (1970).
431. T. Kruck, G. Sylvester and I. P. Kunau, *Angew. Chem. Internat. Edn.*, **10**, 725 (1971).
432. M. A. Bennett and D. J. Patmore, *Inorg. Chem.* **11**, 2387 (1971).
433. M. A. Bennett, R. N. Johnson, G. B. Robertson, T. W. Turney and P. O. Whimp, *J. Amer. Chem. Soc.*, **94**, 6540 (1972).
434. A. Yamamoto, S. Kitazume, L. S. Pu and S. Ikeda, *Chem. Comm.*, 79 (1967).
435. A. Misono, Y. Uchida, M. Hidai and M. Araki, *Chem. Comm.*, 1044 (1968).
436. (a) J. H. Enemark, B. R. Davis, J. A. McGinnety and J. A. Ibers, *Chem. Comm.*, 96 (1968).
(b) B. R. Davis, N. C. Payne and J. A. Ibers, *Inorg. Chem.*, **8**, 2719 (1969).
437. A. Sacco and M. Rossi, *Chem. Comm.*, 316 (1967).
438. A. Yamamoto, S. Kitazume, L. S. Pu and S. Ikeda, *J. Amer. Chem. Soc.*, **93**, 371 (1971).
439. G. Speier and L. Markò, *Inorg. Chim. Acta.*, **3**, 126 (1969).
440. J. Kovàcs, G. Speier and L. Markò, *Inorg. Chim. Acta*, **4**, 412 (1970).
441. M. Aresta, C. F. Nobile, M. Rossi and A. Sacco, *J. Chem. Soc.* (*D*)., 781 (1971).
442. J. Chatt and J. M. Davison, *J. Chem. Soc.*, 843 (1965).
443. U. A. Gregory, S. D. Ibekwe, B. T. Kilbourn and D. R. Russell, *J. Chem. Soc.* (*A*), 1118 (1971).
444. J. T. Mague and J. P. Mitchener, *Inorg. Chem.*, **11**, 2714 (1972).
445. J. R. Sanders, *J. Chem. Soc.* (*A*), 1333 (1972).
446. S. Cenini, unpublished results.
447. P. Meakin, E. L. Muetterties and J. P. Jesson, *J. Amer. Chem. Soc.*, **94**, 5271 (1972).
448. D. L. Williams-Smith, L. R. Wolf and P. S. Skell, *J. Amer. Chem. Soc.*, **94**, 4042 (1972).
449. E. K. von Gustorf, O. Jaenicke and O. E. Polansky, *Angew.Chem. Internat. Edn.*, **11**, 532 (1972).
450. J. D. Warren and R. J. Clark, *Inorg. Chem.*, **9**, 373 (1970).

451. G. Hata, H. Kondo and A. Miyake, *J. Amer. Chem. Soc.*, **90**, 2278 (1968).
452. T. Ito, S. Kitazume, A. Yamamoto and S. Ikeda, *J. Amer. Chem. Soc.*, **92**, 3011 (1970).
453. S. Tyrlik, K. Falkowski and K. Leibler, *Inorg. Chim. Acta.*, **6**, 291 (1972).
454. S. Komiya, A. Yamamoto and S. Ikeda, *J. Organometallic Chem.*, **42**, C65 (1972).
455. C. H. Campbell, A. R. Dias, M. L. H. Green, T. Saito and M. G. Swanwick, *J. Organometallic Chem.*, **14**, 349 (1968).
456. S. T. Wilson and J. A. Osborn, *J. Amer. Chem. Soc.*, **93**, 3068 (1971).
457. C. G. Pierpont and R. Eisenberg, *Inorg. Chem.*, **11**, 1094 (1972).
458. C. G. Pierpont, A. Pucci and R. Eisenberg, *J. Amer. Chem. Soc.*, **93**, 3050 (1971).
459. M. H. B. Stiddard and R. E. Townsend, *Chem. Comm.*, 1372 (1969).
460. K. R. Laing and W. R. Roper, *Chem. Comm.*, 1568 (1968).
461. K. R. Laing and W. R. Roper, *J. Chem. Soc. (A)*, 2149 (1970).
462. B. W. Graham, K. R. Laing, C. J. O'Connor and W. R. Roper, *J. Chem. Soc. (D)* 1272 (1970); *ibid., J.C.S. Dalton*, 1237 (1972).
463. J. Chatt and H. R. Watson, *Proc. Chem. Soc.*, 243 (1960).
464. M. L. H. Green, L. C. Mitchard and W. E. Silverthorn, *J. Chem. Soc. (A)*, 2929 (1971).
465. R. Mathieu and R. Poilblanc, *Inorg. Chem.*, **11**, 1858 (1972).
466. M. Hidai, K. Tominari and Y. Uchida, *J. Amer. Chem. Soc.*, **94**, 110 (1972) and references therein for the preliminary notes.
467. T. Uchida, M. Hidai and T. Kodama, *Bull. Chem. Soc. Japan*, **44**, 2883 (1971).
468. L. K. Atkinson, A. H. Mawby and D. C. Smith, *J. Chem. Soc. (D).*, 157 (1971).
469. J. Chatt and A. G. Weed, *J. Organometallic Chem.*, **27**, C15 (1971).
470. T. A. George and C. D. Seibold, *J. Organometallic Chem.*, **30**, C13 (1971).
471. T. A. George and C. D. Seibold, *Inorg. Nuclear Chem. Letters*, **8**, 465 (1972).
472. B. Bell, J. Chatt and G. J. Leigh, *J.C.S. Dalton*, 2492 (1972).
473. M. L. H. Green and W. E. Silverthorn, *J. Chem. Soc. (D).*, 557 (1971).
474. D. J. Darensbourg, *Inorg. Nuclear Chem. Letters*, **8**, 529 (1972).
475. J. Chatt, R. H. Crabtree and R. L. Richards, *J.C.S. Chem. Comm.*, 534 (1972).
475*bis*. M. Aresta, *Gazzetta*, in press.
476. T. A. George and C. D. Seibold, *J. Amer. Chem. Soc.*, **96**, 6859 (1972).
477. J. Chatt, G. A. Heath and R. L. Richards, *J.C.S. Chem. Comm.*, 1010 (1972).
478. J. Chatt, G. A. Heath and G. J. Leigh, *J.C.S. Chem. Comm.*, 444 (1972).
479. S. Otsuka *et al.* to be published.
480. S. Otsuka, A. Nakamura and T. Yoshida, *J. Amer. Chem. Soc.*, **91**, 7196 (1969).
481. S. Otsuka, Y. Tatsuno and K. Ataka, *J. Amer. Chem. Soc.*, **93**, 6705 (1971).
482. S. Otsuka, A. Nakamura and Y. Tatsuno, *J. Amer. Chem. Soc.*, **91**, 6994 (1969).

483. A. Nakamura, Y. Tatsuno, M. Yamamoto and S. Otsuka, *J. Amer. Chem. Soc.*, **93**, 6052 (1971) and references therein.
484. S. Otsuka, A. Nakamura and Y. Tatsuno, *Chem. Comm.*, 836 (1967).
485. S. Otsuka, A. Nakamura, Y. Tatsuno and M. Miki, *J. Amer. Chem. Soc.*, **94**, 3761 (1972).
486. M. Green, S. K. Shakshooki and F. G. A. Stone, *J. Chem. Soc. (A)*, 2828 (1971).
487. S. Otsuka, T. Yoshida and Y. Tatsuno, *J. Chem. Soc. (D)*., 67 (1971).
488. S. Otsuka, T. Yoshida and Y. Tatsuno, *J. Amer. Chem. Soc.*, **93**, 6462 (1971).
489. (a) R. S. Dickson, J. A. Ibers, S. Otsuka and Y. Tatsuno, *J. Amer. Chem. Soc.*, **93**, 4636 (1971).
(b) R. S. Dickson and J. A. Ibers, *J. Amer. Chem. Soc.*, **94**, 2988 (1972).
490. J. K. Stalick and J. A. Ibers, *J. Amer. Chem. Soc.*, **92**, 5333 (1970).
491. P. Heimbach and R. Traunmüller, *Justus Liebigs Ann. Chem.*, **727**, 208 (1969).
492. Y. Suzuki and T. Takizawa, *J.C.S. Chem. Comm.*, 837 (1972).
493. R. S. Dickson and J. A. Ibers, *J. Organometallic Chem.*, **36**, 191 (1972).
494. S. O. Grim, L. J. Matienzo and W. E. Swartz, Jr. *J. Amer. Chem. Soc.*, **94**, 5116 (1972).
495. Y. Saito, J. Takemoto, B. Hutchinson and K. Nakamoto, *Inorg. Chem.*, **11**, 2003 (1972) and references therein.

Subject Index

R